Hasnain Bokhari

Information and Communication Technologies in Pakistan

History and analysis of electronic public services (2000-2012)

disserta
Verlag

Bokhari, Hasnain: Information and Communication Technologies in Pakistan. History and analysis of electronic public services (2000-2012), Hamburg, disserta Verlag, 2020

Buch-ISBN: 978-3-95935-536-0
PDF-eBook-ISBN: 978-3-95935-537-7
Druck/Herstellung: disserta Verlag, Hamburg, 2020

2014 an der Philosophischen Fakultät der Universität Erfurt angenommene Dissertation

Bibliografische Information der Deutschen Nationalbibliothek:
Die Deutsche Nationalbibliothek verzeichnet diese Publikation in der Deutschen Nationalbibliografie; detaillierte bibliografische Daten sind im Internet über http://dnb.d-nb.de abrufbar.

Table of Contents

List of Figures

Abbreviations

ABC	Automated Border Control
ADBI	Asian Development Bank Institute
ATM	Automated Teller Machines
B2B	Business to Business
B2C	Business to Customer
BB	Branchless Banking
BBS	Bulletin Board Systems
BPS	Basic Pay Scales
CDA	Capital Development Authority
CERS	Computerised Electoral Rolls System
CMS	Content Management System
CNIC	Computerised National Identity Card
CRC	Child Registration Certificate
CSP	Civil Service of Pakistan
CTO	Chief Technology Officer
DCO	District Coordination Officer
DGR	Directorate Generate of Registration
DMG	District Management Group
DoD	Department of Defense
ECAC	Electronic Certification Accreditation Council
ECOSOC	Economic and Social Council
ECP	Election Commission of Pakistan
EGD	Electronic Government Directorate
eGovernment	Electronic Government
eService	Electronic Service
ETO	Electronic Transaction Ordinance
EVM	Electronic Voting Machines
FDI	Foreign Direct Investment
FI	Financial Institution
FIA	Federal Investigation Agency
FRC	Family Registration Certificate
FTTH	Fibre-to-the-Home
G2B	Government to Business
G2C	Government to Citizen
G2E	Government to Employee
G2G	Government to Government

GDP	Gross Domestic Product
GNI	Gross National Income
HDI	Human Development Index
HTTP	Hypertext Transfer Protocol
ICS	Indian Civil Service
ICT4D	Information and Communication Technologies for development
ICTs	Information and Communication Technologies
IFES	International Foundation for Electoral Systems
IGF	Internet Governance Forum
IMF	International Monetary Fund
ISPs	Internet Service Providers
ITLF	Information Technology Law Forum
ITU	International Telecommunication Union
KESC	Karachi Electric Supply Company
KP	Khyber Pakhtunkhwa
LDI	Long Distance and International
mCommerce	Mobile Commerce
MDGs	Millennium Development Goals
MNO	Mobile Network Operator
MoIT	Ministry of Information Technology
MoST	Ministry of Science and Technology
MRP	Machine-Readable Passport
NAB	National Accountability Bureau
NADRA	National Database Registration Authority
NDF	National Data Forms
NDH	National Data Warehouse
NDO	National Database Organization
NEGC	National Electronic Government Council
NITB	National Information Technology Board
NPM	New Public Management
NRA	National Registration Act
OTC	Over-The-Counter
PCB	Pakistan Computer Bureau
PCO	Population Census Organization
PDA	Personal Digital Assistant
PKR	Pakistani Rupee
POS	Point of Sale

PSTN	Public Switched Telephone Network
PTA	Pakistan Telecommunication Authority
PTCL	Pakistan Telecommunication Limited
SAARC	South Asian Association for Regional Cooperation
SBP	State Bank of Pakistan
SCOT	Social Construction of Technology
SCP	Supreme Court of Pakistan
SECP	Security and Exchange Commission of Pakistan
SMS	Short Messaging Service
SNGPL	Sui Northern Gas Pipeline Ltd
SNIC	Smart National Identity Card
STS	Science, Technology and Society
UN	United Nations
UNICTTF	United Nations Information and Communication Technologies Task Force
UP	Uttar Pradesh
USF	Universal Service Fund
USSR	Union of Soviet Socialist Republics
VKC	Village Knowledge Centre
WAPDA	Water and Power Development Authority
WLL	Wireless Local Loop
WSIS	World Summit on the Information Society
WWW	World Wide Web

Acknowledgements

This work is a slightly revised version of the doctoral research written at the Faculty of Humanities, University of Erfurt, supervised by Prof. Dr. Jamal Malik and Prof. Dr. Kai Hafez. I am deeply indebted to my first supervisor, Prof. Malik, for the guidance and support he has lent in this scholarly journey. He provided numerous opportunities for intellectual stimuli in one-on-one sessions and research colloquia, mentoring, and astute observations that shaped this work. His extensive study and experience of South Asia gave me a strong footing to attempt a critical assessment of the cultural and governance system in South Asia by reviewing and (re)analysing problem statements through various academic approaches. I am equally grateful to my second supervisor, Prof. Hafez, whose valuable input and comments have immensely helped in bringing this work to fruition. I will remain thankful for his undivided support and commitment.

A number of people who I came across before and during the process of this research deserve a mention. During the summer of 2005, I met Peter Rave and Geraldine de Bastion as an intern at the Department of Information & Communication Technologies for sustainable development at the GIZ headquarters in Eschborn/Germany. The idea of researching eGovernment was seeded at the GIZ while I worked in their preparatory process for World Summit on Information Technology (WSIS), Tunis. I am thankful to both Peter and Geraldine for giving me this opportunity and also for their inputs and observations from the field whenever requested. I am also thankful to Prof. Dr. Wolfgang Kleinwächter for welcoming me to the highly stimulating world of Internet Governance. Participating in his first summer school on Internet Governance during 2007 and meeting its faculty members - Avri Doria, Dr. William Drake, Bertrand de La-Chapelle and late Dr. Heike Jensen - laid the groundwork for learning how to and how not to approach the subject of ICTs and Internet. Heike's refreshing perspectives on gender and ICTs extolled me to consider this important perspective in my research. Her untimely demise is a loss for Department of Gender Studies at Humboldt University, Berlin and the Internet Governance Forum.

This study relies on the data collected during field visits to Pakistan. Data collection in general can be a challenging exercise especially if the said data in question comes from government sources. I would like to thank the staff members of Electronic Government Directorate at the Ministry of Information Technology in Islamabad, particularly Zubair Saleem, Bilal Manzoor, Faiz Bashir Malik and Muhammad Nasir for patiently understanding my concerns and providing me their inputs. I am especially grateful to Mr. Usman Mobin at National Database Registration Authority (NADRA) in Islamabad. He understood the value of this academic pursuit and provided access for the study of kiosks and eSahulat services. His associate Usman Javaid lent invaluable assistance in helping me explain the context and the frameworks followed in the kiosks, electoral rolls and eCommerce platform. Similarly, Ahsan Mansoor who then worked at Mobilink provided remarkable ushering to understand citizen-centric services initiatives through the mobile operator's Value-Added Services.

I am also thankful to Prof. Richard Heeks at the Centre for Development Informatics for introducing me to Dr. Anita Greenhill. This led to a visiting research fellowship at the Manchester University in 2010. The platform provided by both Prof. Heeks and Dr. Greenhill gave me opportunities of scholarly exchange with fellow researchers working on development studies and ICTs.

I owe a special gratitude to the Heinrich Böll Foundation in Germany which saw potential in my research subject and included me in their Doctoral scholarship programme. It was Böll Foundation's financial support that enabled this study and the chance to present it at various international forums such as Internet Governance Forum (Hyderabad, 2008 and Sharm al-Sheikh, 2009); International Multi-Topic Conference, Mohammad Ali Jinnah University, Islamabad (2009) and International Conference on Theory and Practice of Electronic Governance (ICEGOV2010) in Beijing.

Last but not least I will remain indebted to my alma mater, the University of Erfurt, the Faculty of Humanities, for providing me the opportunity to grow both professionally and personally. It is because of the encouragement and support of various colleagues and fellows at the University of Erfurt that this research has been possible.

Abstract

This study deals with the use of Information and Communication Technologies for public service delivery in Pakistan. Set during the arrival of new ICTs in the period 2000-2012, the research raises the question if and how far have these ICTs affected manual public service delivery mechanisms in Pakistan? The research takes into consideration five initiatives from different public- and private-sector organisations in Pakistan. In this regard, initiatives by Electronic Government Directorate, National Database Registration Authority, Election Commission of Pakistan, and a private mobile network operator, Mobilink are explored and evaluated. Drawing largely on quantitative and qualitative data, this study reveals that the presence and utility of new communication channels in Pakistan have paved the way for a new form of electronic public service delivery which relies on pre-existing sociocultural norms such as oral communication and personal interaction contours. The study suggests that the form and model of eGovernment observed in Pakistan show a blend of manual and automated processes for public service delivery. The study also touches upon the issues of governance and politics in Pakistan that have had a direct and at times an indirect effect on the public sector.

Chapter 1

Introduction

The use of modern information and communication technologies (ICTs) in public service delivery is relatively a new phenomenon in Pakistan. Towards the end of 1990s, there were neither publicly available electronic transaction mechanisms nor abundant automated banking terminals in Pakistan. Internet was in its infancy stages of development and the internet users were in few thousands. There was only one fixed line phone provider, Pakistan Telecommunication Limited (PTCL) which dominated the entire telecommunication infrastructure. Getting a phone connection from PTCL was no less an ordeal. Mobile phone was considered luxury and its subscribers were either wealthy businesspeople or a handful of politicians. The same applied to the broad- and telecast scenario – with an exception of one private television restricted to few metropolitan cities in 1990s, there were two television channels heavily controlled and monitored by the state. It seems difficult to imagine that modern ICTs would find their outreach within such a limited digital mediascape, let alone electronic public service delivery.

On the contrary, within the first decade of the new millennium, Pakistan progressed in the field of digital communication by leaps and bounds and by end of 2010 multiple mechanisms of digital public service delivery became available. How did Pakistan's technological paradigm reach this level of development, presents an interesting example of a country that leapfrogged from a nascent to an advanced stage of technological development? The steps taken towards development of communication technologies and provision of electronic public services are not only of interest to academic curiosity but they also serve as a point of socio-cultural and technical turn in an underdeveloped country like Pakistan.

Such steps included the formation of new institutions as well as the promulgation of new ordinances, acts and policies to provide a legal cover to the ICTs-based services. For the provision of electronic public services, Government of Pakistan established the Electronic Government Directorate (EGD) in 2002 which was mandated to advise government in the uptake of ICTs, develop electronic public services (eServices) and electronic government (eGovernment) solutions and more importantly to formulate eGovernment strategy for Pakistan. Post-2000, mobile phone penetration also rapidly grew in Pakistan. Realising mobile phones' potential and reach, Election Commission of Pakistan (ECP) decided to engage with its citizens in the 2013 electoral process by launching SMS service accessible via ordinary mobile phones[1]. The ECP launched a unique mobile number through which the citizens could receive updated information about their constituency and

---

1 Voter verification: ECP launches SMS service (2012f, February 29), The Express Tribune. Retrieved 04.05.2014 from https://tribune.com.pk/story/343403/voter-verification-ecp-launches-sms-service/

the location of their polling station via mobile phone. The ECP's SMS application became one of the most frequently used electronic public services developed to strengthen the voter registration process.

In addition to the EGD, other public-sector organisations were also formed to develop electronic public services projects. In this context, formation of National Database Registration Authority (NADRA) under the Ministry of Interior heralded a new era for digital identity with its Computerised National Identity Cards (CNIC) project. NADRA went onto develop further eGovernment solutions such as digital kiosk machines and its variation eSahulat; a utility for citizens to access G2C electronic public services. The developments in the public-sector organisation also influenced the private sector, especially the mobile network providers who relied on their expanding customer base in Pakistan to develop innovative citizen-focussed digital services for public consumption. If the number of electronic services produced during 2000 through 2012 are seen in retrospect, there appears to be multiple government and semi-government agencies developing either their individual eGovernment solutions or in collaboration with each other.

Considering the period of 2002-2012 almost a decade after the formation of the EGD, yet Pakistan's experience with electronic public services delivery could still be considered in its formative years. According to 2012's United Nations eGovernment survey, Pakistan is ranked 156th in eGovernment development ranking[2]. The survey focused on the evaluation and analysis of electronic public service delivery mechanisms across the world, in which countries like South Korea and Denmark dominate the ranks. The survey also includes case studies and examples from countries which according to UN standards are considered as developing countries. With eGovernment services and projects being undertaken by several governments the past few years, the literature focusing on eGovernment in developing countries has substantially grown. The question that arises here is how different the phenomenon of eGovernment in developing countries is from technologically advanced countries such as South Korea or Denmark. Similarly, how has Pakistan fared in the area of eGovernment so far. This requires dwelling on a few developments in their chronological and historical perspectives.

1.1. Of ICTs, Internet and the Web

At the beginning of new millennium, the world witnessed notable progresses, among others in two distinct areas, (i) information and communication technologies and (ii) emphasis on human/social development particularly

---

2   E-Government Survey 2012: E-Government for the People. Retrieved 12.12.2013 from https://publicadministration.un.org/egovkb/Portals/egovkb/Documents/un/2012-Survey/Complete-Survey.pdf

focussing on methods to improve and measure people's well-being. During this period, material advances in semiconductor sector saw phenomenal growth. Resultantly, personal computing continuously benefited by increase in the processing power of the computers, affordability and accessibility for a wider public across the world.

At the cusp of the millennium change another development that added a new avenue for digital connectivity was the progress in the telecommunication sector. Post-1990s, telecommunication sector worldwide went through deregulation paving the way for private investment thereby reducing the monopoly of national or government-run telephony and wireless infrastructure[3]. This eventually meant that a number of privately-owned service providers could also provide fixed line telephone connections thus increasing the competition as well as quality of service[4]. Another similar trend during 1990s which was carried over in the new millennium was the commercialisation and popularity of mobile phones[5]. Initially only meant to serve the purpose of always available wireless telephone with an additional functionality of sending text messages, mobile phones morphed shapes, sizes and specifications and compacted into multifunctional smart phones rather than the original feature phone.

Parallel to these technological trends there was another communication technology that used the existing Public Switched Telephone Network (PSTN) for faster transmission of text, audio and video data in comparison to fax machines or telephones. Internet – a brain child of the United States Department of Defense (DoD) originally planned to elevate US hegemony and control over the then USSR at the peak of Cold War[6]. The Soviets' space foray with Sputnik caused the US to invest in a communication technology that would work as a node-based network and continue to work even in the event of disruption at a certain or multiple node[7]. Internet's use in the US military was extended to educational networks across the US and later in Europe[8]. During 70s and 80s communication features such as e-mail and announcements via Bulletin Board Systems (BBS) available on the internet continued to attract the popularity and possibilities this

---

3   In his book 'Telecommunications deregulation and the information economy', Shaw provides useful insights not only about the history of deregulation but also attempts to interpret the impact of telecommunications sector coupled with the wave of globalisation that opened doors for market competition thereby profiting the socio-economic sector. See Shaw, J. K. (2001). Telecommunications deregulation and the information economy. Boston; London: Artech House

4   Holmes, P., Kempton, J. and McGowan, F. (1996). International competition policy and telecommunications. Lessons from the EU and prospects for the WTO. Telecommunications Policy, 20(10), 755-767

5   In just 25 years, the mobile phone has transformed the way we communicate (2010, January 01), The Guardian. Retrieved 04.05.2014 from https://www.theguardian.com/business/2010/jan/01/25-years-phones-transform-communication

6   Abbate, J. (1999). Inventing the Internet. Cambridge, Mass: MIT Press

7   Slotten, H. R. (2002). Satellite communications, globalization, and the Cold War. Technology and Culture, 43(2), 315-350

8   Ryan, J. (2010). A History of the Internet and the Digital Future. London: Reaktion Books

communication medium had to offer[9]. It was only in late 1980s when British scientist Tim Berners-Lee, while working at CERN realised the potential of internet as a two-way communication medium and suggested an idea of providing text documents and images over the internet using hypertext transfer protocol (http)[10]. Lee's scientific contribution, named as World Wide Web (WWW) went onto change the entire communication landscape and continues to redefine the way people work and live in 21st century[11].

Along with a few European countries, Silicon Valley in California, USA served as an epicentre of technological advancements in computer, internet and mobile phones[12]. Realising the potential of internet and WWW, academic institutions across US and Europe started linking, populating and using the web for research and educational purposes[13]. The scholastic prospects offered by the new communication technology also inspired venture capitalists[14]. Progress in semiconductors and computing technology threw rapid waves of devices leading to mass uptake and adoption of devices. In the late 90s, popularity of the internet/web along with devices, enabled new names such as eBay and Amazon to establish their digital footprints offering services that made use of the web and its reach. These were new contenders against the comparatively well-placed technology-based companies such as Microsoft, Apple, IBM or Xerox[15], that had their footing on the hardware side. Despite becoming global names, their business models were simple - eBay focused on providing an online auction platform where people could trade old and new items while Amazon, starting as an online bookstore evolved into a one stop shop for household items[16].

It is important to note that such online shopping stores relied especially on existing communication infrastructure[17]. That is, telephone and internet provided the technological infrastructure while digital payment gateways provided means

---

9   Morris, M. and Ogan, C. (1996). The Internet as mass medium. Journal of Communication, 46(1), 39-50

10   Berners-Lee, T., Cailliau, R., Groff, J.-F. and Pollermann, B. (1992). World-Wide Web: the information universe. Internet Research, 2(1), 52-58

11   Gillies, J. and Cailliau, R. (2000). How the Web was born : the story of the World Wide Web. Oxford: Oxford University Press

12   Archibugi, D. and Michie, J., (Eds.) (1997). Technology, globalisation and economic performance. Cambridge: Cambridge University Press

13   Baer, W. S. (1998). Will the Internet Transform Higher Education? https://www.rand.org/content/dam/rand/pubs/reprints/2005/RP685.pdf; For an informed and critical overview also see Kostopoulos, G. K. (1998). Global delivery of education via the Internet. Internet Research, 8(3), 257-265

14   Kenney, M. and Florida, R. (2000). Venture Capital in Silicon Valley: Fueling New Firm Formation. In Kenney (Ed.), Understanding Silicon Valley : the anatomy of an entrepreneurial region (pp. 98-123). Stanford, Calif.: Stanford Univ. Press

15   Tian, Y. and Stewart, C. (2006). History of E-Commerce. In Khosrow-Pour (Ed.), Encyclopedia of e-commerce, e-government, and mobile commerce (pp. 559-564). Hershey: IGI Global

16   Brandt, R. L. (2011). One click : Jeff Bezos and the rise of Amazon.com. London: Viking

17   Freund, C. and Weinhold, D. (2002). The Internet and international trade in services. American Economic Review, 92(2), 236-240

to close the financial transaction[18]. Today their popularity can be measured by the fact that these tech companies are now listed as publicly traded companies with their shares mostly hinting bullish trends. Shopping on the web or also referred to as online shopping went on to redefine business vocabulary. The terms such as online business, Business to Customer (B2C) and B2B (Business to Business), online service or eServices started to become lingua franca of internet bolstering the service sector and international trade[19]. The launch of such tech companies also redefined organisational structures where most of the day to day office chores were complimented and assisted by computerised or automated modules. Companies which were migrating towards offering online services went through restructuring or process reengineering in order to be relevant and strengthen their customer base[20].

The idea of customer-centred or customer-focused services started to receive more emphasis by the technology companies[21]. Based on the customers' shopping trends and preferences, tech companies devised special computer algorithms with an idea to facilitate their customers for instance by showing them advertisements based on their social, political and cultural preferences. The public sector in the western countries witnessing the uptake of internet, one-stop-shop websites, eServices and frequent use of payment gateways also felt the need to partake and reach out to their stakeholders via the internet[22]. The idea was to make citizen as a centre of attention and create/offer public services via electronic means and in this case via internet[23].

The idea of offering electronic public services however has not been entirely alien in Western countries and it was not developed overnight. Use of computers, spreadsheets, fax and/or electronic banking in the public sector had gradually grown along with the advancements in computer and communication technology[24]. Towards the end of 1990s and early 2000s, public sectors in these mostly developed countries drew inspiration from the popularity of eBusiness and started to develop citizen-focused eServices accessible via government web portals. The principle remained the same as in eBusiness, that is, to use the power

18   Boudreaux, G. and Sloboda, B. (1999). The fast changing world of the internet. Management Quarterly, 40(2), 2; Asokan, N., Janson, P. A., Steiner, M. and Waidner, M. (1997). The state of the art in electronic payment systems. Computer, 30(9), 28-35

19   Shea, T., Ariguzo, G. and White, D. S. (2007). Putting the world in the world wide web: the globalisation of the internet. International Journal of Business Information Systems, 2(1), 75-98

20   Wells, M. G. (2000). Business process re-engineering implementations using Internet technology. Business Process Management Journal, 6(2), 164-184

21   Wigand, R. T. (1997). Electronic commerce: Definition, theory, and context. The information society, 13(1), 1-16

22   McIvor, R., McHugh, M. and Cadden, C. (2002). Internet technologies: supporting transparency in the public sector. International Journal of Public Sector Management, 15(3), 170-187

23   Lai, R. and Haleem, A. (2002). E-governance: an emerging paradigm. Vision, 6(2), 99-109

24   Kraemer, K. L., Andersen, K. V. and Perry, J. L. (1994). Information technology and transitions in the public service: A comparison of Scandinavia and the United States. International Journal of Public Administration, 17(10), 1871-1905

of internet as a communication medium, standardise and make the offering of electronic public services via web efficient, faster, uniform and cheaper[25].

1.2. Of Development Goals and the Millennium Summit

Parallel to the technological growth, predominantly prevalent in the Western hemisphere, another notable development that took place at the dawn of new century was the Millennium Summit chaired by the then United Nations Undersecretary General Kofi Annan and the announcement of Millennium Development Goals (MDGs)[26]. UN's MDGs included a list of eight goals that ranged from alleviating poverty to controlling the spread of diseases such as HIV/AIDS and providing universal primary education – a list of ambitious tasks with measurable targets and a deadline set at 2015[27]. The declaration was signed by the leaders of 189 countries making it a historic millennium declaration. MDGs meant to reduce the disparity between developing and developed countries and aimed to decrease the social/economic divides[28]. The literature that has come out of aid-based or development institutions such as UN, World Bank or International Monetary Fund (IMF) tends to dichotomise the idea of development by distinguishing developing countries with developed countries based on various indicators such as Gross Domestic Product (GDP), Foreign Direct Investment, Gross National Income, Human Development Index (HDI), per capita or other parameters of measuring economic growth[29]. In the following chapters wherever the dichotomy of developing and developed countries is referred to in this research, will be based on a similar inspiration from the above-mentioned economic/human development indices.

However, it is pertinent to question why and how the MDGs are relevant to the ICT debate? Within the MDGs context, there is a reasonably strong relationship between ICTs and human development. MDG's goal number 8 called for an action to develop a global partnership for development among others, explicitly stating that benefits of new ICTs be available for all[30]. This meant that the ICTs were considered as an important instrument that could play a significant role in

25   Tat- Kei Ho, A. (2002). Reinventing local governments and the e- government initiative. Public administration review, 62(4), 434-444

26   Hulme, D. and Scott, J. (2010). The political economy of the MDGs: Retrospect and prospect for the world's biggest promise. New Political Economy, 15(2), 293-306

27   Vandemoortele, J. (2008). Making sense of the MDGs. Development, 51(2), 220-227

28   Vandemoortele, J. (2011). A fresh look at the MDGs. Journal of the Asia Pacific Economy, 16(4), 520-528

29   Avgerou, C., Smith, M. L. and Van den Besselaar, P., (Eds.) (2008). Social Dimensions of Information and Communication Technology Policy. Boston: Springer

30   United Nations Development Programme. (2008). The Role of Information Communication Technologies (ICTs) in achieving the Millennium Development Goals. In Mozambique: National Human Development Report. Retrieved 04.05.2014 from http://hdr.undp.org/sites/default/files/mozambique_nhdr_2008_ict.pdf

achieving the MDGs[31]. As a consequence, a terminology Information and Communication Technologies for development (ICT4D) picked up momentum among aid agencies, academia, development studies literature whereas several academic journals and degree programmes were also started after ICT4D[32]. International development and multilateral organisations such as World Bank, IMF, UNICEF, UNDP, GIZ, DFID, OECD, USAID or CIDA initiated their own chapters on ICT4D and made ICTs as part of their development agendas[33]. One of the important factors of ICT adoption for and by the developing countries was based on the increasing digital divide in the developed and the developing world that played into the income divide between these groups[34].

Meanwhile, the interest in ICT4D grew worldwide and two parallel initiatives were also formalised by the UN to bridge the global digital divide. Upon the request of UN Economic and Social Council (ECOSOC), Secretary General Kofi Annan initiated an ICT Task Force (UNICTTF) in November 2001, initially mandated for three years[35]. UNICTTF was a multi-stakeholder initiative which brought together representatives of governments, the private sector, civil society and international organisations with the sole purpose to exploit ICTs for promoting development[36]. The second parallel initiative to address the global digital divide took place in the form of two summits under the auspices of UN, namely World Summit on the Information Society (WSIS). The idea of WSIS was initially proposed by the government of Tunisia through International Telecommunication Union (ITU) in 1998 and ITU tabled this proposal to the UN.

---

31   Siriginidi, S. R. (2009). Achieving millennium development goals: Role of ICTS innovations in India. Telematics and Informatics, 26(2), 127-143

32   For a list of ICT4D academic journals, graduate and post-graduate courses, see for example: http://groupspaces.com/ipid/pages/ict4d-courses [last accessed 04.05.2015]; Also see Lee, H., Jang, S., Ko, K. and Heeks, R. (2008). Analysing South Korea's ICT for development aid programme. The Electronic Journal of Information Systems in Developing Countries, 35(1), 1-15; Gerster, R. and Zimmermann, S. (2003). Information and Communication Technologies (ICTs) for Poverty Reduction?: Discussion Paper. Retrieved 18.09.2008 from http://www.gersterconsulting.ch/docs/ICT_for_poverty_reduction.pdf

33   See for instance: UNCTAD's section on e-commerce and digital economy: https://unctad.org/en/Pages/DTL/STI_and_ICTs/ICT4D.aspx [last accessed 04.05.2015], World Bank funded programme infoDev's section on ICTs: http://www.infodev.org/articles/ict4d-learning-experience [last accessed 04.05.2015], UNICEF's section on Communication for Development: https://www.unicef.org/cbsc/index_66074.html [last accessed 04.05.2015]

34   Fuchs, C. and Horak, E. (2008). Africa and the digital divide. Telematics and Informatics, 25(2), 99-116

35   Lemos, R. and Varon Ferraz, J. (2013). Information and communication technologies for development. In Currie-Alder, Kanbur, Malone and Medhora (Eds.), International Development: Ideas, Experience, and Prospects (pp. 614-629). Oxford: Oxford University Press

36   Flyverbom, M. (2011). The power of networks: Organizing the global politics of the Internet. Cheltenham: Edward Elgar Publishing

One of the initial tasks UNICTTF was to oversee the WSIS first in Geneva in 2003 and then in Tunis in 2005[37].

These initiatives duly attended by the representatives of national governments, development aid agencies, civil society and academia laid groundwork for the awareness and importance of ICTs. However, the interest in the ICT adoption did not restrict to only address the global digital divide. A number of development initiatives in developing countries also started to focus on offering electronic public services – similar to what technological advanced countries had achieved so far[38]. Consequently, a number of eGovernment initiatives were launched. Some initiatives were indigenously motivated whereas some were promoted by international development organisations such as World Bank or UN in attempts, among others, to overhaul public service sector. While the spread of ICTs would help in lowering the digital gap, the eGovernment initiatives would enable faster, effective and efficient public service delivery via ICTs.

The idea to empower the underprivileged citizens of a developing country via ICTs seemed promising. However, there were a few problems – some digital in nature, a few rooted in governance and others embedded in societal structures. In most of the European countries and North America, Internet was used as a primary medium of communication where link to internet and a computer were required for the electronic public service delivery. On the other hand, digital infrastructure at the beginning of millennium in most of the developing countries was limited in its outreach, not to speak of internet[39]. There were limited landline telephones and since internet was provided via landline connections its outreach was also restricted. If the government information had to be communicated via website, that also meant an accessibility to internet and computers[40].

Most of the developing countries in Asia, Africa or Latin America also carried a post-colonial strain. Their public sector, their laws, bureaucratic structures and regulations were either originated during or derived from their respective colonial periods[41]. The idea of governance in several developing countries is restricted to

37   Kurbalija, J. (2008). The World Summit on Information Society and the Development of Internet Diplomacy. In Cooper, Hocking and Maley (Eds.), Global Governance and Diplomacy (pp. 180-207). London: Palgrave Macmillan

38   Alampay, E. (2006). Beyond access to ICTs: Measuring capabilities in the information society. International journal of education and development using ICT, 2(3), 4-22; Mofleh, S., Wanous, M. and Strachan, P. (2008a). Developing countries and ICT initiatives: Lessons learnt from Jordan's experience. The Electronic Journal of Information Systems in Developing Countries, 34(1), 1-17

39   Lu, M.-t. (2001). Digital Divide in Developing Countries. Journal of Global Information Technology Management, 4(3), 1-4; James, J. (2002). Universal access to information technology in developing countries. Regional Studies, 36(9), 1093-1097

40   Rice, M. F. (2003). Information and communication technologies and the global digital divide: Technology transfer, development, and least developing countries. Comparative Technology Transfer and Society, 1(1), 72-88

41   Englebert, P. (2000). Pre-colonial institutions, post-colonial states, and economic development in tropical Africa. Political Research Quarterly, 53(1), 7-36

centralised command and control instead of serving, interacting and facilitating the citizen[42]. Access to information in several post-colonial countries still remains a major problem. Complicated and complex bureaucratic structures make access to government services not only difficult but also in some cases impossible[43]. Similarly, bureaucratic language adopted in several developing countries was that of their colonial period which differed from the language spoken by common people[44]. Most of the government websites in developing countries during the initial years of new century were in European languages, thereby limiting the access to those who could read and write their adopted bureaucratic language.

Though initially fascinating, ICT adoption in developing countries appeared difficult, partly because the major populace in developing countries did not grow up with the technological advancements[45]. For some it meant a jump from using the limited radio and television outlets to use of and proficiency in handling internet and understanding the world wide web[46]. Also, the adoption of ICTs was not envisaged to be an infrastructural project. It did not mean that investment, whether self or foreign-funded, in state-of-art technology would be laid down in developing countries and from one day to the next eGovernment initiatives would start delivering eServices to the citizens. This also required law making within the ICT sector, understanding technological standards, recognizing the digital potential in the public – or the lack of it, and developing eGovernment strategies[47].

In addition to that, there were other burning issues such as limited funds, lack of domestically manufactured technology or robust infrastructure, IT literacy, digital payment gateways and absence of local content on the government websites to deal with[48]. The demarcation and dichotomy in citizens with regard to access to 'haves' versus 'have nots', urban versus rural, developing versus developed were

---

42   Hönke, J. (2013). Transnational companies and security governance: Hybrid practices in a postcolonial world. Oxon: Routledge

43   Chakravartty provides fresh insight, for instance in case of India where she takes government's telecommunication department as an example to elaborate how the bureaucratic structure developed during British colonial period even after the independence of India continuously relied on pre-independence era and in fact its dominance grew over the years, see Chakravartty, P. (2004). Telecom, national development and the Indian state: a postcolonial critique. Media, Culture & Society, 26(2), 227-249

44   Alhassan, A. and Chakravartty, P. (2011). Postcolonial Media Policy Under the Long Shadow of Empire. In Mansell and Raboy (Eds.), The handbook of global media and communication policy (pp. 366-382). West Sussex: Blackwell Publishing

45   Sahay, S. and Avgerou, C. (2002). Introducing the special issue on information and communication technologies in developing countries. The information society, 18(2), 73-76

46   Alzouma, G. (2005). Myths of digital technology in Africa: Leapfrogging development? Global Media and Communication, 1(3), 339-356

47   Gil-García, J. R. and Pardo, T. A. (2005). E-government success factors: Mapping practical tools to theoretical foundations. Government Information Quarterly, 22(2), 187-216; Janssen, D., Rotthier, S. and Snijkers, K. (2004). If you measure it they will score: An assessment of international eGovernment benchmarking. Information Polity, 9(3, 4), 121-130

48   Heeks, R. (2002c). e-Government in Africa: Promise and practice. Information Polity, 7(2,3), 97-114

also carried over in the eGovernment roll-out[49]. Just like the ICT4D, eGovernment in developing countries became a catch-phrase in development agendas. There are scientific conferences that are only dedicated to the topic of eGovernment in developing countries as are there academic journals which only focus on literature, theories, and case studies on ICTs from developing countries.

1.3. Contextualising ICTs in developing countries

The study of ICTs in government and for public service delivery has attained a considerable focus by academia, governments, international development organisations as well as industry. The academic journals that are dedicated to multi/inter-disciplinary studies and engage academics from varied backgrounds ranging from communication studies, information sciences, political science, management and organisational studies, and sociology[50]. The advanced use of ICTs by the government is very diverse. Such uses can be found either at the policy and institutional level, furthermore it can also be seen in more practical ways such as in the delivery of services, in political and democratic processes and administration within government itself[51]. The use of ICTs for public service delivery is also interchangeably used with eGovernment. Most of the conferences on ICTs and on the theory and use of eGovernment are generally focused on the innovation in technologies and tools used in the development and deployment of eServices. While innovation and use of modern technologies in public service delivery is among one of the tasks of eGovernment services, the focus on cultural and societal aspects, technology and power discourse, governance and citizenship seems to get overshadowed in the presence of eService rankings and indicators[52].

The contribution of eGovernment is not only focused on the innovative eServices, instead it also influences the reconfiguration and reconceptualisation of policy and government's social contract with the citizen[53]. Citizen-focused or citizen-centric services in this regard have gained attention of major eGovernment services in which the services are designed according to the ease and use of the

---

49   Gascó, M. (2005). Exploring the e-government gap in South America. International Journal of Public Administration, 28(7-8), 683-701; Selwyn, N. (2002). 'E-stablishing'an inclusive society? Technology, social exclusion and UK government policy making. Journal of social Policy, 31(1), 1-20

50   Heeks, R. and Bailur, S. (2007). Analyzing e-government research: Perspectives, philosophies, theories, methods, and practice. Government Information Quarterly, 24(2), 243-265

51   West, D. M. (2005). Digital government: Technology and public sector performance. Princeton: Princeton University Press

52   Henman, P. (2010). Governing Electronically: E-Government and the Reconfiguration of Public Administration, Policy and Power. Hampshire: Palgrave Macmillan; Bhatnagar, S. (2004). E-government : from vision to implementation: a practical guide with case studies. New Delhi: Sage publications; United Nations. (2005). Designing e-government for the poor. Retrieved 04.05.2014 from https://www.adb.org/sites/default/files/publication/159381/adbi-e-gov-poor.pdf

53   Nygren, K. G. (2009). E-governmentality: On electronic administration in local government. The Electronic Journal of e-Government, 7(1), 55-64

citizens[54]. The idea of citizen-centric approaches draws inspiration from eCommerce discourse as depicted in the section above. The same also applies to one-stop shop solutions – the provision of government's services via web portals have also placed the focus on the citizen through which the service hours of the government services have been expanded to almost round the clock. With the increased sophistication levels in the web based/browser technologies, it has also become possible to customise the government information according to individual needs[55]. Such format and vision of eServices are enabling governments to administer citizen demands across the whole enterprise[56].

While this may appear to be a (technologically) utopian style of governance, there are issues that have emerged, if not directly in the field of eGovernment but especially in surveillance studies, in power/knowledge discourse and/or network society literature[57]. In the network society literature, the recognition of technology has been enamoured by Castells' perspective that the network society has emerged as a result of developments in ICTs[58]. New networked ICTs have now made it technically possible to create inter-organisational structures that involve a higher degree of communication and joint-working[59]. Part of the proliferation of eGovernment services as mentioned earlier can be credited to the commercialisation of the internet and lowering costs of personal computers. While countries and societies that evolved with the technology in terms of digital literacy and technological growth seem to have taken eGovernment service to the next level, eGovernment in the case of developing countries presents quite a number of challenges[60].

Madon believes that the concept of eGovernment originated in the countries that are categorised as industrial countries[61]. The route to technological evolution that the industrialised countries and their respective administrations have followed is largely missed in the developing countries. There is a need to address these

---

54   Reddick, C. G., (Ed.) (2010). Citizens and e-government: Evaluating policy and management: Evaluating policy and Management. Hershey: IGI Global

55   Lenk, K. and Traunmüller, R. (2002). Electronic Government: Where Are We Heading? In Lenk and Traunmüller (Eds.), Electronic Government: First international conference, EGOV 2002, Aix-en-Provence, France, September 2-6, 2002: Proceedings (pp. 1-9). Berlin: Springer

56   Hagen, M. and Kubicek, H. (2000). One Stop Government in Europe: An Overview. In Hagen (Ed.), One-stop-government in Europe: Results from 11 National Surveys (pp. 1-36). Bremen: University of Bremen

57   Warren, M. (2007). The digital vicious cycle: Links between social disadvantage and digital exclusion in rural areas. Telecommunications Policy, 31(6), 374-388; Dawes, S. S. (2008). The evolution and continuing challenges of e- governance. Public administration review, 68(S1), 86-102

58   Castells, M. (2009). Communication power. Oxford; New York: Oxford University Press

59   Castells, M. (2010a). The power of identity. Oxford: Wiley-Blackwell

60   Warschauer, M. (2003b). Technology and social inclusion rethinking the digital divide. Cambridge: MIT Press; Servon, L. J. (2008). Bridging the digital divide: Technology, community and public policy. Oxford: Blackwell Publishing

61   Madon, S. (2009). E-governance for Development: a focus on rural India. Basingstoke: Palgrave Macmillan

questions for developing countries before considering eGovernment's functionality in different administrative environments. A wide variety of literature on eGovernment in developing countries has mostly referred to the success/failure dichotomy and the discussion on the much-needed grey areas seems to have been ignored[62]. The focus on ICTs has increased at the cost of ignorance about social and organisation factors and institutional and informal aspects that directly influence the way technology is accepted and used[63]. Nevertheless, such an approach, as Madon and Parkinson point out is slowly changing towards the socially inclusive technology[64]. Heeks has also maintained that in the eGovernment debate, socio-technical aspects of an organisation shall be taken into account in order to study eGovernment applications[65]. It is also pertinent to mention that Andrew Feenberg in his book, *Alternative Modernity*, has debated at length on his post-modernist view of technology by adopting a social constructivist approach[66]. Feenberg may not have taken eGovernment application as his frame of reference. However, his point of view about technology is more of a critical theorist trying to employ a constructivist lens to the possibility of reshaping the technical world[67]. Turning back to our original premise of contextualising ICTs and eGovernment in developing countries it would therefore be wrong to assume that development of eGovernment is directly proportional only to the strength of ICT infrastructure that supports and enables the execution of eGovernment.

It might not be difficult for people in developed countries to interact with their government round the clock through the click of a mouse but to accept the same level of efficiency and flexibility for developing countries is going to be more difficult[68]. The argument that many eGovernment projects fail in developing countries has to do with the technological infrastructure and lack of technical know-how, may appear to be a rather simplistic view of the problem. There are of course other and more important factors that have to do with the historical, social and traditional structures on which the public administration in these countries have followed a developmental path[69]. Their dependencies are nurtured,

---

62   Steyn, J., Belle, J.-P. v. and Mansilla, E. V., (Eds.) (2010). ICTs for Global Development and Sustainability: Practice and Applications: Practice and Applications. Hershey: IGI Global

63   Fountain, J. E. (2002). Building a deeper understanding of e-government. In The Future of e-Governance Workshop (September 10, 2002) at the Campbell Public Affairs Institute in Maxwell School of Syracuse University. Retrieved 04.05.2014 from https://www.maxwell.syr.edu/uploadedFiles/campbell/events/fountain.pdf

64   Madon, S.: 2009

65   Heeks, R. (2006a). Implementing and managing eGovernment: An International Text. London: SAGE

66   Feenberg, A. (1995). Alternative modernity : the technical turn in philosophy and social theory. Berkeley: University of California Press

67   ibid.

68   Avgerou, C., et al.: 2008

69   Wescott, C. G. (2001). E-Government in the Asia Pacific region. Asian Journal of Political Science, 9(2), 1-24

for instance, in more face-to-face interaction than the click of a mouse[70]. The approach to develop unified model of eGovernments for both the developed and developing world that is one-size-fits-all eGovernment solutions may continue to fail in developing countries[71]. Schuppan for instance takes the same line of argument by discussing the case of eGovernment for Sub-Saharan Africa by maintaining that the values, attitudes, and rationalities of the population with respect to the political-administrative system play an important role when implementing eGovernment[72].

In supporting the case of developing countries, Bhatnagar calls for conceptualising the eGovernment approach that cater to both urban and rural citizen base in developing countries[73]. In the presence of digital divide, lack of digital literacy, computer and internet affinity, the service provision via websites over the internet in developing countries does not seem to be a reasonable solution[74]. At the same time digitalisation of electronic public services continuously takes place in developing countries albeit without web-based services. There appears to be re-utilisation and re-discovery of existing communication patterns and partial digitisation in several developing countries. In this regard, the rural areas across the world have thus seen a number of projects that offer a mixed approach of manual and automated systems[75]. As a result, a number of projects across the world have dropped the option of eServices via websites and instead started to focus more on alternative approaches. In Rede Govereno, Brazil for example, there are ATM-style kiosk machines that allow citizens to access the government online portal and other services. Similarly, a number of rural kiosks or telecentres have been established in rural parts of India[76]. These telecentres or village kiosks are an attempt toward an inclusive approach to provide mediated access to information and services by employing ICTs especially computers and Internet.

Pakistan's case seems to be following a similar path in the uptake of ICTs for the provision of electronic public services. There are a number of government

---

70   Drüke, H., (Ed.) (2004). Local Electronic Government: A comparative study. London: Routledge

71   Islam, M. S. and Business, S. (2008). Towards a sustainable e-Participation implementation model. European Journal of ePractice, 5(10), 1-12; Dawes, S. S., Cresswell, A. M. and Pardo, T. A. (2009). From "need to know" to "need to share": Tangled problems, information boundaries, and the building of public sector knowledge networks. Public administration review, 69(3), 392-402; Sagheb-Tehrani, M. (2010). A model of successful factors towards e-government implementation. Electronic Government, An International Journal, 7(1), 60-74

72   Schuppan, T. (2009). E-Government in developing countries: Experiences from sub-Saharan Africa. Government Information Quarterly, 26(1), 118-127

73   Bhatnagar, S. C. (2002). E-government: lessons from implementation in developing countries. Regional Development Dialogue, 23(2), 164-175

74   Vassilakis, C., Lepouras, G. and Halatsis, C. (2007). A knowledge-based approach for developing multi-channel e-government services. Electronic Commerce Research and Applications, 6(1), 113-124

75   Rocheleau, B. (2007). Whither E- Government? Public administration review, 67(3), 584-588

76   Mukerji, M. (2008). Telecentres in rural India: emergence and a typology. The Electronic Journal of Information Systems in Developing Countries, 35(1), 1-13

agencies in Pakistan that offer IT-based public services but it is not clear who sets the agenda and who takes the lead. Pakistan had set up an Electronic Government Directorate (EGD) to advise government on eGovernment projects and more importantly policy making and setting up eGovernment strategy. From 2002 and 2012 EGD formulated two eGovernment policy plans and announced several eGovernment projects, however, after 10 years of trying to make sense of eGovernment phenomenon, EGD was dissolved. Nevertheless, eGovernment services have continued to grow in Pakistan but at the same time Pakistan has continued to decline in eGovernment rankings between 2002 to 2012. In this regard, Pakistan presents a highly interesting case for academic enquiry when it comes to the analysis of ICT-based public services. It not only requires a deeper understanding of Pakistan's IT initiatives but also the communication patterns, bureaucratic structures and governance system in Pakistan whose influence can be seen on eGovernment initiatives as well. Similarly, if there are instances of electronic public service delivery in Pakistan, then they could help identify the model of eGovernment that emerges in situations where alternative means of digital communication are used in absence of or as alternative to web-based public services.

1.4. Study Hypotheses

Hereinafter, this study proposes three distinctive hypotheses for the study of ICTs in Pakistan. Taking Pakistan as a case study, this study recognises the rapid emergence of new ICTs and assumes that the ICTs not only foster the public services but they also give rise to new forms of service delivery channels.

H1. The modern information and communication technologies foster the emergence of a new form of public service delivery.

Similarly, this study also presupposes that while Internet has traversed almost all paths and aspects of human life, but considering that there are huge knowledge and digital gaps, the Internet alone as a tool for public service delivery will have limited appeal particularly in developing countries.

H2. The Internet as a tool for electronic public service delivery has only limited effect.

Lastly, this study strongly emphasises the societal norms, communication patterns and social capital that develop over a number of years through mutual exchange of indigenous knowledge and local capacities. Taking these into account, this study assumes that the amalgam of indigenous knowledge and capacities influences the new ICTs and help reshape the technologies according to local cultures.

H3. Social conditions and traditional communication patterns impact the uptake and reshaping of new communication technologies

1.5. Research Questions

The central frame of enquiry in this research revolves around the understanding and analysis of the state of eGovernment in Pakistan. In the presence of new ICTs in Pakistan, this research raises the question as to how the modern ICTs affect the manual public service delivery mechanisms in Pakistan. If the effect(s) are of quantifiable nature, then what developments in ICTs have taken place in Pakistan to facilitate public service delivery? What are the digital patterns available for the provision of service delivery? To be precise, is the web/Internet a primary medium of communication and service delivery or are there other channels for public service delivery? Similarly, how do the existing patterns of government-to-citizen communication in Pakistan correspond to the new service delivery channels? On a more secondary level, this study also raises the question as to how does the government structure built during the British colonial period correspond and conform to the modern and digital means of communication.

1.6. Scope of the study

The time period covered by the present study encompasses the 12 years of the use of ICTs in government-to-citizen services, particularly looking at the developments from 2000 through 2012. The period of 2000-2012 is significant not only for massive investments in the ICT sector in Pakistan but also for this study since most of the developments and the discourses that led to them in electronic public services started in this period. In addition, this time period is also synonymous with the formation of formal governmental institutions such as the Electronic Government Directorate and the National Database Registration Authority that worked extensively in the digitisation drive. Moreover, this is the period when Pakistan went through a major liberalisation policy not only in the mass media but also the telecommunications infrastructure sector.

In structural terms, the scope of the study is primarily limited to the electronic services initiated at the federal level for citizens. This limitation is not particularly constrictive as provincial or local-level eServices initiatives in Pakistan made use of the same data banks centralised at the federal level. Thus, in order to develop an understanding of the state of eServices vis-à-vis eGovernment scenario in Pakistan, the focus is redoubled at the federal level initiatives. Similarly, within the ICT or eGovernment studies, the focus of services can be according to the recipient of these services that are tailored towards business enterprises; for intergovernmental information; and for citizens. The scope of this study is thus particularly centred on the services that are focused towards the provision of electronic services for the citizens.

1.7. Significance of the study

Information and Communication Technologies are rapidly traversing almost all paths of life in Pakistan. By end of 2012 there were reportedly over 80 million mobile subscribers in Pakistan[77]. The government, private entities such as banks and mobile telecom operators are making heavy use of modern technologies to not only stay in constant connection with their customers, but there is also a rise in service delivery and resultant competition among the said entities. ICTs, on one hand, are presenting a good alternative for delivery mechanism while simultaneously the constant digital divide among the digital migrants, digital natives and the digital unknowns -the one without access or opportunity- continues unabated as this study will try to reflect. From 2000 through 2012, there has not only been a rise in terms of public services through electronic means; this period has also witnessed varying trends in digital communication patterns which were either unknown or contrasted with the existing communication patterns in Pakistan.

This research is an attempt to document the changing patterns of digital and traditional communication in Pakistan by assembling fresh empirical evidence. This was achieved by collecting quantitative data from primary digital resources such as NADRA, which has previously been either unknown or less available publicly. One key significance of this study is that it highlights the role of governance in Pakistan's post-colonial state while documenting the history of communication technologies, their uptake and impact when meted against cultural, societal and economic factors in order to arrive at a justification of the current state of digital communication for electronic public service delivery. It attempts to provide concrete evidence for the pattern of citizen use of government data either in the form of digital information services or public service delivery. This study has attempted to reason the resistance or acceptance to digital technologies such as automated kiosk machines considering their compatibility to modern and traditional communication channels. Similarly, this research has intended to study the new means of communication employed by the government to connect with its citizens by underlining the issues of digital literacy, legacy regulations from the period of British colonialism, as well as the traditional forms of communication which are more dominant in Pakistan's predominantly rural and agrarian set-up. Based on the qualitative and quantitative data, this study tries to conceptualise the eGovernment model that has resultantly emerged due to the mix of manual and digital strategies currently being employed in Pakistan.

---

77   International Telecommunication Union. (2013b). ITU ICT Statistics. Retrieved 04.05.2014 from http://www.itu.int/en/ITU-D/Statistics/Pages/stat/default.aspx

1.8. Methodology

The main objective of this research endeavour, as elaborated above, is to investigate the role of modern ICTs being employed in the use of public service delivery. In order to approach this question, it is also necessary to understand the state of ICTs in Pakistan. Similarly, to find out the patterns of digital communication technologies, it is not only needed to conduct an exploratory enquiry but also bolster the results with the help of hard data to establish digital usage patterns. The role of the web/internet in this regard presents fewer possibilities considering the sparse number of internet users with dedicated wired means in Pakistan. There has been phenomenal growth in developing alternative service delivery models to overcome the access issue. While the terrain of eGovernment is quite vast, this study only focuses on the Government to Citizen (G2C) scenario in Pakistan.

The question here is of the kind of technologies currently in use in Pakistan for eServices? Could the Internet be considered as one of the potential modes of communication between the Government and citizens? If so, can the usage pattern of eService delivery be measured? In order to approach these questions, the role of quantitative data, for instance, to gauge the performance indicators or citizen attitude, seems a plausible query to begin with. Similarly, it becomes essential, to supplement the quantitative enquiry with qualitative inputs by reviewing governmental records, policy documents, government websites and expert interviews in order to help interpret or find the impact factor of the eServices. Considering the importance of both the quantitative and qualitative data, this study makes use of a mixed approach to establish usage patterns as well as impact of eServices on Pakistani society.

Use of both quantitative and qualitative strategies has been in use since the 1950s[78]. The blending of both methods in a single investigation has been given several names such as the integrated or combined approach. Sale has termed it as methodological triangulation in which he considers it as the convergence of quantitative and qualitative data[79]. In more recent studies such as the ones by Creswell and Tashakkori et.al, the use of both the approaches has been recognised as mixed method research or mixed methodology[80]. In both of these the central premise is set around the fact that the use of quantitative and qualitative

---

78   Sale, J. E., Lohfeld, L. H. and Brazil, K. (2002). Revisiting the quantitative-qualitative debate: Implications for mixed-methods research. Quality and quantity, 36(1), 43-53; Creswell, J. W. and Plano Clark, V. L. (2007). Designing and conducting mixed methods research. Thousand Oaks: SAGE Publications

79   Sale, J. E., et al.: 2002

80   Creswell, J. W. (2003). Research design: qualitative, quantitative, and mixed method approaches. Thousand Oaks: Sage Publications

approaches in combination can provide a better understanding of a research problem[81].

1.8.1. Fieldwork in Pakistan

In order to begin with the empirical research, the first phase in Pakistan started in 2008. Two visits in April and August were made to identify potential data resources, key persons as well as relevant ministries and/or departments which could facilitate a study of G2C services. In this phase, initial contact was made with the Electronic Government Directorate (EGD) in Islamabad. EGD, which is the official body responsible for the electronic public services, and operates under the Ministry of Information Technology, Government of Pakistan. A number of government officials at the EGD facilitated the information necessary for this research. The primary concern at the EGD remained to find out if there were any projects being planned or implemented in which the Government of Pakistan provides services to the citizens through communication technologies. The emphasis also remained in seeking answers from the officials[82] as to what kinds of strategies were adopted to curb the digital divide. Among other things, the initial results from the empirical evidence collected from the EGD revealed, among other things, that apart from government's official web portal prepared by the EGD, (discussed and analysed in Chapter 4), there were electronic services projects initiated solely for the automation of government departments. Moreover, EGD's data also did not help in figuring out whether alternatives to internet as medium of digital communication were considered.

The scope of this study required quantifiable data that could guide to the type of electronic public services available but also to help understand their communication patterns, socio-demographic usage and frequency of use. To answer these objectives with nuance, a second phase of field work took place from February through April 2009. In this phase, a contact with the Ministry of Interior's technological section, National Database Registration Authority (NADRA), also headquartered in Islamabad, was initiated. The initial response from NADRA was promising in comparison to EGD. Here was a possibility to closely review two G2C projects namely, Kiosk Machine and eSahulat, and also get the data produced by these projects.

Besides reviewing two public-sector organisations, I was also interested if there were G2C projects initiated at the private sector level. In this regard, contact was established with the Value-Added Services (VAS) department of Pakistan's biggest mobile network operators, Mobilink, to review electronic public services initiated by Mobilink. A study was subsequently completed in 2011 of this mobile

---

81   Bergman, M. M., (Ed.) (2008). Advances in mixed methods research: Theories and applications. London: Sage

82   The list of experts, which were contacted and/or interviewed, is provided in the Appendix I.

platform from a private organisation provided valuable insights into communication, uptake patterns and enrich the qualitative analyses with the G2C services.

In addition to a public and private sector initiatives, this research also covers an additional study on a collaborative effort of two public-sector organisations. The deployment, design and functioning of the Election Commission of Pakistan's SMS-based application for voter registration which was developed in collaboration with NADRA provides further insights to strengthen the mixed method approach. This research therefore, covers five case studies in total as depicted in table below.

Table 1: List of Case Studies analysed

Nr.	Case Study	Organisation	Sector	Focus
1	Web Portal Analysis	Electronic Government Directorate	Public Sector	G2C
2	Kiosk	National Database Registration Authority	Public Sector	G2C
3	eSahulat	National Database Registration Authority	Public Sector	G2C
4	Mobile Platform	Mobilink	Private Sector	G2C
5	Voter Verification SMS App	Election Commission of Pakistan	Public Sector	G2C

1.8.2. Quantitative Data

Both NADRA's kiosk machine and eSahulat projects are among the first G2C electronic public services in Pakistan. The kiosk machine, explained in detail in Chapter 5, would be considered as Pakistan's first ATM-styled touchscreen enabled platform deployed in 2005 to provide various services such as citizen verification or payment of utility bills. Similarly, the eSahulat program, an extension of the kiosk machine, targeting a wider citizen base was deployed in 2007. While the number of internet users, mobile phone subscribers and/or fixed line phone subscribers can be obtained through multiple sources such as Internet Service Providers (ISPs), the Pakistan Telecommunication Authority (PTA), and the likes of International Telecommunication Union (ITU), the details of the Kiosk machine and eSahulat are not publicly available. Neither did NADRA release any whitepapers or performance indicators on the basis of which assumptions, trends, behaviours or citizen attitudes could be made. The same also applies to Mobilink, which does not make the statistics public for their mobile commerce platform. The enumerative data from NADRA and Mobilink on citizens' usage patterns could help understand a bigger picture how, where and when are the kiosk, eSahulat and mobile commerce were used.

1.8.3. Data collection procedure

For data collection from NADRA, a formal request was placed in 2008 with the Chief Technology Officer (CTO) at the NADRA's headquarters. The then CTO, Mr. Osman Mobin, understood the need and use of this research and agreed to establish a resource to facilitate this study, which helped in querying the database according to the research needs. Similarly, for data collection from Mobilink, a request was made to the VAS department where Mr. Ahsan Mansoor facilitated the request as well as collection of data.

1.8.4. Nature of data

In order to develop the citizen usage pattern, the data for a number of variables was requested from the NADRA from 2005 through 2009. These variables included the number of Kiosk machines, number of services offered, city of use, frequency of use, purpose of use, gender and the language of technological medium. The idea behind seeking numbers for such variables was not only to compute the frequency, but by including the city and gender, but the idea was to also add socio-demographic factors to the entire result. A number of sessions took place from February 2009 through April 2009 in acquiring this data. Special clearance was sought from NADRA to enter its premises. Similarly, extra meetings were required with the database administrator. Based on the requested questions, he specially had to collaborate in running queries on NADRA's Kiosk and eSahulat databases. The data (see data sample in Appendix III) came from four different NADRA databases. These databases belonged to different kiosk and eSahulat programmes tagged with the city and gender. City data in every spreadsheet was given in the short form which was manually configured in order to be able to generate maps to view the frequency of use of kiosk data. The data received from NADRA was based on anonymity and only for the purpose of exploring the digital trends by reviewing the frequency of use, amount generated at each node and the gender of kiosk user. The data did have some erroneous entries which for the purpose of drawing statistical inferences were first cleaned up. How closely and minutely NADRA itself monitors and scrutinises the kiosk and eSahulat data itself remains beyond the scope of this study. However, within the context of this study, kiosk's data has helped understand the digital trends. More importantly, the data has also been helpful to realise that there is a need to tap into more alternative channels of electronic public service delivery.

1.8.5. Data privacy and ethical consideration

NADRA houses Pakistan's most sensitive database of its citizens. It was, however, made clear in the formal request that the data acquired from NADRA will solely be used for research purposes. In addition, any citizen's or business entity's personal information such as name, address or age were not asked to comply with privacy norms. Gender was the only variable that could help in providing the estimates in usage patterns which NADRA had agreed to share.

1.8.6. Qualitative Interviews

While the quantitative data from NADRA fed into the needs of usage patterns, the qualitative study for this research was primarily focused in understanding the state of eGovernment particularly through the reference of Electronic Government Directorate in Pakistan. The EGD's primary focus has, since its inception, remained the development of ICT infrastructure, policy guidelines as well as to identify possible avenues for eServices in Pakistan. In this regard 15 semi-structured qualitative interviews were held with experts from EGD, NADRA, Pakistan Telecommunication Authority, banking sector, civil society partners, Internet consultants and mobile telecommunication professionals. The questions raised in these interviews were focused on seeking an understanding of Pakistan's eGovernment strategy, its approach, language barriers, media of service delivery and populating of content on these media. In addition to the expert interviews, additional input was also collected while participating in internet or eGovernment conferences with international field experts and academia such as those in GIZ (2010), Internet Governance Forum (2008, 2009) annual congregations and during a visiting fellowship in Manchester University (2010) where I was associated with the Manchester Business School and Centre for Development Informatics. Most of the interview sources were interviewed once. However, there were sources in NADRA as well as in PTA who were contacted for follow-up questions, particularly to seek their opinion on the initial results of quantitative data.

1.8.7. Government Website Analysis

Pakistan's official web portal, www.pakistan.gov.pk falls under the responsibility of the EGD. In order to measure its value for G2C services, its content analysis was done through the modifications in West as well as Rorissa's method of computing eGovernment website index[83]. This was achieved by considering a

---

83   West, D. M. (2007). Global perspectives on e-government. In Mayer-Schönberger and Lazer (Eds.), Governance and information technology: From electronic government to information government (pp. 17-32). Cambridge: The MIT Press; West, D. M. (2004). E- government and

number of factors such as availability of publications, databases, audio/video clips, local/foreign language access, disability access, having privacy policies, security policies or digital signatures on transactions. This method allows to find out the practicality of any website with a focus on G2C eServices. To compute the eGovernment index it involves an objective set of procedures and contents of relevant eGovernment service websites. It involves a two-stage process.

i. Count 18 features (4 points each) on each website (up to maximum of 72 points)
ii. Additional online executable services (up to 28 points)

eGovernment index for site i,

$$e_i = 4f_i + x_i$$

where,

f_i=the number of features present on website i, $0 \leq f_i \leq 18$

x_i=the number of online executable services on website i, $0 \leq x_i \leq 28$

Based on West's method, Pakistan's official federal web portal has been computed and analysed in comparison with its provincial websites. The results of this analysis are presented in Chapter 4.

1.9. Structure of thesis

The research conducted for this study primarily focuses on the key initiatives and major policy directives that have led the development of electronic services for citizens in Pakistan. While the singular interest lies in the period between 2000 and 2012, an attempt has been made to provide a chronological account of ICT deployment for public service framed with the backdrop of the social, economic and political aspects framing the topic. This study is divided into eight chapters.

Chapter 1 introduces the subject and sets up the agenda undertaken in this research. It provides the current state of research and study in the field of ICTs and eGovernment whereby situating the study focus in Pakistan with a developing country context. By introducing a host of primary and related secondary research questions, this chapter also introduces the central theme of the study as to how far has the Pakistani Government employed modern communication technologies for the provision of eServices. Chapter 1 also provides detail about the methodology employed to examine the central and secondary questions on the status of eServices in Pakistan. This chapter elaborates the way quantitative data was collected from the Electronic Government Directorate, Mobilink, and National Database Registration Authority in Pakistan. In particular the sociodemographic

---

the transformation of service delivery and citizen attitudes. Public administration review, 64(1), 15-27

variables that helped guide the understanding of the use and impact of technology in public service delivery are focussed upon.

Chapter 2 takes stock of theoretical background for the study of ICTs. By focusing primarily on communication studies as a possible theoretical lens, this chapter touches upon the subfields of development communication and ICT4D that have influenced current studies and research on eGovernment. It discusses different approaches to understand eGovernment from both a developed and developing country scenario. In doing so, this chapter evaluates and analyses eGovernment in developing countries parsing the topics of digital divide, eReadiness and public-sector reform. This chapter attempts to establish the fact that technological exports can have an impact on culture and society. In addition, it also probes if and how technology reshapes itself according to the social and cultural aspiration, knowledge and capital of a given society.

Chapter 3 outlines the historical patterns of the development and use of communication technologies and traces their origin back to British India with the invention of telegraph and telephone. By briefly highlighting the major technological developments such as television and radio, this chapter also introduces the arrival and reception of the internet in the light of pre-existing communication patterns and means. This chapter also examines the governance structure in Pakistan and explains how Pakistan's colonial legacy is entrenched in its governance and administrative practices. These organisational structures and hierarchies also help understand the formation of new institutions when the implementation of eGovernment was rolled out.

Chapter 4 provides an overview of current trends of ICTs and eServices in Pakistan. It begins by tracing the major policy and practical initiatives taken in the field of ICT and eGovernment. It presents Pakistan's first IT Policy and Action Plan of 2000 and argues how this policy became a starting point for the formation of new institutions such as EGD and ECAC. This chapter then reviews the salient features of Pakistan's first eGovernment strategy initiated in 2005 which helps in understanding EGD's perceived scope and the model of eGovernment that Pakistan envisaged. In addition, this chapter offers an analysis of the Government of Pakistan's web portal and provides a purview of select projects that EGD has undertaken.

Chapter 5 takes a look at the development of NADRA that has been solely developed to undertake eGovernment implementation in Pakistan. This chapter presents the history and scope of NADRA and explains how the computerisation of national identity cards developed by NADRA became the crucial basis of developing G2C eServices in Pakistan. In this context, this chapter reviews four different variations of NADRA's eCommerce platform developed in 2005 in the form of Kiosk machine. It tries to take the analysis of eService delivery further by examining NADRA's kiosk machine and e-Sahulat initiatives in the light of data collected as a part of empirical research by analysing the aforementioned sociodemographic variables and their frequency through kiosk and eSahulat. This chapter aims at not only providing a comparison between two eServices, but also helps understand the changing patterns of development and growth of electronic public service delivery in Pakistan.

Chapter 6 takes a look at two additional eGovernment case studies - projects developed by the Election Commission of Pakistan and a private mobile network operator Mobilink. In doing so this chapter sheds light on how ICTs enable interagency collaboration and digital partnerships between public-sector and private organisations. A qualitative review of the structure and usage of the SMS App developed by the ECP-NADRA for voter verification is offered. This chapter then moves on to the topic of branchless banking in Pakistan that has enabled the uptake of mobile banking in Pakistan on the back of collaborations between private institutions such as mobile network operators and banks with NADRA.

Chapter 7 discusses observations gleaned from all five case studies presented and reviewed earlier. It tries to assess the dominance and role of the then new institutions such as EGD and NADRA that were established for the growth of eGovernment in Pakistan. Taking mobile phone as an emerging communication medium and by analysing the growth of eSahulat in comparison to kiosk machine, this chapter discusses the value of cultural and societal norms and communication patterns. In the light of the five case studies, this chapter tries to evaluate the emerging communication patterns in Pakistan and equates their relevance to technology adoption in comparison to the technological diffusion or spread of ICT infrastructure. This chapter traces the possibility of an eGovernment model emerging and prevailing in Pakistan. Towards the end, this chapter also offers few suggestions on how and which communication patterns can be helpful for the growth of eGovernment implementation in Pakistan.

Chapter 8 concludes the study by presenting a rejoinder of the findings of this research endeavour. This chapter discusses the role of internet and its future in the light of changing patterns of digital communication in Pakistan. It argues that the technological investments made in the first 12 years of the millennium along with the formation of new institutions such as EGD and particularly NADRA has changed the service delivery paradigm thereby giving rise to innovation in and growth of multiple channels. This chapter concludes by pointing the way forward for eGovernment with the advent of ubiquitous computing, big data and artificial intelligence and possible pitfalls the new wave of digitisation may succumb to in Pakistan.

Chapter 2

Theoretical Background

In order to approach the central research question of how to find the use and extent of Information and Communication Technologies for the provision of electronic public service delivery in Pakistan, this study relies on the theories of communication studies. Currently, the literature dealing with studies on electronic public service delivery is either dealt within the studies on eGovernment or in the Information and Communication Technologies for Development (ICT4D) literature[84]. Similarly, for the past ten years, parallel streams have also started on eGovernment in developing countries whose body of knowledge is primarily geared with experiences of digital divide, ICT4D limitations, and to a certain extent the public-sector reform through technologies[85]. Since ICTs have traversed a number of disciplines, an increasing number of studies are continuously illuminating the studies of ICTs, such as Castells on networked society[86], Madon's on eGovernment for development[87], Heek's development informatics perspectives[88], Henman's Foucauldian conceptualisation of eGovernment through studies on governmentality, or critical theorists such as Andrew Feenberg's constructivist approaches[89] on ICTs in alternative modernity[90]. To approach the case of eServices in Pakistan, the main objective of this section would be to develop an understanding of eGovernment in developing countries by highlighting the discourses in ICT4D literature as well as eGovernment.

The research in the field of Information and Communication Technologies had picked up momentum around the mid-1990s, particularly during the commercialisation of the Internet[91]. Although various communication technologies were also used in the field of development communication yet the increased use of the Internet has led to the adoption of not only the Internet but

---

84   Unwin, T. (2009). ICT4D : information and communication technology for development. Cambridge: Cambridge University Press; Prakash, A. and De', R. (2007). Importance of development context in ICT4D projects: A study of computerization of land records in India. Information Technology & People, 20(3), 262-281; Heeks, R. (2006b). Theorizing ICT4D Research. Information Technologies & International Development, 3(3), 1-4

85   Helbig, N., Gil-García, J. R. and Ferro, E. (2009). Understanding the complexity of electronic government: Implications from the digital divide literature. Government Information Quarterly, 26(1), 89-97; Baptista, M. (2005). E-government and state reform: Policy dilemmas for Europe. The Electronic Journal of e-Government, 3(4), 167-174

86   Castells, M. (2010b). The rise of the network society. Chichester, West Sussex: Wiley-Blackwell

87   Madon, S.: 2009

88   Heeks, R. (2010). Do information and communication technologies (ICTs) contribute to development? Journal of International Development, 22(5), 625-640

89   Feenberg, A.: 1995

90   Henman, P.: 2010

91   Steyn, J. and Johanson, G. (2011). ICTs and sustainable solutions for the digital divide: theory and perspectives. Hershey, PA: IGI Global

also several modern ICTs concepts such as network effects, multi-access possibilities and wider roll-out of digitisation in the field of development aid and government[92].

The research on ICT4D is continuously being populated with several theoretical constructs such as ICT4D, ICT4D 1.0, Government 2.0 and/or electronic government[93]. Similarly, the emphasis on the study of ICT4D in developing countries or eGovernment in developing countries has also increased and both the terms ICT4D and eGovernment for development are interchangeably used[94]. With the rise in the use of ICTs for development or eGovernment, there has also been a growth of articles in research journals that particularly focus on them, such as Government Information Quarterly, Information Technologies and International Development, Information Technology for Development, African Journal of Information Systems, The Electronic Journal of Information Systems in Developing Countries, Journal of Emerging Trends in Computing and Information Sciences, Journal of eGovernment, International Journal of Electronic Government Research, and Journal of eGovernance, just to name some of the most cited ones.

Rapid developments and innovation in the field of digital communication technologies that have been brought about, particularly in the past three decades, are bringing profound changes in society[95]. The omnipresent nature of communication technologies that have traversed every walk of human life has left almost no option to be anonymous and disconnected[96]. The social circles and information networks are developed through an inter-layered network of computers that never shut down[97]. The boundaries between being online and offline are getting blurred. Our lives are now governed and, to a possible extent, moulded by the digitally networked society[98]. With or without our consent, our profiles and personal data may be searched in the form of spread sheets in some parts of the world for reasons ranging from market research in the commercial

---

92   Roman, R. and Colle, R. D. (2003). Content creation for ICT development projects: Integrating normative approaches and community demand. Information technology for development, 10(2), 85-94

93   Chun, S. A., Shulman, S., Sandoval, R. and Hovy, E. (2010). Government 2.0: Making connections between citizens, data and government. Information Polity, 15(1), 1-9

94   Venkatesh, V., Sykes, T. A. and Venkatraman, S. (2014). Understanding e- Government portal use in rural India: role of demographic and personality characteristics. Information Systems Journal, 24(3), 249-269; Loudon, M. and Rivett, U. (2013). Enacting openness in ICT4D research. In Smith and Reilly (Eds.), Open development: networked innovations in international development. (pp. 53-78). Cambridge MA: MIT Press

95   Steinmueller, W. E. (2002). Knowledge- based economies and information and communication technologies. International Social Science Journal, 54(171), 141-153

96   Karagiannis, T. (2014). The New face of the Internet. In Harper (Ed.), Trust, Computing and Society (pp. 38-67). New York: Cambridge University Press

97   Raine, L. and Wellman, B. (2012). Networked: The New Social Operating System. Massachussets: Massachusetts Institute of Technology

98   Hanna, N. K. (2010). Transforming government and building the information society: Challenges and opportunities for the developing world. New York: Springer

sector or national governments for revenue, taxes, security and/or demographic data collection, or by international governments in the form of biometric passport control and airline data[99]. It is tempting to say that we live in a truly wired world where our information exists in the form of spreadsheet data continuously being dragged from one cell to another[100].

The Internet has become vital to our social interaction. It is redefining the social contract of citizens with governments[101]. It is rapidly changing the way we interact with our governments and vice versa. This scenario may stand valid for countries with a high penetration of technologies along with increased digital literacy and internet access; however, the situation for developing countries appears to be precarious due to a profound divide that exists in the form of poverty and access to clean water, let alone a weak technology infrastructure and access to computers[102]. In order to bridge these gaps, attempts are made by national governments, international donor agencies, and by non-governmental organisations who have increasingly recognised the potential of information and communication technologies that can not only be used as a tool for community development but also as a possible means to bring together the digital haves and the have-nots[103].

To begin a discussion on the use of communication technologies in the field of development, it needs to be retraced from the use of simple broadcast technologies such as radio and television through several development communication projects and best practice proposed by community organisations, NGOs and/or international development organisations such as UN and World Bank[104].

---

99 Ogura, T. (2006). Electronic government and surveillance-oriented society. In Lyon (Ed.), Theorizing surveillance: The panopticon and beyond (pp. 270-295). London: Routledge

100 The mysterious powers of Microsoft Excel (2013, 21 April), BBC News. Retrieved 04.05.2014 from https://www.bbc.com/news/magazine-22213219

101 Reddick, C. G. (2010). Comparative e-government. New York: Springer

102 Bessette, G. (2004). Involving the community: a guide to participatory development communication. Ottawa: International Development Research Centre (IDRC). Retrieved 29.10.2009 from https://idl-bnc-idrc.dspacedirect.org/bitstream/handle/10625/31476/IDL-31476.pdf?sequence=33&isAllowed=y

103 Reed, T. V. (2014). Digitized lives: culture, power, and social change in the internet era. London: Routledge

104 United Nations Development Programme. (2009). Communication for Development: A glimpse at UNDP's practice. Retrieved 27.08.2011 from https://www.undp.org/content/dam/aplaws/publication/en/publications/democratic-governance/oslo-governance-center/civic-engagement/communication-for-development-a-glimpse-at-undps-practice-/Communication%20for%20Development%20A%20Glimpse%20at%20UNDP%C2%92s%20Practice%20(2009).pdf; World Bank's Chapter on Development Communication contains a repository of several reports, documents and policy papers, see: World Bank (2008b). "Develoment Communication (DevComm)". Retrieved 27.08.2011 from http://web.worldbank.org/archive/website01216/WEB/0__MENUP.HTM.

2.1. Development Communication and ICT4D

The use of technology for development goes back to the post World War II era when major aid interventions were planned under the umbrella of modernising the southern hemisphere by the economic and cultural elites based in the northern hemisphere[105]. The central theme of this modernisation was aimed at liberalisation and development of underdeveloped or post-colonial countries in order for them to follow the same route as many post-war countries[106]. Development in this context was associated with economic and social condition whereas the role of communication was linked to the spread of information that may be used and employed to improve the conditions of regressive countries. Such an approach towards modernisation paradigm however was subjected to disparagement and disapproval particularly during the 1960s for being too West-centric and relied for instance on unfair distribution of social benefits in non-western countries[107]. Alternative theoretical model that propagated in reaction to modernisation theory was named dependency theory.

Dependency theory tried to provide a critical view of modernisation in favour of a political-economic perspective[108]. The advocates of this school of thought have criticised some of the basic postulations of the modernisation paradigm for disregarding the factors such social, historical, and economic that are central and crucial to the growth of any country. Similarly, dependency theorists on the other hand also highlighted the significance and importance of communication and culture as they support diversity and provide richer and meaningful cultural expression[109]. These developments resulted in a demand of a new world information and communication order which could enable stable exchange of communication, information, and cultural programs among developed and underdeveloped countries.

The dependency theory did make a significant impact but from the 1980s onwards it started to lose its relevance to an emerging approach focusing on participation or human participatory factor[110]. The participatory form of development attempts to recognise and acknowledge the cultural realities that are necessary and building blocks of development and unlike its predecessors it tries to look beyond political

---

105   Servaes, J. (2008). Communication for development and social change. New Delhi, India: Sage Publications

106   Chong, A. and Inter-American Development, B. (2011). Development connections : unveiling the impact of new information technologies. New York: Palgrave Macmillan

107   Carleheden, M. and Borch, C. (2010). Editorial: Successive Modernities. Journal of Social Theory, 11(2), 5-8

108   Servaes, J. and Malikhao, P. Development Communication Approaches in an International Perspective. In Servaes (Ed.), Approaches to Development Communication (pp. 214-251). Paris: UNESCO

109   Servaes, J., Jacobson, T. L. and White, S. A., (Eds.) (1996). Participatory communication for social change. New Delhi: Sage

110   Storey, D. (1999). Popular Culture, Discourse and Development. In Jacobson and Servaes (Eds.), Theoretical approaches to participatory communication (pp. 337-358). Cresskill, NJ: Hampton Press

and economic dimensions of development[111]. The development focus has since shifted from economic growth and there has been a gradual incorporation of socio-cultural diversity that can lead to a sustainable and meaningful outcome[112]. It is since then that sustainable development remains the top agenda for major international development agencies particularly in the United Nations Millennium Development Goals[113].

Due to major initiatives in development communication, mass media and communication technologies became a part of the development agenda as a significant tool[114]. The use of media in development was often done through television, radio, print media or even with theatre to ensure participation by giving representation to the local communities. One of the earliest uses of technology can be linked to the radio, which was viewed as a major phenomenon that helped improve and uplift the economic and social conditions of the poor[115]. Lerner in his *The Passing of Traditional Society* primarily focused on the role of media in elaborating actions and models that would allow economic gains by those residing in predominantly agrarian societies[116]. The information he collected about local and international media, particularly the radio, eventually aided locals in learning new ideas and practices. Similarly, in 1969, UNESCO produced a publication *The Impact of Communication on Rural Development* which provides a detailed examination of radio's role and how it can help bring innovation in the agricultural practices by the rural farmers[117].

In several parts of the world, community radio remained a useful medium to enhance civic engagement in marginalised communities that helped in articulating priority issues despite linguistic, ethnic, and literacy barriers[118]. Community radio triggered the participatory approach among communities by engaging the local audience to help address and solve problems. It created its niche by providing listeners with access to information such as health suggestions from medical professionals for treating illnesses or region-specific farming techniques from an

---

111   Servaes, J. and Malikhao, P. (2005). Participatory communication: The new paradigm. Media & global change. Rethinking communication for development, 91-103

112   Servaes, J. (2013). Comparing development communication. In Esser and Hanitzsch (Eds.), The handbook of comparative communication research (pp. 86-102). New York: Routledge

113   Easterly, W. (2009). How the millennium development goals are unfair to Africa. World Development, 37(1), 26-35; Also see for example: Sachs, J. D. and McArthur, J. W. (2005). The millennium project: a plan for meeting the millennium development goals. Lancet, 365(9456), 347-353

114   Sparks, C. (2007). Globalization, development and the mass media. London: Sage

115   Asiedu, C. (2012). Information communication technologies for gender and development in Africa: The case for radio and technological blending. International Communication Gazette, 74(3), 240-257

116   Lerner, D. (1968). The passing of traditional society. New York: Free Press

117   Roy, P., Waisanen, F. B. and Rogers, E. M. (1969). The Impact of communication on rural development : an investigation in Costa Rica and India. Report of a research project initiated by Unesco. Paris: UNESCO

118   Pettit, J., Salazar, J. F. and Dagron, A. G. (2009). Citizens' media and communication. Development in Practice, 19(4-5), 443-452

agricultural specialist[119]. Community radio provided its listeners access to diverse array of sensitive and taboo topics such as information about diseases and substance abuse or confronting human rights violations. In South America, Bolivia and Colombia have present examples of community radio that date back to 1940 where community-based groups made several efforts to broadcast issues and cultural information in local languages[120].

The rise of more sophisticated communication and information technologies in the 1990s, such as the satellite dish or the Internet, presented new possibilities for development aid specialists particularly after the announcement of MDGs[121]. The potential of the new technologies not only enhanced the mass media penetration, for instance through visual means and general proliferation but it produced new opportunities to increase the communication at the local level by employing newer technologies such as the mobile telephones and/or Internet[122]. There is a growing amount of literature that also critiques the use of ICTs in the presence of daunting poverty standards in most of the developing countries[123]. The question arises if ICTs are also able to help alleviate poverty[124]. Can the ICTs ensure the creation of employment opportunities, regular food supplies, health-care centres, shelter, and/or access to information? Poverty in this case shall not only be categorised as lack of resources only, instead it shall be contextualised as also the collective awareness on people's part for exercising their role in the fight against poverty. For instance, unavailability of legal information of a country's land reform legislation can deny rural farmers livelihood and homes their rights and security[125].

119   Sarfo, E. (2011). Offering healthcare through radio: An analysis of radio health talk by medical doctors. Journal of Language, Discourse and Society, 1(1), 104-125

120   McPhail, T. L., (Ed.) (2009). Development communication : reframing the role of the media. Chichester, West Sussex: Wiley-Blackwell

121   Golding, P. (2011). Connected services : a guide to the Internet technologies shaping the future of mobile services and operators. Chichester, West Sussex: Wiley

122   Ibid.

123   Clarke, S., Wylie, G. and Zomer, H. (2013). ICT 4 the MDGs? A perspective on ICTs' role in addressing urban poverty in the context of the Millennium Development Goals. Information Technologies & International Development, 9(4), 55-70

124   Torero, M. and Von Braun, J. (2006). Information and communication technologies for development and poverty reduction: The potential of telecommunications. Baltimore: Johns Hopkins University Press; Flor, A. G. (2001). ICT and poverty: The indisputable link. . Paper presented at the Third Asian Development Forum on "Regional Economic Cooperation in Asia and the Pacific", organised by Asian Development Bank.Retrieved 09.10.2009 from http://ftp.unpad.ac.id/orari/library/library-ref-ind/ref-ind-1/application/poverty-reduction/!%20ICT4PR/ICT%20and%20poverty%20-%20the%20link%20(WB%20Report).pdf

125   Mefalopulos, P. (2008). Development communication sourcebook broadening the boundaries of communication. Retrieved 28.09.2010 from http://siteresources.worldbank.org/EXTDEVCOMMENG/Resources/DevelopmentCommSourcebook.pdf

2.2. ICT4D 1.0 to ICT4D 2.0

Heeks has termed the rise of the Internet and MDGs as the era of ICT4D 1.0[126]. The digital technologies of the 1990s gave revised meanings to the development goals which were in the search for delivery mechanisms. These two domains, as Heeks argues, intersected and gave rise to ICT4D that eventually led to an array of new publications, development programs, and project funding. In this regard, gathering of G8 countries and the eventual setting up of the Digital Opportunities Task Force in 2000 had set an agenda for action on ICT4D[127].

In 2001 the ICT4D agenda moved forward when the United Nations initiated the ICT Task Force with an agenda to focus on topics such as internet governance, enabling environments, and education[128]. As a result, individual ICT4D projects were set up all over the world ranging from small to multi-stakeholder activities between donor agencies, technology companies, governments and NGOs[129]. It is perhaps around the dawn of the millennium that the dichotomisation of "success and failure" and/or "top down and bottom-up" approaches in ICT4D projects started to occupy a permanent place in literature[130]. As Unwin and Kleine have also highlighted, the UN ICT Taskforce trigged a production of a host of project reports that presented the projects largely as success stories[131]. This dichotomy - while one hand started to ignore the grey areas such as those related to indigenous culture and society, the black and white approach as Unwin argued - stood in sharp contrast to some academic observations, which perceived the top-down and supply-driven initiatives with limited or no impact[132]. A host of indigenous projects initiated for instance by local NGOs such as web dissemination of suppressed indigenous cultural knowledge or citizens' journalism websites received lesser attention at international arenas[133].

126   Heeks, R. (2008). ICT4D 2.0: The next phase of applying ICT for international development. Computer, 41(6), 26-33

127   Kirkman, G. S., Cornelius, P., Sachs, J. and Schwab, K., (Eds.) (2002). The global information technology report, 2001-2002: Readiness for the networked world. New York: Oxford University Press

128   Bracey, B. and Culver, T., (Eds.) (2005). Harnessing the potential of ICT for education: a multistakeholder approach: Proceedings from the Dublin Global Forum of the United Nations ICT Task Force. New York: United Nations Publications

129   Unwin, T. (2005). Partnerships in development practice: evidence from multi-stakeholder ICT4D partnership practice in Africa. Retrieved 09.09.2011 from https://unesdoc.unesco.org/ark:/48223/pf0000142982

130   Best, M. L. (2010). Understanding our knowledge gaps: Or, do we have an ICT4D field? And do we want one? Information Technologies & International Development, 6(SE), pp. 49-52

131   Kleine, D. and Unwin, T. (2009). Technological Revolution, Evolution and New Dependencies: what's new about ICT4D? Third World Quarterly, 30(5), 1045-1067

132   Unwin, T.: 2009

133   Schmidt, P. (2003). New model, old barriers: remaining challenges to African civil society participation. Information Technologies and International Development, 1(3-4), 100-103; Hafkin, N. J. (2002). Gender issues in ICT policy in developing countries: An overview. Expert Group Meeting organised by United Nations Division for the Advancement of Women (DAW) on "Information and communication technologies and their impact on and use as an instrument for

Two important developments that took place after the formation of the ICT Task Force was the inauguration of the first phase of the World Summit on the Information Society (WSIS) in 2003[134]. Organised by the ITU in Geneva, it gathered 11,000 delegates from governments, donor agencies, companies and civil society from 175 countries. Although the WSIS created a debate on crucial topics pertaining to information society by incorporating multiple actors to discuss the use of ICTs, it has equally been criticised for its excessive technological orientation[135]. The second phase of WSIS, which took place in Tunis, in 2005 saw bigger participation of 19,401 members[136]. Similarly, another development that took place prior to the Tunis summit was the formation of the Internet Governance Forum (IGF) which was first recommended in the report of the Working Group on Internet Governance[137]. This report became one of the inputs to the Tunis summit. As a multi-stakeholder and cross-sectoral forum, the IGF brings together representatives from government, business, development co-operation, international bodies, regulatory agencies, industry organisations and unions, producers of ICT, NGOs, media, foundations, academics, and individual practitioners[138]. This chain of events from the announcement of MDGs to Digital Opportunity Taskforce, UN ICT Task Force and WSIS to IGF are an enabling force for streamlining the agenda for eGovernment and also as a tool for development. The following hereinafter will focus on the discussion on eGovernment as well as eGovernment in developing countries.

2.3. From ICT4D to eGovernment

With the commercialisation of the Internet and the boom in mobile telecommunication, ICT4D initiatives have inspired many governments to undertake ICTs in the administrative processes to improve the efficiency of public service delivery[139]. By the dawn of the new millennium most of the ICT4D literature started to introduce and establish the term *electronic government* or *eGovernment*. While the debate on ICTs continues, there is a growing body of

the advancement and empowerment of women".Retrieved 03.05.2014 from citeseerx.ist.psu.edu/viewdoc/download?doi=10.1.1.463.3492&rep=rep1&type=pdf

134 International Telecommunication Union (2003). "Building the Information Society: a global challenge in the new Millennium". Retrieved 03.02.2008 from https://www.itu.int/net/wsis/docs/geneva/official/dop.html.

135 Servaes, J.: 2008

136 International Telecommunication Union (2005). "Tunis Agenda for the Information Society". Retrieved 03.03.2008 from https://www.itu.int/net/wsis/documents/doc_multi.asp?lang=en&id=2266|2267.

137 Internet Governance Forum (2006). "About the IGF". Retrieved 03.03.2008 from https://www.intgovforum.org/multilingual/tags/about.

138 Mathiason, J. (2008). Internet governance: the new frontier of global institutions. Oxon: Routledge

139 Al Ajeeli, A. and Al-Bastaki, Y. A. L., (Eds.) (2011). Handbook of research on E-services in the public sector : E-government strategies and advancements. Hershey: Information Science Reference

literature that now focuses on eGovernment in developing countries. Over the last ten years eGovernment definitions are continuously evolving, however, a common understanding is that eGovernment refers to the use of information and communication technologies (ICTs), particularly internet-based applications, to provide faster, cheaper, easier, and more efficient access to and delivery of information/services to the public, businesses, other agencies and government entities[140]. During the last ten years several ideas of providing modern administration and democracy in public service delivery have emerged[141].

Bhatnagar refers to some of the earlier definitions of eGovernment by the use of communication technologies in three areas of public action: (1) relations between the public authorities and civil society, (2) functioning of the public authorities at all stages of the democratic process and (3) the provision of public services (electronic public services)[142]. International development organisations such as UN, World Bank or OECD, which have been actively advocating the use of ICTs in government, also share a similar definition. Governments are increasingly employing the use of communication technologies to deliver services online and to improve public administrative processes by redesigning[143]. Reorganisation of the public sector resultantly helps reduce costs and improves the efficiency of services offered to citizens, government agencies and businesses[144]. Similarly, ICT-enabled interaction encourages the development of (new) democratic spaces in which relationships among the public sector, citizens, and enterprises are redefined according to a participatory perspective[145].

eGovernment initiatives have rapidly increased in the past few years and the emphasis on delivering government services through electronic medium has become a focus for national governments[146]. Examples of such public-sector services may include among others tax filing, issuance and renewal of identity

---

140 Grönlund, Å. (2010). Ten Years of E-Government: The 'End of History' and New Beginning. In Wimmer, Chappelet, Janssen and Scholl (Eds.), Electronic Government: 9th IFIP WG 8.5 International Conference, EGOV 2010, Lausanne, Switzerland, August 29 - September 2, 2010. Proceedings (pp. 13-24). Berlin: Springer

141 Hameed, T. (2007). ICT as an enabler of socio-economic development. Retrieved December 17, 2011 from http://unpan1.un.org/intradoc/groups/public/documents/un-dpadm/unpan043799.pdf

142 Bhatnagar, S.: 2004

143 United Nations. (2003). Global E-government Survey 2003. Retrieved 04.05.2014 from https://publicadministration.un.org/egovkb/portals/egovkb/Documents/un/2003-Survey/unpan016066.pdf

144 Organisation for Economic Co-operation and Development. (2003). OECD E-Government Flagship Report "The E-Government Imperative". Retrieved 04.05.2014 from http://www.oecd.org/officialdocuments/publicdisplaydocumentpdf/?cote=GOV/PUMA(2003)6/ANN&docLanguage=En

145 infoDev and Center for Democracy & Technology. (2002). The E-Government Handbook For Developing Countries: A Project of InfoDev and The Center for Democracy & Technology. Retrieved 04.05.2014 from http://unpan1.un.org/intradoc/groups/public/documents/apcity/unpan007462.pdf

146 Weerakkody, V., El-Haddadeh, R., Sabol, T., Ghoneim, A. and Dzupka, P. (2012). E-government implementation strategies in developed and transition economies: A comparative study. International Journal of Information Management, 32(1), 66-74

cards, drivers' licenses and passports, online application for government jobs, obtaining birth or marriage certificates, just to name a few[147]. The entire premise is set in giving citizens an option or the convenience of accessing information via government websites thus making their queries easier by avoiding the hassle of phone calls or physical visit to government[148]. Here, while dealing with the impact of eGovernment on public service delivery, another important aspect is to assure information availability[149]. The idea of such an eGovernment scenario is to help citizens provide government information.

However, receiving information services via websites has also gone through major transformations due to the developments in mobile phone technology[150].Governments are now increasingly offering services on smart phones in the form of SMS and Apps (short for applications)[151]. The nature of government service delivery is thus facing rapid transformation[152]. Public service delivery is now being offered by multiple digital channels making delivery of services faster and cheaper.[153]

2.4. Citizen-centricism & power relationships in a networked society

The idea behind efficiency and effectiveness in the public services is to understand the requirements and necessities of the citizens and build services around citizen's stipulations[154]. In this context, the citizen becomes the focus of the service being provided by the governments which employ different tools, ICTs being one of them. Within an eGovernment scenario, this concept is referred to as

147   Chen, H., Brandt, L., Gregg, V., R, T., Dawes, S., Hovy, E., Macintosh, A. and Larson, C. A., (Eds.) (2008). Digital government e-government research, case studies and implementation. Berlin: Springer

148   Welch, E. W., Hinnant, C. C. and Moon, M. J. (2004). Linking citizen satisfaction with e-government and trust in government. Journal of public administration research and theory, 15(3), 371-391

149   Brown, M. M. (2007). Understanding e-government benefits: An examination of leading-edge local governments. The American Review of Public Administration, 37(2), 178-197

150   Wang, Y.-J. (2009). Rethinking e-government services : user-centred approaches. Paris: OECD; Reddick, C. G.: 2010

151   Susanto, T. D. and Goodwin, R. (2011b). An SMS-Based e-Government Model: What Public Services can be Delivered through SMS? In Al Ajeeli and Al-Bastaki (Eds.), Handbook of Research on e-Services in the Public Sector: eGovernment Strategies and Advancements (pp. 137-146). Hershey PA: Information Science Reference

152   Heeks, R. (1999). Reinventing government in the information age : international practice in IT-enabled public sector reform. London: Routledge; Papadopoulos, T., (Ed.) (2012). Public sector reform using information technologies transforming policy into practice. Hershey, Pa.: IGI Global

153   Al-Hakim, L., (Ed.) (2007). Global e-government theory, applications and benchmarking. Hershey, PA: Idea Group Publishing

154   van Velsen, L., van der Geest, T., ter Hedde, M. and Derks, W. (2009). Requirements engineering for e-Government services: A citizen-centric approach and case study. Government Information Quarterly, 26(3), 477-486

citizen-centric or citizen focussed eServices[155]. However, such sort of citizen-centricism also prompts additional set of corollaries. By making the citizen as the focus point in eGovernment scenarios, the notions of power and citizenship in a paper-based government scenario seems to be contested[156]. The question here arises as to how far the citizen has assumed a central position in its relationships with the government. More precisely, in what way(s) and to what extent the citizen encounters changes in her/his role and relationships with government by moving to an eGovernment environment.

In the manual or the paper-based environment, personal and physical identification of the citizen and the verification of identity through authentication processes reside at the heart of government service provision[157]. Authentication processes associated with paper-based authentication systems have largely been a ration/social card, passport or personal identity document. The identity card holder would physically present the document in office where the document carrier would be verified through the information asked, coupled with the photograph, included in the document. These identity checks may be supplemented by a face-to-face assessment of the person, or subsumed by official's knowledge of the citizen within the local community[158].

With the emerging digital era, there are three main procedures of personal identification that may include accepting a self-declared statement of identity that could consist of information such as username, registration number, address, or password[159]. Another possible way would be accepting an item of identity the person physically possesses for instance a smartcard, electronic tag, or mobile phone. In more complex situations, this verification may involve biometric scan. This development particularly marks the citizen as a unique customer of the government. Personalisation of this level makes the government approach the citizen based on his/her individual preferences or online behaviour recorded with the government, something that may become laborious in a paper-based environment[160].

155   King, S. and Cotterill, S. (2007). Transformational government? The role of information technology in delivering citizen-centric local public services. Local Government Studies, 33(3), 333-354

156   Humes, L. L. and Reinhard, N. (2009). E-Government and the Influence of Power in the Development of Information Infrastructure. Journal of Global Information Technology Management, 12(2), 61-79

157   Davison, R. M., Wagner, C. and Ma, L. C. (2005). From government to e-government: a transition model. Information Technology & People, 18(3), 280-299

158   Anderson, B. (2007). Information and communication technologies in society : e-living in a digital Europe. London: Routledge

159   Kolsaker, A. and Lee-Kelley, L. (2007). G2C e-government: modernisation or transformation? Electronic Government, An International Journal, 4(1), 68-75

160   Warkentin, M., Gefen, D., Pavlou, P. A. and Rose, G. M. (2002). Encouraging citizen adoption of e-government by building trust. Electronic markets, 12(3), 157-162

Henman argues that the digital society and eGovernment is reconfiguring the social contract between the citizen and government[161]. He bases his argument on the lines of Foucault's idea of governmentality to claim that eGovernment has the ability to produce citizens that can act or govern themselves to particular ends[162]. The entire computerisation drive and use of technologies in government redefines the relationship between technology and power. On one hand, technologies are considered to be neutral yet on the other hand their use can bring about centralisation, control and power in select hands. Henman argues that within the eGovernment context, technologies and humans together form a complex network where their interactions takes places at several levels. To understand the relationship within technology and power, it is necessary to understand the way human/technology network evolves and operates and to analyse the outcome of such interactions. Therefore, to understand the impact of eGovernment technologies have on reconfiguration of power equally requires acceptance as well as investigation of an array of available technologies that eventually make up eGovernment[163].

Here, emphasis on the relevance of Foucault's notions of power and governmentality in the current eGovernment scenario is needed[164]. Foucault had already developed his understanding of power before he had coined the term governmentality and on the idea of conduct[165]. If Foucault's understanding of power is seen within an eGovernment perspective, we notice that ICTs and their use have the capability to encompass several areas of human activities. For instance, eHealth scenario presents the mobilisation of bio-power by eGovernment given that how states feel concerned about the health and wellbeing of their subjects and intervene when necessary in several areas of health care be it prevention or recovery[166]. A good example of these scenarios could be of an active role of health departments, especially in European countries in keeping a check on citizens and their interaction with vulnerable citizens. Another form of power is sovereign power whose applicability can be seen in welfare or tax systems. The idea of tax declaration provides a good example of how people keep track of their expenses and expenditures. Due to interconnectivity of ICTs at various regional, national and international level, people feel compelled to declare

---

161   Henman, P.: 2010

162   Henman, P. (2013). Governmentalities of Gov 2.0. Information, Communication & Society, 16(9), 1397-1418

163   Ogura, T.: 2006

164   Gordon, C. (1991). Governmental Rationality: An Introduction. In Burchell, Gordon and Miller (Eds.), The Foucault effect : studies in governmentality - with two lectures by and an interview with Michel Foucault (pp. 1-51). Chicago: University of Chicago Press

165   Bratich, J. Z., Packer, J. and McCarthy, C., (Eds.) (2003). Foucault, cultural studies, and governmentality. Albany: State University of New York Press

166   Gaby, S. and Henman, P. (2004). E-Health: transforming doctor-patient relationships with a dose of technology. In Paper presented at: Aus- tralian Electronic Governance Conference; 14th and 15th April 2004. Centre for Public Policy, University of Melbourne Victoria. Retrieved 08.10.2008 from http://www.public-policy.unimelb.edu.au/egovernance/ConferenceContent.html

and pay taxes or even face punishments in case of tax evasion[167]. Similarly, another form of power in Foucauldian sense is disciplinary power which functions through surveillance or the threat of surveillance[168]. Its basic objective is to internalise self-discipline and self-regulation so that individuals govern themselves in accordance with the objectives of governing authorities[169]. A reasonable eGovernment example in this regard is the installation of speed camera on motor highways, which forces the drivers to control their speed in order to avoid penalties. Similarly, the presence of electronic databases, online payments, digital thumb impressions, swipe cards scenarios forces one to consider tax returns in the fear of being watched[170]. Therefore, we see that the entire ICT/eGovernment discourse is not just technical/technological rather it traverses other disciplines such as political science, communication studies and/or social sciences.

2.5. Stages of eGovernment

In order to understand eGovernment as a concept, it is essential to understand different constituents of this emerging discipline as well as to answer, for instance, what requires to call eGovernment as completely workable? Similarly, what are the stages that governments need to consider while developing or adopting an eGovernment programme? One of the earliest models in this regard that was floated about the stages of eGovernment was Layne and Lee's idea of four stages of eGovernment development which included (1) catalogue stage, (2) transaction stage, (3) vertical integration, and (4) a horizontal integration stage[171].

In the catalogue stage, governments set up their initial online presence via websites. This generally includes online presentation of government information, its functioning, its organisational procedures. This is a somewhat one-way communication stage where government and citizens cannot interact with each other, instead the focus is given to the internal collection government information and its presentation on the web. The advantage of this stage is that if the government has provided forms/pro-forma related to matters such as tax, birth or

---

167 Lips, M. (2010). Rethinking citizen–government relationships in the age of digital identity: Insights from research. Information Polity, 15(4), 273-289

168 Chini, I. (2012). Governmentality and the information society: ICT policy practices in Greece under the influence of the European Union (Doctoral Dissertation, London School of Economics and Political Science, UK). Retrieved 04.05.2014 from http://etheses.lse.ac.uk/847/1/Chini_Governmentality_and_the_information_society.pdf

169 Lips, A. M. B., Taylor, J. A. and Organ, J. (2009). Managing citizen identity information in E-government service relationships in the UK: the emergence of a surveillance state or a service state? Public Management Review, 11(6), 833-856

170 Dutton, W., Guerra, G. A., Zizzo, D. J. and Peltu, M. (2005). The cyber trust tension in E-government: Balancing identity, privacy, security. Information Polity, 10(1, 2), 13-23

171 Layne, K. and Lee, J. (2001). Developing fully functional E-government: A four stage model. Government Information Quarterly, 18(2), 122-136

licensing, then such forms could be downloaded and printed by the citizen which s/he could later either submit in person or send by post.

The second stage of transaction offers more interactivity and two-way communication between the government and the citizen. It includes more government services and access to online forms. Citizens and businesses are able to engage in electronic transactions with their governments. Possible examples of this stage may include the use of interactive forms for registering a business, applying for a drivers' license, or filing for unemployment benefits. The transaction stage would of course require more work on the governments' part as it would have to increase its efforts in linking and connecting the internal procedures between two or multiple government machineries to the online world.

While the cataloguing and transaction stages could be considered as one of the initial phases in eGovernment. The vertical and horizontal integration stages are somewhat sophisticated forms of eGovernment which can be characterised by integration within each level of government, such as local, state/provincial and/or federal. In the vertical stage, a government agency or ministry is supposed to be fully automated and all its procedures are digitised whereas its website is able to deliver information from ministry/agency's database. The stage of horizontal integration goes one step further and it means that not only different government departments are digitised in their individual capacities, but these departments and their respective database are able to communicate and digitally respond to each other.

This stage is the most complex and does not only require higher level of digital sophistication but it also requires advanced level of security. An example of horizontal integration can be a citizen applying for the renewal of her/his identity card via government website via Ministry A's website. As soon the citizen submits an online query regarding the renewal, the Ministry A's database contacts the databases of taxation, law and order departments to check if the citizen in question is being flagged by any fellow agency. If the responses from the fellow agency databases are in positive, the citizen's request by the Ministry's A website gets accepted (see Figure 1).

In a study conducted by the United Nations on eGovernment, it expanded the Layne and Lee model further by identifying five stages of eGovernment evolution which include: (1) emerging, (2) enhanced, (3) interactive, (4) transaction, and (5) connected. The role of the emerging state is the same as of the "catalogue" stage in which the government makes sure of its online presence in the form of a web page with links to ministries or departments[172].

---

172   United Nations. (2008). Global E-Government Survey 2008: From E-Government to Connected
      Governance   Retrieved   04.05.2014   from   https://publicadministration.un.org/egovkb/en-
      us/Reports/UN-E-Government-Survey-2008

Figure 1 : Four stage eGovernment model

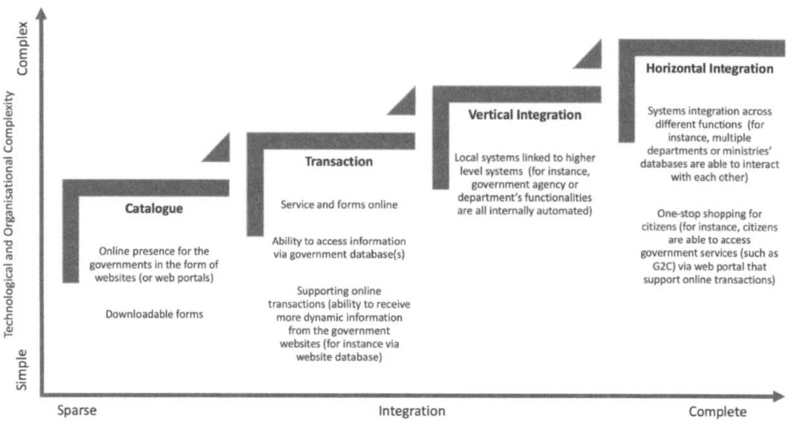

Source: Own illustration, adopted from Layne & Lee's four stage model[173]

Much of the information stays static with little possibility for the citizens to interact. Similarly, at the "enhanced" stage governments provide more information on policy and governance matters. They have links to archived information such as documents, forms, reports, laws and regulations. Stage three, "interactive", delivers online services such as downloadable forms for tax payments or applications for license renewals.

According to the UN's eGovernment evolutionary model it is only possible at stage four of "transaction" stage for governments to start to transform themselves by introducing two-way interactions between the citizen and government. This stage could include options for paying taxes online, applying for ID cards without needing to send printouts, birth certificates or passports. The major benefit of this stage is that all transactions are conducted online. The stage five of "connected" is somewhat similar to the idea of Layne and Lee's vertical and horizontal stage, where governments transform themselves into one connected entity that has all its departments and ministries connected by an integrated back office infrastructure. This would probably reflect the most sophisticated level of online eGovernment.

While comparing the description of both the models that emerged over the last decade, it is noticeable that the basic understanding of an eGovernment's evolution remains similar. While ICTs make it possible to develop eGovernment services, there is a varied degree of differences, reasons and the use of eGovernment in different countries. Countries with well-established technological infrastructures such as United States and many in the European

---

173   Layne, K., et al.: 2001

Union have well-structured eGovernment services[174]. However, developing countries still face problems pertaining to legacy ICT infrastructure and people's ability to use eServices[175]. Many developing countries, which started with the implementation of eGovernment, faced difficulties in executing it the way other countries had experienced. Their difficulties and failure to cope with the widely accepted eGovernment model started a new debate focusing on the failure of eGovernment[176]. There has now been a continuous emphasis on eGovernment for developing countries, and new channels of communication are being researched to deliver eServices[177]. The following section discusses different aspects of eGovernment in developing countries and highlights some of the issues and challenges that are currently a part of the debate over eGovernment in developing countries.

2.6. eGovernment and developing countries

One of the major objectives of eGovernment, regardless of developed or developing country, remains the increase of efficiency and productivity of public and private sector by incorporating modern ICTs[178]. There has been considerable increase in the national budgets for development of ICT infrastructure and eGovernment applications. The private sector in this regard has also taken advantage of these developments and provided services such as e-commerce. It has perhaps been the popularity of e-commerce activity that has given rise to the notion of citizen-centric eServices. A similar trend of citizen-centrism is also being followed in developing eGovernment applications for the citizens. However, for many developing countries the concept of citizen-centric or citizen focus is relatively new[179].

In a developing country context, citizen has been an entity who has for a long time been ignored or been co-opted in the struggle for basic necessities of life. Post decolonisation, most of these developing countries have skewed administrations that banked on the colonisers' older ways and means of centrist command and directions[180]. Thus, providing citizen-focused services in developing countries

---

174 Weerakkody, V. and Reddick, C. G. (2012). Public sector transformation through e-government: experiences from Europe and North America. New York: Routledge

175 Ndou, V. (2004). E-government for developing countries: opportunities and challenges. The Electronic Journal of Information Systems in Developing Countries, 18

176 Heeks, R. (2005a). e-Government as a Carrier of Context. Journal of Public Policy, 25(1), 51-74

177 Reddick, C. G., Abdelsalam, H. M. and Elkadi, H. A. (2012). Channel choice and the digital divide in e-government: the case of Egypt. Information technology for development, 18(3), 226-246

178 Gotoh, R. (2009). Critical factors increasing user satisfaction with e-government services. Electronic Government, An International Journal, 6(3), 252-264

179 Mofleh, S., Wanous, M. and Strachan, P. (2008b). The gap between citizens and e-government projects: the case for Jordan. Electronic Government, An International Journal, 5(3), 275-287

180 Dwivedi, O. P. and Nef, J. (2004). From Development Administration to New Public Management in Postcolonial Settings. In Mudacumura and Shamsul (Eds.), Handbook of Development Policy Studies (pp. 153-176). New York: Marcel Dekker

still sounds far-fetched. There are issues related to design and/or implementation of eGovernment that poses a big challenge for its uptake in developing countries. In one of the earlier studies quoted by Basu and conducted by the UN, the eGovernment level was identified for 190 nations[181]. The study highlighted five stages of eGovernment ranging from emergence to integration. Considering it was the earlier phases of eGovernment at the world level, none of the surveyed nations had achieved integration at the time of the survey. There were only 17 countries that had achieved the transaction stage. Most developing nations were either at the emergence or at the broadcast stage, thereby allowing few interactive services for the citizens. Basu noted that countries that had performed well in this ranking with extensive and considerable IT and government infrastructure were regarded as generally well-funded. The first few concerns raised about the functioning of eGovernment were if its workability and solutions depended on resources or certain specific knowledge management essentially translating into provision of considerable financial and IT resources.

Additionally, the feasibility of having a workable eGovernment application is also largely dependent on the government's overall ability, willingness and readiness to invest in information technology. Basu has listed some of the symptoms and consequences that were identified in the UN study. These symptoms and consequences, as it can be seen from Table 2, are largely based on issues that range from inadequate design systems; high maintenance costs; underestimated project costs; or inappropriate choice of software[182]. Similarly, he points out the importance of institutional weakness of unclear objectives and insufficient planning that together with the local environment makes it difficult for eGovernment projects to finish, let alone function.

Heeks, who has extensively researched this subject, believes that most of the eGovernment applications in developing countries have failed[183]. He argues that 50% of eGovernment applications undertaken until 2002 were partial failures. Most of these projects had undesirable outcomes and in most cases the goals were not reached. Similarly, 35% of applications were either completely abandoned or not implemented at all. Most of the failures are attributed among other reasons, to the context gaps. Heeks points towards the technology transfer mechanisms that were adopted in a hurried approach towards implementing eGovernment applications[184]. The technology that many developing countries purchased or received as a donation resulted in design-reality gaps because the imported technologies did not seem to conform to the socio-cultural conditions and proved incompatible due to varied skill sets. Similarly, IT infrastructure was also a major

181   Basu, S. (2004). E-government and developing countries: an overview. International Review of Law, Computers & Technology, 18(1), 109-132
182   Ibid.
183   Heeks, R. (2002b). Information systems and developing countries: Failure, success, and local improvisations. The information society, 18(2), 101-112
184   Heeks, R.: 2006a

hindrance. The infrastructure does not only include the internet access but it also banks on reliable telecommunication facilities, and constant electricity access[185].

Table 2: Factors influencing the eGovernment in developing countries

Core Factors	Symptoms	Consequences
Institutional Weakness	Insufficient planning Unclear objectives	Inadequately designed systems Cost over-runs
Human resources	Shortage of qualified personnel Lack of professional training	Insufficient support Isolation from sources of technology
Funding arrangements	Underestimated project costs Lack of recurring expenditure	Unfinished projects Higher maintenance costs
Local environment	Lack of vendor representation	Lack of qualified technical support
Technology and information changes	Lack of backup systems/parts Limited hardware/software Inappropriate software	Implementation problems System incompatibility
Legal inadequacy	Complex legislative procedure	Over reliance on customer application Lack of legal framework

Source: Basu[186]

The World Bank's infoDev programme on eGovernment for developing countries, in this regard, presents a very detailed plan of action for developing eGovernment applications - what it calls as the five elements of implementing eGovernment transformation rests on process reform, leadership, strategic investment, collaboration and civic engagement[187]. All the five elements constitute important steps towards the preparatory phase of eGovernment applications. "Process Reform" suggests carefully planning, streamlining and consolidating offline processes before putting them online. It further suggests responding to local needs by relying on local knowledge capacities. "Leadership" requires a need for an office and to designate a senior official as a focal point for eGovernment innovation, planning and oversight. The "strategic investment" highlights the need of defining clear goals and to catalogue available resources, ranging from funding to personnel by making short- and long-term plans. One of the most important phases is that of "collaboration", which requires the establishment of a consultative process in the planning phase that would include possibilities to hear from and speak with local business entities, NGOs and other

---

185 Hawari, A. a. and Heeks, R. (2010). Explaining ERP failure in a developing country: a Jordanian case study. Journal of Enterprise Information Management, 23(2), 135-160
186 Basu, S.: 2004
187 infoDev and Center for Democracy & Technology: 2002

government agencies. The last aspect of "civic engagement" points to consultations for designing applications that are focused around citizens.

Apart from the suggestions made by various practitioners, there is a need to understand and deconstruct the issues that hinder the development and growth of eGovernment in developing countries. These issues include, but are not limited to, digital divide, eReadiness, multilingual and local content, public sector reform and/or channels of electronic public service delivery[188]. In the light of these issues, this section revisits the Layne & Lee model of eGovernment which has so far been the blueprint for any eGovernment service in the developed world and attempts to review its applicability for developing countries.

2.6.1. Digital Divide & localising ICTs

One of the most crucial factors necessary for a workable eGovernment scenario in developing countries remains the ongoing debate of persistent digital divide. In its study, the OECD defined digital divide as "the gap between individuals, households, businesses and geographic areas at different socio-economic levels with regard both to their opportunities to access ICT and to their use of the Internet for a wide variety of activities"[189]. Digital divide then distinguishes between the digital haves and have-nots, that is, the ones with access to ICT and those without[190]. There is perhaps a need to deconstruct this binary notion of digital divide and to include different social, cultural and political inequalities along with technological disparities that eventually contribute to its prevalence [191]. It would also be inappropriate to consider that the digital divide can be overcome by investing in the digital equipment. Availability of digital infrastructure alone does not guarantee the alleviation of digital divide, there are other factors such as digital access which is even more challenging. What happens if a considerable chunk of the population is denied access to information as well as technology – access and information that is ought to be public and widely available, say in libraries or community centres[192].

---

188   Fonseca, C. (2010). The digital divide and the cognitive divide: Reflections on the challenge of human development in the digital age. Information Technologies & International Development, 6(SE), pp. 25-30

189   Organisation for Economic Co-operation and Development. (2001). Understanding the Digital Divide. Retrieved 04.05.2014 from http://dx.doi.org/10.1787/236405667766

190   Gunkel, D. J. (2003). Second thoughts: toward a critique of the digital divide. New media & society, 5(4), 499-522

191   O'Hara, K. S. D. (2006). inequality.com : power, poverty and the digital divide. Oxford: Oneworld Publicationss

192   Helbig, N., et al.: 2009

Similarly, the digital divide also reflects various other patterns of socioeconomic inequalities persistent in our society[193]. Norris, for example in his study of digital divide, argues that the digital divide should be seen as a phenomenon that has three differing facets, namely, global divide, social divide and a democratic divide[194]. The global divide in Norris's view is a varied Internet access between developed and developing nations. Similarly, he views a social divide as a gap between those having access to information and those without it and lastly, a democratic divide where there is a disparity between those who do or do not have the digital means to engage in public life. For any country undergoing an eGovernment implementation, not only requires technological diffusion but may also require a lot of socio-political reengineering of society and culture[195].

Warschauer seems to be in consensus with Norris on digital divide. In his intellectually stimulating study on the digital divide, he states that it is not just the technology diffusion, technology adoption and uninterrupted internet access on which the scope of technological application sits upon[196]. It goes beyond the questions of technology and places culture in a central position. Warschauer refers to the studies in critical theory of technology such as ones by Andrew Feenberg, Bruno Latour, Maria Bakardijeva and James Sleven who support the idea of social constructivism and social embeddedness of technology according to the local cultures[197]. This argument seems to go in the direction of technological determinism, that is, if technology shall adapt to cultures, or if cultures shall adjust themselves to technology[198].

The debate on technological determinism and social constructivism has also inspired many eGovernment studies and it seems pertinent to review these concepts that may help towards the concluding parts of this study. Bijker argues that since our (Western) culture is becoming highly prevalent with modern communication networks where computers are increasingly taking over the manual tasks, there is perhaps a need to pause, look back, review and above all, understand a culture in which these technological advancements are taking

---

193   Yu, L. (2006). Understanding information inequality: Making sense of the literature of the information and digital divides. Journal of Librarianship and Information Science, 38(4), 229-252
194   Norris, P. (2001). Digital divide : civic engagement, information poverty, and the Internet worldwide. Cambridge: Cambridge University Press
195   Cruz-Jesus, F., Oliveira, T. and Bacao, F. (2012). Digital divide across the European Union. Information & Management, 49(6), 278-291
196   Warschauer, M.: 2003b
197   Bakardjieva, M. (2005). Internet society : the Internet in everyday life. London: SAGE; The debate on technological determinism has dominated the decade of 90s and as Gunkel has put it that at the turn of the millennium, technological determinism has developed into two categories, called 'hard' and 'soft determinism'. Hard determinism has the ability to make technology the sufficient or necessary condition for social change, whereas soft determinism takes technology to be a primary factor that may or can facilitate change. For more detailed works on technological determinism see for example, Gunkel, D. J.: 2003

place[199]. He believes that a technological culture with a standard approach towards Science and Technology appears simplistic, in which technological development follows a linear path. Technology constructed this way, he believes, is conceived by a number of experts who from inception until its end hold the decisive power until it is made operational[200]. Such a technological development that assumes an autonomous approach where the society is kept at arm's length is bound to have direct effects on the society in terms of hesitation to use, or even rejection[201]. This, for example, can be seen in terms of eServices being launched with modern technology without having the necessary capacity building, as we shall see in the case of NADRA's kiosk machine in Chapter. This approach will eventually pave the way for technology to "determine" society instead of it being the other way around. It remains to be seen if users can also influence the development of technology which might make it evolve beyond its presupposed utility. Could there be ways in which users of technology may become an agent or influencing force in shaping its development?

The lack of social influence in the technological development marked the beginning of alternative thinking in studies of Science, Technology and Society (STS) during 1980s[202]. In contrast to the standard or determinist image of technology, a constructivist analysis of technology started to appear. The constructivist approach views the technical and political domains as highly intertwined. It perceives the technological development as a social process, which restricts the political and ethical issues to be dealt by, or left to, the scientists or technologists. Instead of technologically deterministic view, constructivist approach of analysing technology suggests a development of technology that is socially driven or constructed[203]. Technology, in which social shaping should be supplemented by the technical building of society[204]. In comparison to determinist view of technology, which presupposes the society's ability to deal with the final product, the constructivist approach suggests the development process to be influenced by the technology's economic, social, cultural or ecological effects and its implications may eventually help shape up the evolutionary styled technological solution[205].

---

199   Bijker, W. E., Hughes, T. P. and Pinch, T., (Eds.) (2012). Social Construction of Technological Systems: New Directions in the Sociology and History of Technology. Cambridge: MIT Press

200   Hoff, J., Horrocks, I. and Tops, P. (2003). Technology and Social Change. Democratic governance and new technology (pp. 28-48). London: Routledge

201   Williams, R. and Edge, D. (1996). The social shaping of technology. Research policy, 25(6), 865-899

202   Fuglsang, L. (2001). Three perspectives in STS in the policy context. In Cutcliffe and Mitcham (Eds.), Visions of STS: Counterpoints in science, technology, and society studies (pp. 33-49). Albany: State University of New York Press

203   Servaes, J. (2014). Technological determinism and social change: Communication in a Tech-Mad World. London: Lexington Books

204   Verdegem, P. and Verleye, G. (2009). User-centered E-Government in practice: A comprehensive model for measuring user satisfaction. Government Information Quarterly, 26(3), 487-497

205   Hardy, C. A. and Williams, S. P. (2008). E-government policy and practice: A theoretical and empirical exploration of public e-procurement. Government Information Quarterly, 25(2), 155-

62

The advocates of the constructivist view, such as Trevor Pinch, Wiebe Bijker or Thomas Hughes, suggest constructivist reasoning with the help of the concept of relevant social groups as a starting point in what they called as Social Construction of Technology or SCOT strategy[206]. These social groups comprise various members of society such as civil society, businesses, institutions, organisations and industries. The motivation behind the involvement of relevant social groups in the technological process is to have the technical artefact designed through numerous eyes and social reiterations[207]. Their interactions would lead to the different meanings of the same technology. Their divergent views result in negotiations, contestation and interpretation for multiple solutions to the problems associated with the technological design. Thus, the final shape of such a solution would assume consensus for the universal acceptability[208]. The SCOT researchers attribute this phenomenon to 'interpretative flexibility'. Technology that can be flexibly interpreted by majority members of society has a higher tendency to be understood at large[209].

Based on the interpretative flexibility of the technological solution, the constructivist approach, therefore, rather forces the technological route to avoid autonomous development style. It carries the design that can represent contingent products for the activities of social actors. Two aspects can be deduced from such technical design. A consensus-based technical design can on the one hand have the ability to exhibit the democratisation value attached to technology but, on the other hand, it also hints at the power nexus that is being exercised by the relevant

---

180 For more elaborate study on the social shaping of technology, Doctor and Dutton's work on The Social Shaping of the Democracy Network (DNet) provides a very comprehensive read. Hague, B. N. and Loader, B. D., (Eds.) (2005). Digital democracy: Discourse and decision making in the information age. London: Routledge

206 SCOT or Social Construction of Technology has been influencing many eGovernment studies and inspired by technology adoption modules use references to SCOT approaches while developing eGovernment applications. See for example, Bijker, W. E., et al.: 2012; Pigg, K. E. and Crank, L. D. (2004). Building community social capital: The potential and promise of information and communications technologies. The Journal of Community Informatics, 1(1), 58-73; Gronlund, A. (2005). What's In a Field - Exploring the eGoverment Domain. In Proceedings of the 38th Annual Hawaii International Conference on System Sciences. (pp. 3-6). New York: IEEE; Shin, D.-H. (2007). A critique of Korean National Information Strategy: Case of national information infrastructures. Government Information Quarterly, 24(3), 624-645; Thompson, M. (2008). ICT and development studies: Towards development 2.0. Journal of International Development, 20(6), 821-835

207 Pigg, K. E., et al.: 2004

208 Sawyer, S. and Eschenfelder, K. R. (2002). Social informatics: Perspectives, examples, and trends. Annual review of information science and technology, 36(1), 427-465

209 Nygren, K. G. (2010). "Monotonized administrators" and "personalized bureaucrats" in the everyday practice of e-government: Ideal-typical occupations and processes of closure and stabilization in a Swedish municipality. Transforming Government: People, Process and Policy, 4(4), 322-337; Ilshammar, L., Bjurström, A. and Grönlund, Å. (2005). Public E-Services in Sweden. Scandinavian Journal of Information Systems, 17(2), 11–40; Meijer, A. and Lofgren, K. (2010). Selling technology to the policy sciences: Marketing strategies for specialized scholars. Conference: Internet, Politics, Policy 2010: An Impact Assessment.Retrieved 04.05.2014 from http://blogs.oii.ox.ac.uk/ipp-conference/sites/ipp/files/documents/IPP2010_Meijer_Lofgren_Paper.pdf

social groups for the contested design[210]. There are two more stages in SCOT approach that are important for an artefact's wider appeal, namely, 'closure' and 'stabilisation'[211]. The "closure" stage of an artefact corresponds to the multiple interpretations that influence the technical design. "Closure" is reached once the interpretations achieve wide approval and the interpretative flexibility starts to diminish. In terms of shape and function, the artefact stabilises thus reaching the final stage of "stabilisation".

The constructivist approach, such as SCOT, exhibits that the way new technological systems could develop through a rather tedious process of negotiation and renegotiation, contestation and re-contestation. This iterative process produces a technological artefact whose meaning and use is influenced by innumerable societal forces and whose feedback eventually determine the final outcome of the artefact in use. Within the context of eGovernment, such a consensus-based approach attempts to dilute the discouragement of the powerless user of technologically determinist eService that may also help in curbing the digital divide[212]. Such an approach tries to place the human agency at the centre of technological development, albeit, to a limited extent[213].

Andrew Feenberg, one of the proponents of the Critical Theory of Technology, steps in with his constructivist approach by offering to view technology's politicised nature[214]. He shares a more or less similar constructivist view that it is not the technical principles alone that determine the shape of technology. Rather there are also social forces that drive the technological development. He reminds us of the powerful legislative authority of technology that is shaping up our society. That is, what happens when it is technology that decides on our behalf as to how we communicate, how we live and how we travel. Technology in such situations takes the central stage in our lives. However, if technology is so powerful, as Feenberg notes, then will it not be justified to gauge it with the same set of democratic standards as other political institutions[215]. If the technical designs were to be democratised, then social injustice prevailing in the form of digital divide, they may well be tackled with more diligence. Deliberations over the technical change would not only support the largely agreed principles of democracy such as voting on political issues, but it would also make technology a part of broader political spectrum and also help enhance participation and agency. The democratic interventions influencing the design of technology

---

210 Lassinantti, J., Bergvall-Kåreborn, B. and Ståhlbröst, A. (2014). Shaping Local Open Data Initiatives: Politics and Implications. Journal of theoretical and applied electronic commerce research, 9(2), 17-33
211 Bijker, W. E., et al.: 2012
212 Yu, L.: 2006
213 Parvez, Z. and Ahmed, P. (2006). Towards building an integrated perspective on e-democracy. Information, Community & Society, 9(5), 612-632
214 Feenberg, A. (2002). Transforming technology: a critical theory revisited. New York: Oxford University Press
215 Ferneding, K. A. (2003). Questioning technology : electronic technologies and educational reform. New York: Peter Lang

become necessary in order to combat the centralised and elitist control of design. Consensus or even dissensus- based interventions form a necessary prelude to shaping the more localised version(s) of technology[216].

Could there be other possible concerns held by the constructivist view(s) for democratising technological design for its meaning and design? The determinist and constructivist approaches for studying technology are the results of increased interest in technological controversies created in the Western and post-industrial scenario. The issue of socially shaped technology becomes more problematic when the technology ties a knot with the post-colonial society in, what David Lyon calls, a marriage of convergence[217]. If we consider examples from the contemporary world, then technological developments such as computer and internet -to which the human reliance is becoming incredibly worrisome- both evolved in the West. They represent and symbolise a particular view or model of the world[218]. Its inception and design approaches heavily inspired the socio-political, linguistic and/or cultural values known to the West[219]. The way governments today are rapidly employing (borrowed and adopted) technologies (from the West) for their governmental and public utilities, such a design, then have all the possibilities to come into conflict with the society that has entirely different social, political and economic traditions from the West[220].There are no quick answers; however, technology and culture both influence each other to a certain degree[221]. There are instances where a technology changes its shape according to the local culture, as we shall see in the case of the kiosk machines in Chapter. A similar instance is of an increasing number of service centres and knowledge kiosks that have been built all across India[222]. This reshaping and/or social embeddedness is paving the way to many innovative solutions that are currently being taken up in developing countries via alternative media of service delivery than the internet[223].

216   Taylor, J. and Lips, A. (2008). The citizen in the information polity: Exposing the limits of the e-government paradigm. Information Polity, 13(3), 139-152

217   Lyon, D. (2002). Surveillance society : Monitoring everyday life. Buckingham: Open University Press

218   Ciborra, C. and Navarra, D. D. (2005). Good governance, development theory, and aid policy: Risks and challenges of e-government in Jordan. Information technology for development, 11(2), 141-159

219   Heeks, R.: 2005a

220   Thompson, M. (2004). Discourse, 'Development' & the 'digital divide': ICT & the World Bank. Review of African Political Economy, 31(99), 103-123

221   For a detailed study on societal discourse on and through Internet, see for example, Slevin, J. (2000). Internet and society. Cambridge: Polity Press

222   Thomas, P. (2009). Bhoomi, Gyan Ganga, e-governance and the right to information: ICTs and development in India. Telematics and Informatics, 26(1), 20-31

223   Avgerou, C. (2008). Information systems in developing countries: a critical research review. Journal of information Technology, 23(3), 133-146; Chen, W. and Wellman, B. (2005). Minding the cyber-gap: the Internet and social inequality. In Romero and Margolis (Eds.), The Blackwell companion to social inequalities (pp. 523-545). Malden, MA: Blackwell Publishing

Thus, when digital divides continue, there can be solutions that may come from within the societies and cultures based on their indigenous experiences and capabilities. Such indigenous solutions have the potential to impact the technologies to undergo a process of (re)negotiation, (re)iteration and/or (re)design based on the feedback from the local users of technology[224]. One of the promising examples of eGovernment in developing countries comes from India's Gyandoot project, which was initiated in January 2000[225]. The project used information kiosks by establishing low cost Intranet connected with 20 government-owned information kiosks in five blocks of the district. 17 privately owned kiosks were included at the later point. All the information kiosks included a computer, a modem, a printer, a power generator and necessary furniture. All the kiosks were located in government offices or in a business sector. Prior to the initiation of Gyandoot, the procedure to access land records to meet the local records-keeper (a patwari)[226]. With a kiosk, it is made possible to file the request electronically. Every kiosk provided services that cover around 20 to 30 villages by serving a population of about 30,000.

One of the key aspects of choosing the owner of the kiosk was based on the voting system. The village committees and the local community jointly elected the owners of the kiosk. With the help of these kiosks, the villagers could apply for different government services such as to ask for a copy of a land record, old age pension, or birth certificate. Similarly, Gyandoot made it possible for the villagers to file complaints via kiosks[227]. Digital divide when seen through the lens of social constructivism tries to help us understand and take an in-depth view of socio-political forces that are crucial to the design of technology, particularly in the developing countries. SCOT on the other hand with its attributes of interpretative flexibility and social groups tries to highlight the important role that culture can play in the development and application of technology. Apart from the socio-cultural perspectives, there is a need for relevant preparedness that is crucial to the alleviation of digital divide. Within ICT and eGovernment context this is referred to as eReadiness.

---

224 Bakardjieva, M. and Feenberg, A. (2002). Community technology and democratic rationalization. The information society, 18(3), 181-192

225 Cecchini, S. and Raina, M. (2004). Electronic government and the rural poor: The case of Gyandoot. Information Technologies and International Development, 2(2), 65-76

226 For more detailed and critical study on the politics of Gyandoot project see for example: Sreekumar, T. T. (2007). Decrypting e-governance: Narratives, power play and participation in the Gyandoot intranet. The Electronic Journal of Information Systems in Developing Countries, 32(1), 1-24

227 Bhatnagar, S. (2009). Unlocking E-Government potential : concepts, cases and practical insights. Los Angeles: Sage Publications

2.6.2. eReadiness

One of the measures to assess and quantify the digital divide is eReadiness or electronic readiness of the country undergoing an eGovernment implementation. When discussing the access to knowledge via ICTs, eReadiness becomes an important instrument to gauge the country's preparedness to opt for eGovernment projects. Simply put, eReadiness refers to the degree to which a country would be equipped or ready to conduct eServices[228].

E-readiness is, as Heeks in his influential study on Africa points out is based on six crucial factors that provides a starting point towards and before implementation of eGovernment programme(s)[229]. These include:

- data systems infrastructure.
- legal infrastructure
- institutional infrastructure
- human infrastructure
- technological infrastructure
- leadership and strategic thinking

These factors, as also underscored by Dada, go beyond the technological infrastructure and look into, for example, laws and regulations that are required to support the eGovernment initiatives[230]. This, of course, leads to the legislative debate that a country opting to develop eServices, must have set up legislative bodies that can facilitate such efforts at the policy level. The legal infrastructure, as Heeks argues, must be supplemented by the necessary human infrastructure. This shall not be limited to only the knowledge and skill-set required at the development level but it shall equally involve the recipients' training as well. The recipient in this context is the citizen for whom the eServices are being built[231].

As Kuntelj and Vintar argue that such eServices must match the citizen's needs and problems[232]. In the case of developing countries, where there can be a varied range of e-literacy, technological and economic standards, an eService has to be capable to provide services taking limitations into consideration[233]. Coupled with human, technological and legal infrastructure, e-readiness for a country also

---

228   Ojo, A., Janowski, T. and Estevez, E. (2005). Determining Progress Towards e-Government-What are the Core Indicators? In 5th European Conference on on e-Government (ECEG2005). (pp. 313-322). Retrieved 09.11.2009 from http://collections.unu.edu/eserv/UNU:3052/Adegboyega-ECEG2005.pdf

229   Heeks, R. (2002a). e-Government in Africa: Promise and practice. Information Polity, 7(2), 97-114

230   Dada, D. (2006). E-readiness for developing countries: moving the focus from the environment to the users. The Electronic Journal of Information Systems in Developing Countries, 27(1), 1-14

231   Heeks, R.: 2002a

232   Kunstelj, M. and Vintar, M. (2004). Evaluating the progress of e-government development: A critical analysis. Information Polity, 9(3), 131-148

233   Taylor, J., et al.: 2008; ibid.

requires institutions that not only have the ability to understand the ICT developments but are also able to facilitate eGovernment projects and services[234].

2.6.3. Local Content and multilingualism on the web

Another important factor, which is crucial to the functioning of eGovernment in developing countries, is the language access-divide due to prevalence of English as a dominant language of the Internet[235]. In the early years of internet development, the content for the web was mostly written in English language. Since the Internet was originally developed in the United States, it was obvious that the medium of communication remained English. Similarly, after the invention of the World Wide Web by Sir Tim Berners Lee in the United Kingdom, most of the content that developed had an obvious inclination towards western culture and values[236]. This phenomenon has resulted into a web environment that is heavily tilted towards English-speaking users in the developed world with advanced educational skills. In other words, a huge disparity in access to the web and the Internet is compounded in terms of the number of websites in developing countries, amount of local language content, and the use of online content.[237]

In 2003, the UN Global eGovernment Survey reported that English was available to some extent on 125 websites out of 173, either as the default site language or in addition to the native language[238]. In 2012 survey, the United Nations eGovernment report cited more than 80 per cent of all websites were using English as its primary means of communication. In this regard, support for the local content production and the multilingual approach towards eServices is of paramount importance for a wider reach and acceptability of eGovernment which otherwise could create another vicious circle of information/digital haves and have- nots[239].

Similarly, the top four languages according to Internet World Stats between 2004 through 2012 decade have been English, Chinese, Spanish and Japanese[240]. The web content as well as the language speakers of this content have seen phenomenal growth and these speakers have grown in millions. However, over the course of eight years, that is from 2004 through 2012, English has maintained its dominance in all these years. In 2004 alone there were 289 million English

234   Gichoya, D. (2005). Factors affecting the successful implementation of ICT projects in government. The Electronic Journal of e-Government, 3(4), 175-184

235   Becker, S. A. (2005). E-government usability for older adults. Communications of the ACM, 48(2), 102-104

236   Berners-Lee, T., et al.: 1992

237   Bwalya, K. J. (2009). Factors affecting adoption of e-government in Zambia. The Electronic Journal of Information Systems in Developing Countries, 38(1), 1-13

238   United Nations: 2003

239   United Nations: 2012b

240   Internet World Stats (2012). "Internet World Users by Language". Retrieved 04.05.2014 from https://www.internetworldstats.com/stats7.htm.

language web content users that increased by 64% in 2012 amounting to 565 million users.

Figure 2: English language domination of the web/internet content

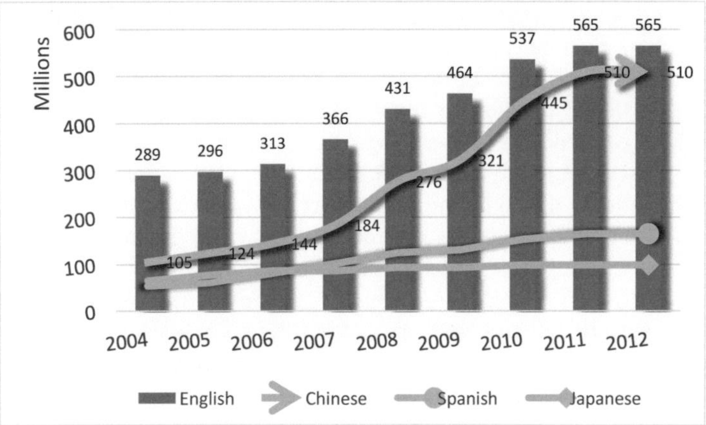

Source: Own illustration based on Internet World Stats[241]

Chinese web content users are the second to follow to English with 510 million web content users. The dominance of the English language for government websites severely affects the progress of eGovernment projects. These websites are legible to people who have the English language proficiency. The remaining population, which though literate, may be deprived of the information on the government websites[242].

2.6.4.eGovernment & Public-Sector Reform

For eGovernment in developing countries to work, the aspects of digital divide, eReadiness and local or multilingual content serve as building blocks, however, does eGovernment mean the automation or digitisation of manual processes?[243] It is often argued that digitising the records and bringing in more technological solutions in public administration can create efficient and effective way of public service delivery. Saxena opines that since eGovernment is predominantly a western approach, therefore the implications of eGovernment may well be different in developing countries the way it has been perceived, designed and

---

241 ibid.

242 Mutula, S. M. (2005). Peculiarities of the digital divide in sub-Saharan Africa. Program: electronic library and information systems, 39(2), 122-138

243 Asgarkhani, M. (2005). Digital government and its effectiveness in public management reform: A local government perspective. Public Management Review, 7(3), 465-487

69

experienced in the western countries[244]. In addition to that, prior to the use of ICTs for the public service delivery, western governments have over the period of time made several largely successful attempts in improving the efficiency and responsiveness of their governments by bringing public sector reform.

Such a reform was coupled with the wave of "New Public Management" (NPM) that started from Australia and New Zealand during 90s and inspired Europe and North America[245]. Saxena, points out that western countries have already gone through the reform process during the NPM phase before opting for the eGovernment systems[246]. Most of these reforms, therefore, came from within these countries based on their own socio-political and administrative approaches[247]. In comparison, most of these reforms in developing countries have been brought about through the assistance of international development organisations or at least blueprints offered by them. The dilemma, if it may be termed, with the developing countries remained that these public-sector reforms had limited appeal and impact due to the top-down approaches employed. Consequently, despite all attempts at restructuring the economies in various developing countries, the public administrations in developing countries remain highly centralised and extremely bureaucratic[248].

In addition to the centralised and complicated bureaucratic structures, several developing countries in Asia or Africa still adopt procedures and structures from their colonial past[249]. Resultantly the governance and administrative system that emerge is based on hybrid approaches with their roots in the legacy/old systems or at times in their colonial past and some driven by the donor agenda[250]. Within the eGovernment perspective, several developing countries with stale and post-colonial structures eventually became recipients of the eGovernment practices. On the other hand, eGovernment advocates the overhauling and/or a process re-engineering of governance structures before opting for digitisation[251]. As a

244   Saxena, K. (2005). Towards excellence in e-governance. International Journal of Public Sector Management, 18(6), 498-513

245   Torres, L., Pina, V. and Royo, S. (2005). E-government and the transformation of public administrations in EU countries: Beyond NPM or just a second wave of reforms? Online Information Review, 29(5), 531-553

246   Homburg, V. (2004). E-government and NPM: a perfect marriage? In Proceedings of the 6th international conference on Electronic commerce. (pp. 547-555). Delft: ACM Press.

247   Cordella, A. (2007). E-government: towards the e-bureaucratic form? Journal of information Technology, 22(3), 265-274

248   Barima, A. K. and Farhad, A. (2010). Challenges of making donor-driven public sector reform in sub-Saharan Africa sustainable: Some experiences from Ghana. International Journal of Public Administration, 33(12-13), 635-647

249   Guma, P. (2012). Public Sector Reform, E-Government and the Search for Excellence in Africa: Experiences from Uganda. Electronic journal of e-government, 11(2), 241-253

250   Huque, A. S. (2005). Explaining the myth of public sector reform in South Asia: de-linking cause and effect. Policy and Society, 24(3), 97-121

251   Fuchs, C., et al.: 2008; Coleman, S. (2005). African e-governance-Opportunities and Challenges. Oxford University: Oxford Internet Institute

consequence, the kind of eGovernment system that emerges from the developing countries is half adopted which also leads to potential collapse.

2.6.5.Revisiting Layne & Lee model

Based on the four key factors that are key to the development and growth of eGovernment in developing countries, it seems necessary to revisit the earlier premise, based on the stages of eGovernment developed by Layne & Lee. This sort of an eGovernment model serves as a blueprint to initiation of any eGovernment application. Considering the digital divide factors coupled by the lack of local content and e-readiness approaches, it is pertinent to review the Layne & Lee model of eGovernment and assess its application in developing countries.

A closer look at this model indicates that the web along with the internet as a communication medium is at the heart of every stage. This may eventually mean that every single citizen trying to interact with the government for any possible activity should qualify the prerequisites of the digital literacy as well as the possession of required gadgetry. Among other things, it also needs to be seen if the Internet, as the major carrier of digital information, has the inclusive potential to deliver eServices/information to a wider spectrum of audience? This raises a whole lot of new questions, for instance, did all the countries around the globe pursuing eGovernment ensure not just the adequate diffusion of communication technologies but also the readiness aspects discussed earlier in this chapter. The diffusion in this context means the availability and presence of all internet-related technologies[252]. This obviously includes uninterrupted electricity supply and subscription to the telephone service in a conventional Internet scenario. A reasonable access for the financially challenged could be Internet cafes, but that probably might still be for people with low e-literacy to handle the internet information at the internet cafes[253].

According to the United Nations Report, the internet subscription per 1000 users among the OECD countries in comparison to the least developing countries was 562 whereas for the least developed countries it was low as 17[254]. Considering that communication technologies have developed and evolved in most of the OECD countries, their figure of 562 internet users at least project that 50% of population is actually equipped with the internet. What can be deduced from these statistics? Internet, which is supposed to be the sole carrier of eService data, does not even have 5% of subscribers in the countries where information technology revolution has not had a timely reach or probably has not yet reached. Similarly,

---

252   Reddick, C. G. (2004). A two-stage model of e-government growth: Theories and empirical evidence for US cities. Government Information Quarterly, 21(1), 51-64
253   Schuppan, T.: 2009
254   United Nations Conference on Trade and Development. (2007). The Least Developed Countries Report. Retrieved 04.05.2014 from https://unctad.org/en/Docs/ldc2007_en.pdf

71

according to another survey, among the world's least developed nations, 32 countries were found to be in the emerging phase that is at the cataloguing stage of eGovernment. The government websites or portals are characterised by static and insufficient information. These portals are infrequently updated and offer very few interactive features and non-existent online services. Similarly, several cases from developing countries were found at the junction of an enhanced and interactive stage, that is the transaction and vertical integration stages.

With the Internet alone and together with these statistics the applicability and functioning of Layne & Lee model arguably becomes questionable[255]. It may well serve to the future of eGovernment but it certainly disqualifies the societies and the groups, which as discussed earlier are at the peripheries in the technological age[256]. Their utilisation of the internet as a core carrier of eService data may still seem a far cry. Therefore, the Layne & Lee model is perhaps not an ideal or global solution to eGovernment in developing countries. For a more inclusive approach, particularly for the marginalised groups, the demand for another eGovernment model(s) seem inevitable[257]. These models may well use the core of Layne & Lee's approach of developing eServices but the end product, the one at the citizen's end, may have to incorporate all the societal and cultural norms and communication patterns with which s/he has been acquainted to. In other words, while technology adoption may continue to happen amidst technological advancements, a less technologically-deterministic approach in developing a citizen-centric eService may be necessary for eGovernment to work in developing countries[258].

2.6.6.ICTs and changing service delivery channels

According to the United Nations' eGovernment development index, the top 20 countries that have been ranked high on eGovernment index do not include a single developing country. While some of the reasons have been given and explained in the sections above, India's experience with Gayandoot tries to hint at one of the problems. Most of the developing countries have higher rural population that remain underserved not only in teledensity but also in social and human development sectors. The example from India has shown that Gyandoot has attempted to cater to the rural sector. Similarly, if the critique of the Internet being the sole medium of communication is taken into consideration, as it was

---

255 Yildiz, M. (2007). E-government research: Reviewing the literature, limitations, and ways forward. Government Information Quarterly, 24(3), 646-665

256 Baqir, M. N. and Iyer, L. (2010). E-government maturity over 10 Years: A comparative analysis of e-government maturity in select countries around the world. Comparative E-Government (pp. 3-22). New York: Springer

257 Heeks, R.: 2002b

258 Davison, R., Vogel, D., Harris, R. and Jones, N. (2000). Technology Leapfrogging in Developing Countries-An Inevitable Luxury? The Electronic Journal of Information Systems in Developing Countries, 1(1), 1-10; Warschauer, M. (2003a). Demystifying the digital divide. Scientific American, 289(2), 42-47

discussed in the section above, then shall that mean that the developing countries will not be able to partake in eGovernment drive till a certain level of eReadiness, digital literacy and means to access the internet are attained universally or at least broadly?

In this regard, we see from the example of the Gyandoot initiative that the service delivery mechanism is rapidly changing. Due to the rise in mobile phone penetration, a number of countries are now relying on eServices that can be offered through mobile phones, interactive voice response systems, digital television, and self-service terminals[259]. Village kiosks and telecentres for the rural areas are also offering an array of government services.

The United Nations' 2012 study on eGovernment has also given a special and detailed emphasis on the use of multi-channels for public service delivery and also tried to emphasise why the need of multichannel public service delivery for developing countries is becoming important[260]. With growing convergence in technology and the increasing pervasion of the Internet, the web has become the most used eGovernment delivery mechanism. This, however, does not tackle the problem of accessibility for users that do not have stable internet connections or lack the cognizance of or means to learn the use of a certain technology. Traditionally, there were multiple sources of delivery of information like post, face-to-face contact or telephone[261]. The oral factor in these means ensured that messages reached even those who could not read or lived in distant areas.

To make these services inclusive for every citizen requires multiple channels that can be used by them at their own convenience and ease[262]. Mobile phones represent one such alternative delivery mechanism. As statistics from ITU show, more and more people are available on mobiles than fixed line telephony and some are in fact accessing the internet through it[263]. Mobile phones extend outreach to usually ignored groups such as rural poor, seniors and the physically challenged as the relative economical technology allows every user to keep a personal device[264].

---

259   Germanakos, P., Samaras, G. and Christodoulou, E. (2007). Multi-channel delivery of e-services in the light of m-government challenge. In Kushchu (Ed.), Mobile Government: An Emerging Direction in e-Government (pp. 292-317). Hershey: IGI Global

260   United Nations: 2012b

261   AlAwadhi, S. and Morris, A. (2009). Factors influencing the adoption of e-government services. Journal of Software, 4(6), 584-590

262   Gorla, N. (2008). Hurdles in rural e-government projects in India: lessons for developing countries. Electronic Government, An International Journal, 5(1), 91-102

263   International Telecommunication Union. (2010). ICT Statistics. Retrieved 12.09.2011 from https://www.itu.int/en/ITU-D/Statistics/Pages/stat/default.aspx

264   Porter, G. (2012). Mobile phones, livelihoods and the poor in Sub- Saharan Africa: Review and prospect. Geography Compass, 6(5), 241-259

Figure 3 : United Nations 2012 eGovernment development index

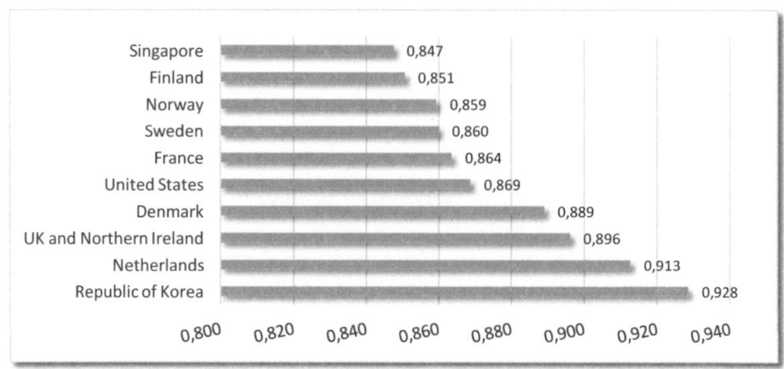

Source: Own illustration based on UN eReadiness report[265]

Development of mobile websites allows specially condensed versions of information to be accessible for a smaller screen with slow speeds. Additionally, a mobile phone allows the employment of location-based services thus using resources effectively and efficiently for the benefit of the user and the service provider. Mobile phones are also being increasingly used to distribute social benefits and provide timely message services that allow mobiles to act as identification mechanisms to facilitate banking. In Turkey, for instance, there is a UYAP SMS information service developed by the Ministry of Justice[266]. This service relies on SMS as a communication channel (instead of web/internet) to inform respondents in a legal case about updates or requests from the court which earlier had to be collected in person. This SMS service is an augmentation to the official delivery mechanism but IT allows clients to prepare for court requests and helps reduce response rates and improve the output of the courts[267].

Kiosks are emerging as another way of extending public service where it can be used on a shared-basis from public areas, thereby ensuring participation of everyone and independence from a self-owned device[268]. Kiosks being dedicated units are built with customised software that allows for easier and user-friendly interfaces besides having an in-built and secure payment mechanism. Another mode is employing public-private partnerships to insert a human-element for facilitating users wherein trained staff assist the user or perform the required

---

265 United Nations: 2012b

266 Ministry of Justice (Republic of Turkey) (2008). "UYAP SMS Information System". Retrieved 10.12.2012 from http://sms.uyap.gov.tr/smseng/engnedir.html.

267 Çam, A. R. (2007). SMS information system: Mobile access to justice. European Journal of ePractice, 4(10), 35-41

268 Ni, A. Y. and Ho, A. T.-K. (2005). Challenges in e-government development: Lessons from two information kiosk projects. Government Information Quarterly, 22(1), 58-74

functions on their behalf[269]. This is especially durable in the case of low-income or technically-illiterate target groups and also serves to be a source of employment for local people and a business opportunity in a market which is competitive enough to allow the private sector to move in as the last mile of delivery.

Telecentres are an example of such an arrangement and allow a one-stop solution for the provision of multiple services[270]. The case of e-seva in the Indian state of Andhra Pradesh is an example of such an arrangement where diverse functions such as applications for driver's license, passports, registration of birth and marriage certificates and filing of taxes and utility bills can be achieved through any of the 400 available counters[271]. Similarly, in Italy Reti Amiche uses alternative delivery mechanism for requesting residence permits, tax payments, certified electronic messages to public administration or payment transaction[272]. All of these services are accessible through tobacconists or post-offices that already exist and are used by the people. Introduction of these services not only facilitates the end user but also these organisations by providing an alternative source of income.

The key to multichannel delivery tries to ensure that digitisation or advances in technology do not miss out the less-engaged or underprivileged sections of the society[273]. The omnipresence of these services allows better delivery of public services while not only ensuring a collection of standardised data and records but also an opportunity to centralise information and keep it updated to the last transaction irrespective of the mode employed. According to the United Nations eGovernment Survey 2012, most the member states employ the web channels to deploy eGovernment services[274]. Additionally, 71 countries utilise public-private partnerships, 32 employ kiosks and 60 provide mobile-based services. In a regional comparison, Asia is closely behind Europe, which leads multichannel service delivery mechanism, in mobile-based and public-private partnerships. The Asian average is higher due to the strong eGovernment campaigns of Gulf countries which have relatively smaller populations and are financially better off than their South and East-Asian counterparts. Choice of multichannel delivery

269 Sharma, V. and Seth, P. (2011). Effective Public Private Partnership through E-Governance Facilitation. Computer & Communications Sciences, 34(1), 15-25

270 Durrant, F. (2002). e-Government and the Internet in the Caribbean: An initial assessment. In Traunmüller and Lenk (Eds.), Electronic Government. First International Conference, EGOV 2002 Aix-en-Provence, France, September 2–6, 2002 Proceedings (pp. 101-104). Berlin: Springer; Rao, S. S. (2008). Social development in Indian rural communities: Adoption of telecentres. International Journal of Information Management, 28(6), 474-482

271 India Filings (2015). "Government Initiatives: eSeva". Retrieved 04.05.2015 from https://www.indiafilings.com/learn/eseva/.

272 Paoli, A. D. and Leone, S. (2015). Challenging conceptual and empirical definition of e-government toward effective e-governance. International Journal of Social Science and Humanity, 5(2), 186

273 Ahmed, N. (2006). An Overview of e-Participation Models. A publication of the UN Department of Economic and Social Affairs (UNDESA). Retrieved 04.05.2014 from http://unpan1.un.org/intradoc/groups/public/documents/UN/UNPAN023622.pdf

274 United Nations: 2012b

mechanism depends heavily on the usage patterns, per capita income, presence of content in local language, technical literacy and adoption of the users[275]. Similarly, centralising information across different mechanisms needs to ensure that information flows across departments and is not bogged down by bureaucratic policies. Other challenges include budgetary constraints and lack of technical knowledge, however, in the long run it appears that multichannel service delivery mechanisms will continue to grow as an alternative to the web/internet service delivery for effective and transformational eGovernment.

2.7. Concluding Remarks

ICTs along with the eGovernment as a strategy do not aim to replace the administrative or government processes. On the contrary, employment of latest technological tools and use of digital transmission mechanism helps improve the governments to provide the public service in more efficient and effective manner. Therefore, ICTs only provide an alternative to improve such governance mechanisms whereas the access to and delivery of public services through traditional and manual mechanism continue to function.

eGovernment is, therefore, a reliance on ICTs and computer technologies for governments to perform in a more proficient manner with its citizens and also within itself. The employment of ICTs and computing systems when adopting eGovernment has the ability to also improve the manual procedures and functioning of the public sector. These diverse aspects of the use of technologies in government, as this chapter has tried to underscore, does not only remain a question of management science or information technology. Adoption of eGovernment leaves its impact on a much broader scale of political, social, economic, linguistic, cultural and technical level(s). The experiences of developing countries while employing eGovernment have, thus, yielded varying results as this chapter has also tried to highlight. These experiences have led to new and revised debates for the role of socio-cultural factors that assume importance while discussing the case of eGovernment in developing countries.

Internet which is the key communication medium in development of eGovernment applications is rapidly redefining the concept of social contract within eGovernment scenario. However, there are several impeding considerations for developing countries that include unequal access to the internet, digital literacies, electronic preparedness or eReadiness which do not only hinder the growth of eGovernment but also question the importance of internet as a central communication medium. As a consequence, the debates on technological determinism within the ICTs and digital divide context continue to populate and present contesting opinions within the eGovernment discourse.

---

275 Gagnon, Y.-C., Posada, E., Bourgault, M. and Naud, A. (2010). Multichannel delivery of public services: A new and complex management challenge. International Journal of Public Administration, 33(5), 213-222

Social constructivist theorists, as this chapter has tried to argue, advocate for more social input in development process of technologies and cultural embedding within technological artefacts.

These approaches carry the ability and potential to result into more responsive eGovernment applications, particularly within the developing countries context. The changing delivery channels of and the access points for the G2C services (such as kiosks, multipurpose telecentres) can be considered as one of the outcomes of technologies that correspond to socio-cultural norms and traditional communication channels prevalent in (respective) society. Pakistan's journey in eGovernment seems to have traversed the same route as that of other developing countries. There has been initial reliance on web service mechanism but given the factors of digital divide, eReadiness, local content, elitist nature of technologies initially produced an eGovernment as highly exclusive domain. However, before dwelling on Pakistan's experience with eGovernment, it is essential to understand the cultural and historical context in which eGovernment is being applied.

Chapter 3

Governance, Society and Media in Pakistan

In order to approach the central question of communication technologies as agents of change in public service delivery, the theoretical background discussed in Chapter 2 has tried to contextualise technologies within a framework of governance and society. This chapter attempts to explore and debate these ideas within a Pakistani context. The aim of this chapter is thus to briefly understand the issues of governance, formation of society, the current mass media and technological investments in a Pakistani scenario. This also includes some of the key factors that can or may directly influence the debate on the use of ICTs in Pakistan. Before discussing the ICT scenario in Pakistan, it is pertinent to mention that the communication structures that exist today in Pakistan are because of reasons that lie in Pakistan's social and political history. These structures are the result of communication patterns that evolved with the development of Pakistani society. Thus, to understand the public services scenario in Pakistan, it is also important to briefly overview the governing structures that have shaped Pakistan's current administrative and bureaucratic set-up and procedures. It is, however, to be noted that it is not within the boundaries of this Chapter to present an analysis of these rather difficult philosophical and, at times, politically intertwined terrains of society, culture and governance.

Pakistani society is largely carved out of what the British left in 1947 with their partition plan with two Pakistans separated by thousands of miles of India dividing the two[276]. Pakistan as a nation-state emerged after India's partition and eventual independence from the British in 1947. The society did not follow a coherent structure: neither did it pre-exist nor does Pakistan have one homogenous culture[277]. It rather has held together people who speak different languages, follow different religious persuasions, social norms, race and ethnicities[278]. These ethnicities also happen to be closely associated with provincial societies that are further differentiated into tribes, clans, dialects or languages and/or traditions[279]. The majority of Pakistani population grows up with one or two extra languages in addition to their own regional ones, the major ones

---

276   Cohen, S. P. (2011). The future of Pakistan. Washington, D.C.: Brookings Institution Press; Cohen, S. P. (2006). The idea of Pakistan. Washington (D.C.): Brookings Institution Press

277   Akbar, M. J. (2012). Tinderbox : the past and future of Pakistan. New York: Harper Perennial

278   Shaikh, F. (2009). Making sense of Pakistan. New York: Columbia University Press

279   Alavi, H. (1986). Ethnicity, Muslim society, and the Pakistan ideology. In Weiss (Ed.), Islamic reassertion in Pakistan: The application of Islamic laws in a modern state (pp. 21-47). Syracuse: Syracuse University Press; Majeed, G. (2010). Ethnicity and ethnic conflict in Pakistan. Journal of Political Studies, 17(2), 51

include Punjabi, Balochi, Sairki, Sindhi, Pushto or Urdu[280]. The largest linguistic group in Pakistan is comprised of Punjabis which make up for over 44% of the total population, followed by Pushtoon (15.4%), Sindhis (14.1%), Saraiki (10.1), those speaking Urdu (7.6%) and Balochis (3.6%)[281]. Despite the obvious disparity among the languages in terms of speakers and nativity, Urdu was declared as the national language, considering that it was the language of the Muslim renaissance in the province of Uttar Pradesh (UP) India[282]. English on the other hand received a status of an official and constitutional language[283].

At the time of partition, East and West Pakistan's combined population was 75 million. Over a period of 70 years Pakistani population has seen a rapid growth[284]. In 1951 the population was 33 million but over the past seven decades population has exponentially grown to 207 million[285]. As the population of Pakistan has grown, so have its problems. The present-day Pakistan, even 70 years after its independence, faces a plethora of social, political and economic challenges[286]. These range from lack of access to basic necessities of health and public education, from acute power and energy crisis to political instability, and from corruption to lack of government's writ[287]. As a young nation, Pakistan's progress has seen several social, political and cultural transformations which included having a rift with India on Kashmir and other issues, lack of politically elected leadership which resulted in experiments with forms of government required to run Pakistan and the fall of East Pakistan to form Bangladesh[288].

280   Rahman, T. (2008). Language policy and education in Pakistan. In Hornberger (Ed.), Encyclopedia of language and education (pp. 383-392). Boston, MA: Springer

281   Qadeer, M. A. (2006). Pakistan : social and culural transformations in a Muslim nation. London: Routledge

282   Daechsel, M. (2006). The politics of self-expression the Urdu middle-class milieu in mid-twentieth century India and Pakistan. London; New York: Routledge

283   Rahman, T. (2003). Language policy, multilingualism and language vitality in Pakistan. In Saxena and Borin (Eds.), Lesser-Known Languages of South Asia: Status and Policies, Case Studies, and Applications of Information Technology (pp. 73-104). Berlin: Walter de Gruyter

284   Hussain, S., Malik, S. and Hayat, M. K. (2009). Demographic transition and economic growth in Pakistan. European Journal of Scientific Research, 31(3), 491-499

285   Pakistan Bureau of Statistics (2017). Province Wise Provisional Results of Census - 2017. http://www.pbs.gov.pk/sites/default/files/PAKISTAN%20TEHSIL%20WISE%20FOR%20WEB%20CENSUS_2017.pdf

286   Burki, J. (2008). Changing perceptions and altered reality : Pakistan's economy under Musharraf, 1999-2007. Oxford: Oxford University Press; McCartney, M. (2011). Pakistan--the political economy of growth, stagnation and the state, 1951-2009. Abingdon, Oxon ; New York: Routledge

287   Imam, A. and Dar, E. A. (2014). Democracy and Public Administration in Pakistan. New York: CRC Press; Asif, M. (2011). Energy crisis in Pakistan : origins, challenges and sustainable solutions. Karachi: Oxford University Press

288   Jaffrelot, C. (2004). A history of Pakistan and its origins. London: Anthem

3.1. Governance and the formation of civil administration in Pakistan

Pakistan, in its seven plus decades of existence, has shared an estranged relationship with democracy with a few interludes of democratic rule occasionally occurring in the 70s and 90s, with a more recent stint from 2008 onwards[289]. Parliamentary democracy in Pakistan has failed four times partially due to military's dominance in the state of affairs at international, national, and local levels[290]. Within first three decades of its independence, Pakistan already experimented with three constitutions with the most recent one being implemented in 1973[291]. Continuous amendments, either in the name of reform and strengthening of democracy, have often led to military rulers prolonging their stay in power[292]. A sustained political stability and lasting democratic setup, however, still seems a far cry for Pakistan's democracy enthusiasts[293].

Pakistan's post-colonial origin and its experiments with democracy have had a direct effect on its governance structure[294]. Pakistan's civil administrative structure and a number of public sector reforms trace their roots in Pakistan's colonial legacy which also needs to be taken into perspective as it has shaped most of Pakistan's initial bureaucratic setup[295]. Prior to the partition of India in 1947, British Raj had established administrative structure that comprised of civil servants. These civil administrators were particularly trained to maintain the control mechanism in the representative institutions[296]. The civil administrative structure which post-partition Pakistan received and adopted carried a similar understanding. It emphasised more on bureaucracy and control and showed less focus on governance and public service[297].

---

289   Siddiqa-Agha, A. (2007). Military Inc. : inside Pakistan's military economy. London: Pluto Press

290   Hussain, E. (2010). Military agency, politics and the state the case of Pakistan (Doctoral Dissertation. Faculty of Economics and Social Sciences, University of Heidelberg, Germany). Retrieved 23.08.2012 from http://archiv.ub.uni-heidelberg.de/volltextserver/10947/1/Thesis.Online_Pub.5.8.10.pdf

291   Shah, A. (2002). South Asia faces the future: Democracy on hold in Pakistan. Journal of Democracy, 13(1), 67-75; Ali, S. A. (1995). Unicameralism in United Pakistan: Why and How? Pakistan Horizon, 48(3), 69-80

292   Talbot, I. (2012). Pakistan : a new history. New York: Columbia University Press

293   Aziz, M. (2008). Military control in Pakistan : the parallel state. London; New York: Routledge

294   Amin-Khan, T. (2012). The post-colonial state in the era of capitalist globalization: Historical, political and theoretical approaches to state formation. New York: Routledge

295   Niaz, I. (2011). Advising the State: Bureaucratic Leadership and the Crisis of Governance in Pakistan, 1952–2000. Journal of the Royal Asiatic Society, 21(1), 41-53; Jones, G. N. (1997). Pakistan: A civil service in an obsolescing imperial tradition. Asian journal of public Administration, 19(2), 321-364

296   Talbot, I. (2010). India and Pakistan. In Brass (Ed.), Routledge Handbook of South Asian Politics (pp. 43-56). Oxon: Routledge

297   Wilder, A. (2009). The politics of civil service reform in Pakistan. Journal of International Affairs, 63(1), 19-37; Khan, T. H., Khan, H. A. and Wariach, M. A. (2014). Governance challenges in Public Sector: an Analysis of Governance and Efficiency Issues in Public Sector Institutions in Pakistan. Governance, 4(9), 58-67

Right from the beginning, Pakistan was overpowered by the civil bureaucracy along with the army, which sociologist Hamza Alavi referred to as the over-developed state structure[298]. The nexus of civil bureaucracy and the army grew over a number of years[299]. The echelons of elite cadre comprising civil and military bureaucracy put more stress on the centre-periphery power configuration[300]. By ensuring a strong power centre at the cost of the provincial and regional actors, the Pakistani establishment eventually pushed these actors further to the margins[301]. Pakistan's initial capital city remained Karachi from 1947 until 1960s. By carving out a new city close to the Kashmir Valley and the Himalayan range, the capital shifted from Karachi to Islamabad in 1963. The federal capital surrounded by the Punjab province had soon started to accommodate more representation of Punjab in its bureaucratic structure[302]. The dominance of Punjab in the civil and military institutions started to overshadow the other three provinces of Sindh, Balochistan and NWFP (now Khyber Pakhtunkhwa)[303]. Such a (re)configuration has led to serious backlashes and provincial squabbling over political autonomy between the Punjab and the poorer and smaller provinces[304].

The administrative structure of Pakistan has thus remained very centralised for the past 70 years and the supremacy of civil bureaucracy can be observed at several levels[305]. According to its constitution, Pakistan follows the Westminster style of parliamentary democracy where there is a separation between bureaucracy and the representative political executive, nevertheless the civil bureaucratic system nurtured in the colonial period has largely remained intact. Whenever the Pakistani political elite tried to establish its writ and reorganise the administrative structure, a strong resistance came its way, sometimes also in the form of premature dissolution of parliaments[306]. Thus, the political representatives

298   Misra, A. (2010). India-Pakistan : coming to terms. New York: Palgrave Macmillan; Alavi, H. (1975). India and the colonial mode of production. Economic and Political Weekly, 10(1975), 1235-1262

299   Aziz, M.: 2008

300   Talbot, I. (2002). Does the Army Shape Pakistan's Foreign Policy? In Jaffrelot (Ed.), Pakistan, nationalism without a nation (pp. 311-333). London: Zed Books

301   Zaidi, S. A. (2005). State, Military and Social Transition: Improbable Future of Democracy in Pakistan. Economic and Political Weekly, 40(49), 5173-5181

302   Jalal, A. (1993). The State and Political Privilege in Pakistan. In Weiner and Banuazizi (Eds.), The Politics of social transformation in Afghanistan, Iran, and Pakistan (pp. 152-184). Syracuse, N.Y.: Syracuse University Press

303   Waseem, M. (2011). Pakistan:A Majority-Constraining Federalism. India Quarterly, 67(3), 213-228

304   Jalal, A. (2014). The struggle for Pakistan : a Muslim homeland and global politics. Cambridge (Mass.); London: The Belknap Press of Harvard University Press

305   Hussain, M. and Kokab, R. U. (2013). Institutional influence in Pakistan: Bureaucracy, cabinet and parliament. Asian Social Science, 9(7), 173-178

306   Ahmed, S. (2001). The Fragile Base of Democracy in Pakistan. In Shastri and Wilson (Eds.), The Post-Colonial States of South Asia: Democracy, Development and Identity (pp. 41-68). New York: Palgrave Macmillan

resultantly formed alliances but at the same time also remained largely dependent upon the establishment[307].

The bureaucracy or establishment has been considered a consistent source of 'stability' and has always tried to serve as a counterweight to political upheaval and government instability[308]. The bureaucracy or the civil service was initially given the name of Civil Service of Pakistan (CSP) through which the business of state got carried on. The roots of CSP go back to a similar institution called the Indian Civil Service (ICS) during the pre-partition period[309]. The ICS officers themselves were a result of the training at the College of Fort William in Calcutta. This college was originally set up for the training of East India Company officials but later it also started to recruit the locals who could attain advancement and position in British-controlled India[310]. At the time of the partition, there were approximately 1,100 Indian Civil Service officers, which also included 100 Muslims officers.[311] Out of this configuration, eighty-three of them opted to go to Pakistan. In a vast country like Pakistan, particularly when its two wings, East and West Pakistan were a thousand kilometres apart, found itself without indigenous administration after the British declared independence in 1947. The CSP officers arrived in a country that had almost no administration to begin with and Pakistan relied on the administrative system that was established/left by the British[312]. Since there was an urgent need of administrators, the CSP officers who left India for Pakistan received quicker promotions.

Access to basic necessities and provision of public services remained the responsibility of civil administration set-up. In order to maintain a balance between politics and civil administration the founder figure of Pakistan, Mohammad Ali Jinnah (1876-1948), underscored that CSP officers shall concentrate on day-to-day administrative affairs and avoid any meddling with the state politics. However, after Jinnah's death in 1948, CSP officers developed the administrative as well as the political clout by eventually designing and carrying out implementation of national policies[313]. The CSP was disbanded in 1973 and various services were combined into one administrative system grouped into 22

---

307   Malik, I. H. (1997). The Supremacy of the Bureaucracy and the Military in Pakistan. State and Civil Society in Pakistan: Politics of Authority, Ideology and Ethnicity (pp. 57-80). London: Palgrave Macmillan

308   Ziring, L. (1997). Pakistan in the twentieth century : a political history. Karachi: Oxford Univ. Press

309   Chaudry, A. (2011). Political Administrators: The Story of the Civil Service of Pakistan. Oxford: Oxford University Press

310   Mohabbat Khan, M. (1999). Civil service reforms in British India and united Pakistan. International Journal of Public Administration, 22(6), 947-954

311   Niaz, I. (2010). The Culture of Power and Governance of Pakistan: 1947 - 2008. Oxford: Oxford University Press

312   Bhagwandas, R. (2015). Pakistan: Federal Public Service Commission and its Functions. In Chalam (Ed.), Governance in South Asia : State of the Civil Services (pp. 239-256). New Delhi: SAGE Publications

313   Barany, Z. (2009). "Authoritarianism in Pakistan". In Policy Review. Retrieved 02.06.2014 from https://www.hoover.org/research/authoritarianism-pakistan.

Basic Pay Scales (BPS 1-22)[314], though the former CSP officers retained important positions in the administrative apparatus[315]. This reform led to the creation of District Management Group (DMG) as well as Secretariat Group which plays a key part in policy-level decision making and falls in the higher BPS category (20-22)[316]. Although the elite character of the bureaucracy was partially attenuated but over the years the historical dominance that was synonymous during British period in administrative system has been increasing. Policy planning generally takes place in secretariat divisions whereas their implementations are carried out by the directorates, also referred to as attached departments[317]. The fact that two similar bureaucratic institutions are attached to each other for policy planning and implementation has often led to conflicts[318]. An example of such an arrangement will be witnessed in the later chapters particularly with reference to the formation of Electronic Government Directorate as an attached department under Ministry of Information Technology (see Chapter 4).

3.2. Service Delivery and Public-Sector Reform

Experiments with civil administrative system, political system and military-civil dichotomy and their tussle in Pakistan continued to affect access to public services and efficient service delivery system in past 70 years[319]. With time, the elements of bribery and corruption have also become part and parcel of the administrative culture in Pakistan[320]. Stanley Kochanek believes that the cultural norms in India provided a conducive environment for corruption to grow[321]. A pre-partitioned India already experienced institutionalised form of delinquencies at lower levels of administration. This involved paying small amounts of money to public servants to process documents or receive service that would otherwise be a basic right[322]. These practices continued to follow in Pakistan as well and over past seven decades such wrongdoings have become an integrated part of

---

314   Khan, M. M. (2002). Resistance to administrative reforms in South Asian civil bureaucracies. In Farazmand (Ed.), Administrative reform in developing nations (pp. 73-88). Westport, Conn.: Praeger

315   Braibanti, R. (1966). Research on the bureaucracy of Pakistan: a critique of sources, conditions, and issues, with appended documents. Durham: Duke University Press

316   Rizvi, H. A. (2000). Military, State and Society in Pakistan. London: Macmillan Press

317   Kennedy, C. H. (1987). Bureaucracy in Pakistan. Karachi: Oxford University Press

318   Husain, I. (1973). Mechanics of Development Planning in Pakistan: A Suggested Framework. Pakistan Economic and Social Review, 11(4), 454-462

319   Kazi, S. (1995). Rural Women, Poverty and Development in Pakistan. Asia-Pacific Journal of Rural Development, 5(1), 78-92

320   Maheshwari, S. (1974). Administrative Reforms in Pakistan. The Indian Journal of Political Science, 35(2), 144-156; Syed, A. H. (1971). Bureaucratic Ethic & Ethos in Pakistan. Polity, 4(2), 159-194

321   Kochanek, S. A. (1974). Business and politics in India. Berkeley: Univ. of California Press

322   Islam, N. (1989). Colonial legacy, administrative reform and politics: Pakistan 1947-1987. Public Administration & Development, 9(3), 271-285

administrative system[323]. Getting a passport, driver license or even children's admission to educational institutions require either strong socio-political affiliations or bribes[324]. Service delivery in this regard has been poorly affected by factors that were either adopted from pre-partitioned India or the ones that developed over a period of time[325].

Post 1971, the remaining (West) Pakistan predominantly comprised a rural society. With a huge influx of rural populace into the urban centres, the society has been reshaping itself by becoming more urban every year in demographic and cultural sense. Pakistan's rural population on the other hand also increased from 27.7 to 89.3 million between 1951 and 1998. If the population of high density rural districts and areas near cities to the urban population are also added, the rural–urban balance would tilt toward a rural majority. Therefore, rural communities have remained central to Pakistani society. However, the developments between the rural and urban Pakistan have remained uneven and access to public service for rural centres in comparison to urban areas has been a far larger challenge for the government[326]. There have been eras in Pakistan's history, particularly during General Ayub's (1907-1974) period and also during other dictatorships, that saw continuous foreign aid when Pakistan showed steady economic growth[327].

Despite some considerable achievements in technology and commerce, Pakistan is still challenged with problems similar to those faced at the time of independence. For an exponentially increasing population, the sustained provision of public service delivery has remained a major problem for the Government[328]. The rising population growth rate has also made it difficult for the Government to address the problems of poverty. On one hand, as was earlier argued, most of the public sector is populated by the strong bureaucracy, while on the other hand it has been less proactive in ensuring even public service delivery to both urban and rural areas of Pakistan[329]. Governance at the grassroots level in Pakistan is still very weak and in several areas the government is unable or limited

---

323 Maniruzzaman, T. (1971). Crises in Political Development" and the Collapse of the Ayub Regime in Pakistan. The Journal of Developing Areas, 5(2), 221-238; Khan, F. (2007). Corruption and the Decline of the State in Pakistan. Asian Journal of Political Science, 15(2), 219-247

324 Islam, N. (2004). Sifarish, sycophants, power and collectivism: administrative culture in Pakistan. International Review of Administrative Sciences, 70(2), 311–330

325 Hasnain, Z. (2010). Devolution, accountability, and service delivery in Pakistan. The Pakistan Development Review, 49(2), 129-152

326 Mustafa, D. and Sawas, A. (2013). Urbanisation and political change in Pakistan: Exploring the known unknowns. Third World Quarterly, 34(7), 1293-1304

327 Mahmood, S. (2009). Reform of the public services in Pakistan. New York: Nova Science Publishers

328 Kamran, T. (2008). Democracy and governance in Pakistan. Lahore: South Asia Partnership-Pakistan. Retrieved 30.04.2019 from http://sappk.org/wp-content/uploads/publications/eng_publications/Democracy_and_Governance.pdf

329 Hasnain, Z. (2008). "Devolution, Accountability, and Service Delivery: Some Insights from Pakistan". World Bank Policy Research Working Paper (4610). Retrieved 02.04.2011 from https://openknowledge.worldbank.org/bitstream/handle/10986/6713/wps4610.pdf;sequence=1.

in its capacity to provide basic services such as health, water, education or sanitation[330]. The basic infrastructure facilities are not yet as widespread as it is in its urban centres. According to World Bank indicators, Pakistan is categorised as a low-income country where poverty is widespread[331]. At the dawn of the new millennium, roughly 40% of the population still lived at national poverty line of \$1.59[332]. In 2011, its per capita GNI (gross national income) was recorded as \$1,1213 placing it as a poor nation[333]. Similarly, there is a stark contrast in the living standards of Pakistani households, which are divided into elite upper class, middle class, lower middle class and poor, and thus trigger wide social disparities[334].

A number of reform initiatives taken up by either the military and civilian governments to improve the service delivery appeared to have failed due to political and provincial politics, Pakistan's strong bureaucracy and/or use/lack of inadequate resources[335]. Such state of affairs has resultantly encouraged non-governmental organisations or international donor agencies to step in the role of assisting the government[336]. Similarly, there is an array of civil society organisations that have assumed the role of public service providers for education and health services; essential obligations for which government should be responsible and accounted for[337]. The apathy on the Government's part has also led to the traditional institutions taking charge of the system wherever the government writ has remained weak[338]. In several pockets of Pakistan, the landholders have assumed a quasi-governmental role and often serve as a go-between during times of dispute. Similarly, many religious functionaries mobilise

---

330　Still far from the target (2011c, October 05), Daily Dawn. Retrieved 02.06.2012 from https://www.dawn.com/news/664142

331　International Monetary Fund. (2004). Pakistan: Poverty Reduction Strategy Paper - IMF Country Report No. 04/24. Retrieved 02.02.2009 from https://www.imf.org/external/pubs/ft/scr/2004/cr0424.pdf

332　World Bank. (2015). Pakistan Development Update. Retrieved 05.06.2015 from http://documents.worldbank.org/curated/en/531691468098388002/pdf/957510WP00PUBL0pril 20150edited0fina.pdf

333　World Bank. (2011). The World Bank Annual Report 2011: Year in Review. Retrieved 12.11.2012 from http://siteresources.worldbank.org/EXTANNREP2011/Resources/8070616-1315496634380/WBAR11_YearInReview.pdf

334　Durr-e-Nayab (2011). Estimating the middle class in Pakistan. The Pakistan Development Review, 50(1), 1-28

335　Ishrat, H. (1999). Pakistan-The Economy of an Elitist State. Oxford: Oxford University Press

336　Khan, M. A. and Ahmed, A. (2007). Foreign aid—blessing or curse: Evidence from Pakistan. The Pakistan Development Review, 46(3), 215-240; Bano, M. (2008). Dangerous correlations: Aid's impact on NGOs' performance and ability to mobilize members in Pakistan. World Development, 36(11), 2297-2313

337　Khan, I. A. (2010). Public sector institutions, politics and outsourcing: Reforming the provision of primary healthcare in Punjab, Pakistan. Journal of International Development, 22(4), 424-440

338　Lamb, R. D. and Hameed, S. (2012). Subnational Governance, Service Delivery, and Militancy in Pakistan: A Report of the CSIS Program on Crisis, Conflict, and Cooperation. Retrieved 04.05.2014 from https://csis-prod.s3.amazonaws.com/s3fs-public/legacy_files/files/publication/120610_Lamb_SubnatGovernPakistan_web.pdf

people through madrasas and mosques and offer poorest of the poor food, shelter and (religious) education[339].

Weak grip on the state authority and experimentation with the governance system has continuously weakened civilian rule and eventually provided chances for the military to take over the state of affairs[340]. With a promise of devolution of power, decentralisation, to lessen bureaucracy's control and perhaps also to strengthen his stay in power, the then ruler General Musharraf introduced a set of reforms in the civil administration system which saw the abolition of deputy commissioner's office and a creation of District Coordination Officer (DCO)[341]. DCOs aimed to provide and direct executive level actions and take measures for instance in improving public service delivery and their efficiency. The reforms introduced by General Musharraf for the induction of DCO system also created a room for an additional system of union councils headed by elected district administrator called "Nazim"[342]. Routine public service tasks from legal attestations to birth/death registration or property disputes fall under district administration. Musharraf's devolution of power plan was a radically different one from that of his predecessors (such as Ayub and Zia) as in this case devolution complemented reforms in other public-sector areas such as taxation, civil service, electoral and police department[343]. The current system of governance in Pakistan appears to be an amalgam of Musharraf's decentralisation plan and of the two democratic tenures of the Pakistan People's Party (2008-2013) and Pakistan Muslim League-N (2013-2018)[344].

We see here that a number of factors have contributed to Pakistan's patchy growth, unsustainable economic models, civil-military quest for power and bureaucratic system inspired by the British colonial period that has gone through a number of reform processes yet remain authoritative in nature and less service-oriented in approach[345]. These historical references are essential to understand the context in which electronic public services have been initiated. Some of the

---

339   Malik, J. (2008). Madrasas in South Asia : teaching terror? London: Routledge
340   Siddiqa-Agha, A.: 2007; Waseem, M. (2002). Causes of Democratic Downslide. Economic and Political Weekly, 37(44-45), 4532-4538
341   Shah, S. A. H., Khalid, M. and Shah, T. (2006). Convergence Model of Governance: A Case Study of the Local Government System of Pakistan. The Pakistan Development Review, 45(4), 855-871
342   Cyan, M. R. (2006). Main issues for setting the civil service reform agenda in Pakistan. The Pakistan Development Review, 45(4), 1241-1254
343   Alam, M. (2013). Pakistan's Devolution of Power Plan 2001: A Brief Dawn for Local Democracy? In Sansom and McKinlay (Eds.), New Century Local Government: Commonwealth Perspectives (pp. 44-57). London: Commonwealth Secretariat
344   Taj, A. and Baker, K. (2018). Multi-level Governance and Local Government Reform in Pakistan. Progress in Development Studies, 18(4), 267-281
345   Lange, M. K. (2004). British colonial legacies and political development. World Development, 32(6), 905-922; Jhatial, A. A., Cornelius, N. and Wallace, J. (2014). Rhetorics and realities of management practices in Pakistan: Colonial, post-colonial and post-9/11 influences. Business History, 56(3), 456-484; Iqbal, M. and Ahmad, E. (2006). Is Good Governance an Approach to Civil Service Reforms? The Pakistan Development Review, 45(4), 621-637

electronic services by NADRA (as we will see in the following chapters), for instance, Family Registration Certificate (FRC) and Verisys (citizen verification system) during their formation cited historical references to corruption, fraud and deceit that remain rampant in the Pakistan. The following sections briefly look at the developments of ICTs in Pakistan as well as the other related advances in digital media landscape.

3.3. Developments in IT in Pakistan (British Colonial Period)

As with the case of civil administration sector, development in the ICTs sector in Pakistan can also be traced back to Pakistan's colonial legacy under the British when it first developed the telegraph and later the telephone[346]. The British Empire exercised its control in India by developing and dominating numerous communication channels. These channels included modes of physical communication such as the development of an extensive railway network, steamboats and a postal system. While on the other hand at the level of digital communication, the development of the telegraph is attributed to Dr William Brook O'Shaughnessy who worked under the East India Company in building a telegraphic system between 1836 and 1839 which he eventually managed to finish by 1856[347].

The telegraph system built by O'Shaughnessy connected important cities of Peshawar, Agra, Bombay, Madras and Calcutta to the garrison cities and other important British settlements through overhead telegraph lines[348]. Between 1852 through 1856, the East India Company massively but strategically invested in building up an extensive network of telegraph system that spanned over 4000 miles. This involved construction of high poles that traversed the entire countryside[349].

It is interesting to notice how the geographic topology of telegraph nodes was constructed. It appeared that telegraph seemed to the Company more as tool of control and dominance over India[350]. The Company at that point of time was operated under the administration of Lord Dalhousie (1812-1860), who remained India's Governor General from 1848 to 1856[351]. He will remain synonymous with not only initiating telegraph system in India but also introducing the Post Office

---

346 Pinkerton, A. (2008). Radio and the Raj: broadcasting in British India (1920–1940). Journal of the Royal Asiatic Society, 18(2), 167-191

347 Choudhury, D. K. L. (2000). 'Beyond the reach of monkeys and men'? O'Shaughnessy and the telegraph in India c. 1836-56. The Indian Economic & Social History Review, 37(3), 331-359

348 Choudhury, D. K. L. (2010b). The Telegraph and the Uprisings of 1857. Telegraphic Imperialism: Crisis and Panic in the Indian Empire, c. 1830 (pp. 31-49). Basingstoke: Palgrave Macmillan

349 Gorman, M. (1971). Sir William O'Shaughnessy, Lord Dalhousie, and the Establishment of the Telegraph System in India. Technology and Culture, 12(4), 581-601

350 Solymar, L. (2000). The effect of the telegraph on law and order, war, diplomacy, and power politics. Interdisciplinary Science Reviews, 25(3), 203-210

351 Headrick, D. (2010). A double-edged sword: communications and imperial control in British India. Historical Social Research, 35(1), 51-65

Act[352]. In terms of geographical typology of telegraph poles, Dalhousie thought of a telegraph route that also provided the Company administrative, political and military advantage[353]. In this regard, the telegraph poles were spread all across India and their route included from Calcutta (now Kolkata) to Agra, then south to Bombay (now Mumbai) on the west coast thereby crossing Madras (now Chennai) on the east coast. Similarly, the route from Delhi across entire northwest crossing Ambala, Jhelum, Rawalpindi and Peshawar (see Figure 4).

Figure 4 : Telegraph system developed by the British in 1856

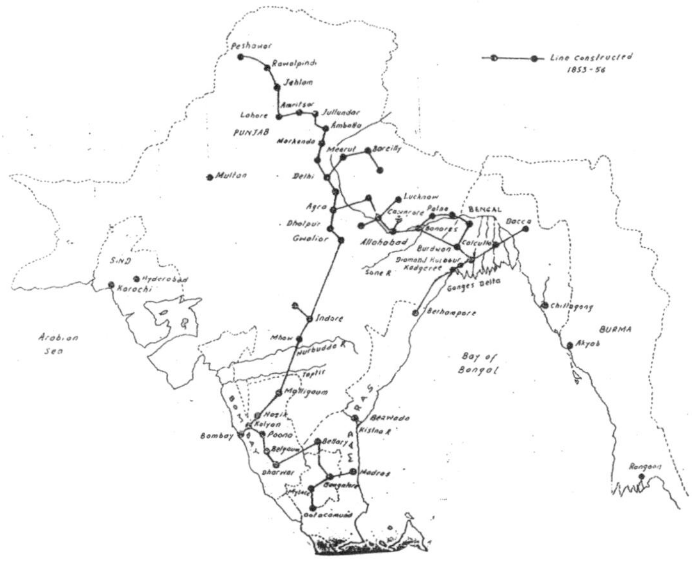

Source: Reza[354]

For development and laydown of telegraph system, the construction system took place in two different stages. The first stage involved in setting up overhead lines that were made up of bamboo poles. The purpose of these overhead lines was to establish a communication system as soon as possible in order to provide the first electricity-based communication system to the British military[355]. Such a system would not only have served the military but would also have provided faster communications to allow intelligence sharing, curb and contain local uprisings by

---

352 Wolpert, S. (2004). A new history of India. New York: Oxford University Press
353 Kerr, I. J. (2007). Engines of Change: The Railroads That Made India. Westport: Praeger
354 Reza, Y. (1999). History of Pakistan Telecommunications. Islamabad: Pakistan Telecommunication Authority
355 Gorman, M.: 1971, p. 589

Indians, for instance[356]. Dalhousie managed to conquer the Sikh territory in the northwest, elsewhere he continuously and considerably spread the British frontiers[357]. Thus, it was only natural for the Company to speed up completion of the telegraph system in order to procure major British possessions in India[358].

In the second stage the temporary bamboo poles were replaced by more permanent type of rods. Once the telegraphic system was established, duly tested and in place, it was used by more than just the military[359]. The Company opened telegraph services to both the English as well as Indian trade associations for commercial and financial purposes. Similarly, the telegraph service was also extended to the print media when India-based two English language newspapers, Delhi Gazette and Lahore Chronicle, were also given a licence to utilise the telegraph services for press communication[360].

It is interesting to note how the telegraph played a crucial role in legitimising the Empire's dominance in colonial India through technical means[361]. On one hand, the British managed to penetrate into the social fabric of Indian society by spreading the telegraph for commercial and journalistic purposes. On the other hand, the telegraph served as the first digital communication technology that gave the Empire an edge, for example, in the 1857 uprising[362]. The uprising that began in the city of Meerut on May 10, 1857 and spread to Delhi had most of the Indian soldiers (sepoys) who were serving in the British army. It did not take long for the sepoys to take over and the following day Delhi had fallen[363].

It was then that telegraph pronounced itself as the sole informer for the British Raj to all stations in north and south. Based on the instant information received via telegraph, the British army had the upper hand over freedom fighters in crushing their rebellion. Although there certainly were numerous social, political, and moral factors that contributed to the English 'success', it was also telegraph, as new technology, that lent superiority to the British[364]. It is interesting to note how rebels remembered telegraph. Sepoys of the 1857 revolt realised that that the

356 Bayly, C. A. (2000). Empire and Information: Intelligence gathering and social communication in India, 1780-1870. Cambridge: Cambridge University Press

357 Llewellyn-Jones, R. (2007). The great uprising in India, 1857-58 : untold stories, Indian and British. Woodbridge: Boydell & Brewer

358 Sarkar, S. (2010). Technological momentum: Bengal in the nineteenth century. Indian Historical Review, 37(1), 89-109

359 Pradhan, Q. (2007). Empire in the hills: the making of hill stations in colonial India. Studies in History, 23(1), 33-91

360 Kuppuram, G. and Kumudamani, K. (1990). History of science and technology in India. Delhi: Sundeep Prakashan; Choudhury, D. K. L.: 2010a

361 Baber, Z. (2001). Colonizing nature: scientific knowledge, colonial power and the incorporation of India into the modern world- system. The British journal of sociology, 52(1), 37-58

362 Choudhury, D. K. L. (2009). The Telegraph, Censorship, and 'Clemency': Canning during 1857. Contemporary Perspectives, 3(1), 34-54

363 Putnis, P. (2013). International press and the Indian uprising. In Carter and Bates (Eds.), Mutiny at the Margins: New Perspectives on the Indian Uprising of 1857: Vol 3: Global Perspectives (pp. 1-17). London, New Delhi: Sage Publications

364 Knayar, P. (2007). The great uprising: India, 1857. New Delhi: Penguin Books

British mobilised their forces to contain the rebellion with quick responses by channelling the communication via telegraph system. Thus, the rebels perceived this technology as hostile as it brought their rebellion down[365]. Perhaps a celebratory attitude was shown by chief commissioner of Punjab, Sir John Lawrence (1811-1879) who declared telegraph as saviour of the British Empire[366].

The invention of telegraph established itself as a digital tool for the Empire to expand its hold over Indian territories through digital means by exchanging data at lightning speed[367]. This immensely helped the British in strategic and military coordination as well as in having a tighter grip over India[368]. On economic front, English and Indian business communities overwhelmingly adopted telegraph as rapid information-sharing device, for instance, for exchange of the prices of commodities as well as information about shipping or weather[369]. In terms of its social effects, Gorman claims that the literate class of Indians made an extensive use of telegraph for exchanging personal and family messages[370]. Thus, telegraph wire also provided further legitimacy to India's imperial government. It significantly helped the British to exercise not just its territorial, political and economic influence but also the cultural authority.[371].

A couple of corollaries need to be drawn here. The invention of digital communication mechanisms that took place at the hands of colonial rulers and the governmentalisation of the entire geographic and socio-political strata of India created a sense of permanent dependency on the British[372]. Such a dependency continued in India and Pakistan from independence in 1947 to date. The bureaucratic structure, language and format of command and control are highly prevalent in both India and Pakistan. This also further clarifies the arguments stated in earlier passages about the colonial influence in the bureaucracy in Pakistan since 1947 that follows the similar governance structures left by the British. Their approach in dealing with the country and its citizens is loosely

---

365   Lahiri, N. (2003). Commemorating and remembering 1857: The revolt in Delhi and its afterlife. World Archaeology, 35(1), 35-60

366   David, S. (2003). The Indian Mutiny: 1857. London: Penguin Books

367   Worth, A. (2014). Imperial media : colonial networks and information technologies in the British literary imagination, 1857-1918. Columbus: The Ohio State University Press

368   Rand, G. (2013). Reconstructing the Imeprial Military after Rebellion. In Rand and Bates (Eds.), Mutiny at the Margins: New Perspectives on the Indian Uprising of 1857: Volume IV: Military Aspects of the Indian Uprising (pp. 93-112). London, New Delhi: Sage Publications

369   Though military in conception, Internet also met such an overwhelming response when it was made public in 1990s especially by the introduction of eCommerce which eventually led to the beginning of securitisation of states through digital means. See for example: Henman, P.: 2010

370   Gorman, M.: 1971

371   Sangwan, S. (1988). Indian response to European science and technology 1757–1857. The British Journal for the History of Science, 21(2), 211-232

372   Subrahmanyam, G. (2006). Ruling continuities: Colonial rule, social forces and path dependence in British India and Africa. Commonwealth & Comparative Politics, 44(1), 84-117

90

inspired by the British colonial rule that is based on more control over its subjects[373].

Similarly, we see that telegraph, as a communication technology, had its background in command and conquer mechanism. It does not come as a big surprise that 100 years after the invention of telegraph, when the USA first experimented in developing a communication technology in 1960s, the Internetwork of computers or the Internet was also a military project[374]. Its main focus, as argued in first and second chapter, was to develop a network that is robust and is able to perform even if some nodes are disconnected. Both telegraph and Internet that started off as a military tool later went on to become commercial product with the latter being more pervasive[375].

3.4. Diffusion of radio, television, and telephone (Post 1947)

The history of mass communication in Pakistan is a story of the gradual build-up to a point that radio, television and access to fixed line telephone was slowly spread out to the entire country. At the time of independence, Pakistan inherited two radio stations, Lahore and Peshawar, and a countrywide network of telegraph services[376]. New broadcasting stations later started from Karachi, Rawalpindi and Pakistan-administered Kashmir. The service was later spread to Quetta and Hyderabad in the 1950s[377].

During the 1970s, the radio service became popular particularly across the rural sector for its particular coverage of information about and for villages and farmers. With arrival of television, the radio maintained its popularity in the rural Pakistan. Towards the end of 1990s, approximately three million television sets were owned by Pakistanis. Similarly, around 30 to 40% population had access to television which increased every year due to the arrival of satellite dish and cable television[378]. Post 2000, the relaxation and privatisation in media opened door to many new television channels which are accessible on local cable networks[379].

---

373   Ziring, L. and Robert LaPorte, J. (1974). The Pakistan Bureaucracy: Two Views. Asian Survey, 14(12), 1086-1103

374   Schell, B. H. (2007). The Internet and society : a reference handbook. Santa Barbara: ABC-CLIO

375   Hoffmann, B. (2004). The politics of the Internet in Third World development : challenges in contrasting regimes with case studies of Costa Rica and Cuba. New York: Routledge

376   Elahi, M. and Zia, A. (2008). Pakistan. In Banerjee and Logan (Eds.), Asian Communication Handbook 2008 (pp. 369-404). Singapore: Asian Media Information and Communication Centre

377   Radio Pakistan (2014). "Chronology of Pakistan Broadcasting Corporation (PBC): Radio Pakistan in the light of history". Retrieved 22.05.2014 from http://www.radio.gov.pk/chronology-of-pbc.

378   Yusuf, H. (2013). Mapping Digital Media: Pakistan: A Report by the Open Society Foundations. Retrieved 02.01.2014 from https://www.opensocietyfoundations.org/uploads/4d1a6626-d6d7-41b5-befc-d8f5a78d545c/mapping-digital-media-pakistan-20130902.pdf

379   Incentives to private TV channels stressed (2003e, August 02), Daily Dawn. Retrieved 01.09.2008 from https://www.dawn.com/news/133754/karachi-incentives-to-private-tv-channels-stressed; Seminar calls for freedom of electronic media (2002a, February 01), Daily

On the other hand, until the 1960s, telephone was considered a privilege or a rarity. Its access was restricted to public officials and businessmen, that too only in bigger (urban) centres[380]. The telephone network did not initially include small villages and towns. In the 1970s the telephone network slowly started to expand at the national level. Use of telephone became common in cities. However, population in smaller towns and villages remained only sparsely connected with telephone networks[381].

In the 1990s, the introduction of satellite and wireless technology slightly extended access to telephones. By end of the 1990s when mobile phones slowly spread across social strata of Pakistan, rural areas started to physically map into the telephone network[382].

3.4.1. ICT Developments and Digital Divide in Pakistan (1990 onwards)

Pakistan's population, as stated earlier, mostly comprises rural population. Socio-economic development in Pakistan has been mostly urban- oriented[383]. This disparity has started to become more obvious in terms of public amenities, developmental budgets and modern ICTs such as fixed-line telephony and internet. Increase in population occurred in rural areas whereas development (infrastructural or social) focused more on urban centres[384]. Ratio of population has so far remained 44% (urban) and 56% (rural)[385]. From 1990 to 2012, the total population of Pakistan grew from 107 million to 187 million[386]. Telephone network had not covered even 6 million population. In terms of statistics, the expansion of telephone network in 1990 showed 0.8 million subscribers and by

Dawn. Retrieved 30.08.2008 from https://www.dawn.com/news/17381/karachi-seminar-calls-for-freedom-of-electronic-media

380   Aziz, M. A. (1979). A history of Pakistan : past and present. Lahore: Sang-e-Meel Publications

381   World Bank. (2013a). DataBank: World Development Indicators. Retrieved 02.06.2013 from https://databank.worldbank.org

382   Batool, S. H. and Mahmood, K. (2010). Entertainment, communication or academic use? A survey of Internet cafe users in Lahore, Pakistan. Information Development, 26(2), 141-147

383   Ahmed, F. (1996). Pakistan: Ethnic Fragmentation or National Integration? The Pakistan Development Review, 35(4), 631-645

384   Malik, S. J. (1993). Poverty in Pakistan, 1984-85 to 1987-88. In Lipton and Gaag (Eds.), Including the Poor: Proceedings of a Symposium Organized by the World Bank and the International Food Policy Research Institute. (pp. 487-519). Retrieved 02.07.2011 from http://documents.worldbank.org/curated/en/649751468764364351/pdf/multi-page.pdf

385   Hasan, A. and Raza, M. (2009). Migration and small towns in Pakistan: Working Paper Series on Rural-Urban Interactions and Livelihood Strategies (Working Paper 15). Retrieved 13.01.2012 from https://pubs.iied.org/pdfs/10570IIED.pdf

386   World Bank: 2013a; National Institute of Population Studies. (2013). Pakistan Demographic and Health Survey. Retrieved 30.01.2014 from https://www.nips.org.pk/abstract_files/PDHS%20Final%20Report%20as%20of%20Jan%2022-2014.pdf

2012 this figure only increased to 5,86 million fixed line subscribers (see Figure 5).

Figure 5: Fixed Line Telephone Subscribers from 1990 through 2012

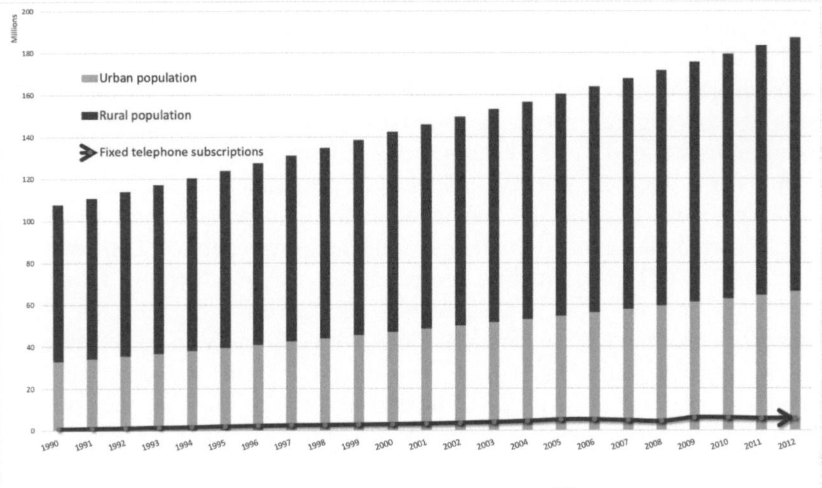

Source: Own Illustration based on World Development Indicators[387]

This uneven growth has left the rural centres mostly disconnected from rest of Pakistan[388]. Developments in other ICTs, particularly during the post-1990s, such as wireless networks and internet have been speedier than the traditional wired network in Pakistan[389]. In 1996, the government established Pakistan Telecommunication Authority (PTA) to regulate, maintain and oversee the provision of telecom services[390]. The teledensity started to increase towards the end of the 1990s particularly. This occurred partially due to deregulation in the telecommunication sector, arrival of internet and mobile network operators[391]. Due to the deregulation policy of the government, new licenses were issued for Long Distance and International (LDI) and Local and Wireless Local Loop (LL, including WLL) services[392]. In the wake of deregulation, PTA started issuing new

---

387   World Bank: 2013a

388   The great digital divide (2009b, September 21), Daily Dawn. Retrieved 07.10.2011 from https://www.dawn.com/news/838989

389   Pakistan Telecommunication Authority. (2007a). Industry Analysis Report. Retrieved 17.12.2010 from https://www.pta.gov.pk/media/industry_report_2007_1.pdf

390   Pakistan Telecommunication Authority (PTA). (1996). The Gazette of Pakistan. Act No. XVII OF 1996. Retrieved 08.03.2009 from https://www.pta.gov.pk/media/telecom_act_170510.pdf

391   Looney, R. E. (1998). Telecommunications policy in Pakistan. Telematics and Informatics, 15(1-2), 11-33

392   Ansari, S. and Saleem, S. (2009). .pk: Pakistan. In Akhtar and Arinto (Eds.). Digital Review of Asia Pacific 2009-2010. (pp. 294-301). Ottawa: International Development Research Centre.

licenses for data network operators and Internet Service Providers (ISPs)[393]. However, as seen in Figure 5, these developments were still not sufficient to help bridge the digital gap (at least in terms of fixed line telephone) between rural and urban Pakistan.

In order to address the digital divide, government of Pakistan under the Ministry of Information Technology launched Universal Service Fund (USF) towards the end of 2006[394]. USF works in collaboration with public and private telecommunication companies which contribute 1.5% of their adjusted revenues whereas the government acts as a facilitator and does not pay any financial contribution in the fund[395]. The idea of the USF particularly focuses on the rural areas in order to increase the level of telecom penetration in the unserved and under-served areas through the provision of local loops and rural voice and broadband communications[396]. Under its rural telecom programme, USF launched development of ICT infrastructure project in 26 various districts. To expand USF's scope and infrastructural network, public sector organisation PTCL (Pakistan Telecommunication Limited) along with other private mobile network operators including Mobilink, Ufone, Telenor and Warid Telecom have been continuously participating in ICT infrastructure projects all across rural Pakistan[397]. On the other hand, to generate a trickle-down effect of USF initiative for the rural communities, PTA also proposed a rural telecentre project named Rabta Ghar that aimed to establish 400 telecentres throughout the country[398]. The telecentres are multipurpose which are run by a private individual and include the computer access, fax, printing, telephone and broadband internet services[399].

In terms of investment, during the decade of 2000-2010, in years 2006 and 2007 Pakistan saw a strong foreign investment of USD 1,905 and 1,824 million

https://idl-bnc-idrc.dspacedirect.org/bitstream/handle/10625/38550/IDL-38550.pdf?sequence=1&isAllowed=y

393　Pakistan Telecommunication Authority. (2003). Annual Report 2002-03. Retrieved 03.10.2008 from https://www.pta.gov.pk/media/annual_report.pdf

394　Universal Service Fund: A success story (2012b, May 18), Business Recorder. Retrieved 02.02.2013 from https://fp.brecorder.com/2012/05/201205181190823/

395　Universal Service Fund (USF) is not an indigenous idea that only works in Pakistan. It is a global practice undertaken by several countries to ensure that telecommunication services are in the reach of wider audience on the principle of availability, affordability and accessibility. For an informed study, see International Telecomunication Union. (2013). Universal Service Fund and Digital Inclusion for All Retrieved 15.12.2013 from https://www.itu.int/en/ITU-D/Digital-Inclusion/Documents/USF_final-en.pdf

396　Universal Service Fund. (2009). Universal Service Fund: Telecom for All. Annual Report 2008-09. Retrieved 25.05.2014 from https://usf.org.pk/assets/publication-pdf/annual-report2008-09.pdf

397　Universal Service Fund (USF). (2013). Universal Service Fund: Telecom for All. Annual Report 2012-13. Retrieved 03.04.2010 from https://usf.org.pk/assets/publication-pdf/annual-report.pdf

398　PTA launches 'Rabta Ghar' (2007b, October 30), Business Recorder. Retrieved March 29, 2010 from https://fp.brecorder.com/2007/10/20071030645727

399　400 telecentres for villages (2007g, February 16), Daily Dawn. Retrieved 09.03.2009 from https://www.dawn.com/news/233033

respectively in the telecom sector (the highest during that decade)[400]. Exponential growth in the telecom sector eventually resulted in the creation of 80,000 to 500,000 jobs directly or indirectly. The telecom sector constituted 2 percent of Pakistan's GDP increase every year (with an expectation of 2009-10). The deregulation in telecommunication sector paved way to multiple mobile network operators to apply for licenses and these also included foreign companies such as Warid Telecom and Telenor.

Table 3: Foreign Direct Investment (FDI) and Telecommunication Share

	2001-02	2002-03	2003-04	2004-05	2005-06	2006-07	2007-08	2008-09	2009-10	2010-11
Total FDI*	484.7	798	949.4	1524	3521	5140	5410	3720	1574	813
FDI in Telecom Sector*	-	-	207.1	494.4	1905	1824	1439	815	79	361
Telecom (%) Share	6.1	13.5	21.8	32.4	54.1	35.5	26.6	22	5	44

Source: Own Illustration based on the Annual Reports of PTA[401]

In US$, Million

Pakistan's political situation also impacted (both positively and negatively) its foreign trade and foreign direct investment during this period. It was a time when Musharraf's military rule ended in 2008 and the civil government was formed after a gap of 10 years. During transition from military to civilian rule, foreign direct investments momentarily saw fluctuations which also impacted the telecom sector (see Table 3). With an exception of 2009, between 2004 through 2011, the telecom share in foreign direct investment in Pakistan has remained 22 plus % with the highest being in 2005-06 (see Table 3). As a result of these investments, mobile phone penetration has been growing at exponential rates.

From two mobile network operators in 1990s to as many as five operators in 2005 brought massive modern digital infrastructure projects to Pakistan. This also triggered a tough competition among operators[402]. These factors directly impacted

---

400   Pakistan Telecommunication Authority. (2011). Annual Report 2010-11. Retrieved 03.07.2012 from https://www.pta.gov.pk/annual-reports/pta_ann_rep_11.pdf
401   Pakistan Telecommunication Authority (2012). Annual Reports. Retrieved 20.01.2012 from https://www.pta.gov.pk/en/data-&-research/publications/annual-reports
402   International Telecommunication Union. (2007). Foreign Direct Investment in Pakistan Telecommunication Sector. Paper presented by PTA staff member Mr. Asif Inam at Regional Seminar on Costs and Tariffs for Member Countries of the Tariff Group for Asia and Oceania (TAS) in Seol, South Korea. Retrieved 03.08.2011 from https://www.itu.int/ITU-D/finance/work-cost-tariffs/events/tariff-seminars/Korea-07/presentations/FDI_Aasif_Inam.pdf

the price of per minute call and per SMS[403]. As a consequence, the lower prices of call/SMS factor started to take the mobile phone market outside the urban centres into the rural belts as the affordability of mobile phones eased it into lives of rural Pakistan[404]. This resulted in massive digital leapfrogging as mobile phone penetration grew much more compared to the fixed line phone subscription in Pakistan[405]. If the Figure 5 is reanalysed we see that the commercialisation of internet did not hugely impact the fixed-line telephone subscriptions and these only saw a meagre increase between 2000 to 2004. Between 2004 through 2012 the total number of fixed-line phone subscribers have not seen any significant increase or decrease.

However, if the Figure 5 is reviewed with an additional variable, that is, with an addition of mobile phone subscription, we see that a dramatic impact of mobile phones usage in Pakistan's both rural and urban market. Fluctuating between 0.5 million to 1.0 mobile subscribers in 2001 to 2003, we see that the mobile subscription was less than fixed phone line subscription. However, within a span of one year, mobile subscription starts to take a steep curve beginning 2004. The competition between local and foreign mobile network operators further helped this geographical outreach[406]. It took three years from 2004 through 2007 to expand the mobile phone coverage to the urban centres. By 2012, that is, after five more years the mobile network operators managed to reach almost all of rural Pakistan (see Figure 6).

Government backed USF initiative and mobile phone companies own financial interests were the major factors of this enormous digital outreach[407]. This scenario is eventually helping the entire telecommunication landscape in Pakistan not only in terms of connectivity but also the electronic services that can be provided via mobile phones as we shall see in the following chapters.

403 Cartels in telecom sector (2012c, October 11), Daily Dawn. Retrieved 15.12.2012 from https://www.dawn.com/news/755950

404 Pakistan Telecommunication Authority. (2008). Annual Report. Retrieved 17.12.2010 from https://www.pta.gov.pk/annual-reports/annrep0708/ch_03.pdf

405 Cellphone users jump to 52m from 1.2m in 2002 (2007d, May 13), Daily Dawn. Retrieved 02.09.2009 from https://www.dawn.com/news/246666

406 Jafri, S. K. (2012). 3G Mobile Spectrum – Issues & Prospects. In SBP (State Bank of Pakistan) Research Bulletin Volume 8, Number 1.Retrieved 13.01.2013 from http://www.sbp.org.pk/research/bulletin/2012/Vol-8-1/opinion2sabina.pdf

407 United Nations. (2015). Universal Service in Pakistan-The Broadband Perspective. Country Perspective presented by Mr. Mudassar Hussain (Ministry of Information Technology, Pakistan) at the Capacity building workshop on improving broadband connectivity in Republic of Korea (1-2 September 2015). Retrieved 02.12.2015 from https://www.unescap.org/sites/default/files/Item%207%20Universal%20Service%20in%20Pakistan.pdf

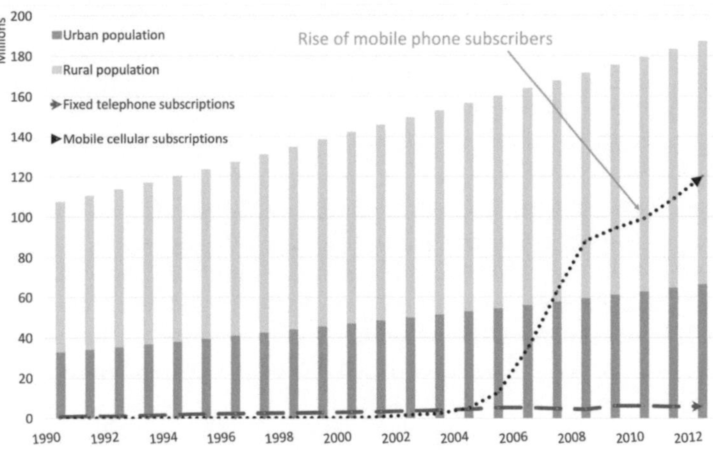

Source: Own Illustration based on the World Development Indicators[408]

In further addressing the digital divide, a number of other digital infrastructure-related projects initiated by private digital communication companies including the ones by the internet service providers had also been started between 2000 and 2010[409]. From a copper wire telephone line, Pakistan has been quickly shifting towards faster networks such as those run on fibre optic cables. For instance, in 2006 a private company named Nayatel introduced Pakistan's first high speed internet network, fibre-to-the-home (FTTH) network in the capital Islamabad with plans to further expand to other cities[410]. Similarly, mobile phone networks have also quickly shifted from 2G to 3G networks and with the introduction of 4G networks voice, video, and data services are more readily available than ever via mobile phones[411]. In order to expand the telecommunication growth, government also reduced prices of internet as well in order to promote the broadband internet access[412]. On the other hand, post 2000, cable operators have also started offering

---

408 World Bank: 2013a

409 Pakistan Telecommunication Authority. (2010c). Paradigm Technologies - Broadband Subscribers Survey: Estimating Broadband End-Users and their Experience of Service and its Performance. Retrieved 17.10.2011 from https://www.pta.gov.pk/media/bb_sub_sur_report_10.pdf

410 25 fastest growing companies announced (2011b, January 20), The Express Tribune. Retrieved 26.02.2012 from https://tribune.com.pk/story/106200/allworld-network-announces-pakistan-fast-growth-25-winners/

411 Ministry of Finance (2015). Pakistan Economic Survey (2014-15): Transport and Communications. Retrieved 15.10.2015 from http://www.finance.gov.pk/survey/chapters_15/13_Transport.pdf

412 PTA cuts bandwidth rates (2006d, June 24), Daily Dawn. Retrieved 06.04.2009 from https://www.dawn.com/news/198300

voice, video and data services for which Pakistan Telecommunication Authority (PTA) has been issuing license acquisitions[413].

Developments in ICT sector have also directly influenced the banking industry in Pakistan. Historically, Pakistan has not been a cashless and card-based society[414]. Due to uneven growth in rural and urban sectors as well as the presence of informal financial sector has modelled the consumer habits on a cash-based economy in Pakistan[415]. Till 2000, there was not even a provision of electronic bank transfers in Pakistan. In 2002, the government announced the first Electronic Transaction Ordinance that enabled banks and electronic commerce activities to take place in Pakistan[416]. This let Pakistan's electronic banking sector make a late entry but there has been a steady progress as banks and local internet websites have started developing different products and services[417]. The State Bank of Pakistan's figures from 2007-08 revealed that e-banking transactions in Pakistan had reached 30.1 million amounting to PKR 3.4 trillion[418]. Similarly, the provision of e-banking also promoted the use of cashless banking and by 2012, the total number of plastic cards (for electronic transactions) reached up to 16.6 million. This also had an impact on the growth of ATM machines. The total number of ATM machines by 2012 had reached over 5,548. The number of Point of Sale (POS) terminals which makes debit card payments possible were recorded as well over 4 million amounting to Rs. 21.05 billion. Similarly, around 7 million transactions on other e-banking channels such as through POS, the Internet, call centre/Interactive Voice Response or IVR, and mobiles were recorded[419].

3.4.2. Social Media in Pakistan

In comparison with mobile phone subscription, Pakistan's internet penetration has remained relatively very low[420]. Despite investments made in ICT sector in the

---

413 Frequency Allocation Board (2004). Broadband Policy. Retrieved 14.06.2009 from http://www.fab.gov.pk/images/pdf/Broadband%20Policy.pdf

414 Alternate channels for E-Billing (2004b, January 15), Business Recorder. Retrieved 10.10.2009 from https://fp.brecorder.com/2004/01/20040115191221

415 State Bank of Pakistan. (2010). The Size of Informal Economy in Pakistan. In SBP Working Paper Series. No. 33.Retrieved 23.08.2011 from http://www.sbp.org.pk/repec/sbp/wpaper/wp33.pdf

416 Electronic Transactions Ordinance promulgated (2002h, September 12), Daily Dawn. Retrieved 02.03.2009 from https://www.dawn.com/news/56846; State Bank of Pakistan (2002b). Electronic Transactions Ordinance. Retrieved 15.03.2009 from http://www.sbp.org.pk/about/act/ETC202.pdf

417 Arshad, A. (2014, November 01). Banking by app. Aurora. Dawn Media Group. Retrieved 15.01.2015 from https://aurora.dawn.com/news/1140685

418 State Bank of Pakistan (2008b). Pakistan 10 Year Strategy Paper for the Banking Sector Reforms. Retrieved 23.05.2011 from http://www.sbp.org.pk/bsd/10YearStrategyPaper.pdf

419 State Bank of Pakistan. (2012a). Payment Systems Quarterly Reports (2012 - 2013). Retrieved 02.12.2012 from http://www.sbp.org.pk/psd/reports/2012/Third-Quarterly-Review-FY11-12.pdf

420 30m internet users in Pakistan, half on mobile: Report (2013a, June 24), The Express Tribune. Retrieved 05.07.2013 from https://tribune.com.pk/story/567649/30m-internet-users-in-pakistan-half-on-mobile-report/

post-2000 scenario, internet growth in Pakistan in first five years of the decade was under five users per 100 inhabitants[421]. Between 2005 and 2012, the internet users have only increased by five users making it about 10% of its total population. Within the South Asian region, Pakistan trails behind Maldives, Sri Lanka, India and Bhutan, making it fifth among rest of the South Asian countries (see Figure 7)

Given internet scenario in Pakistan, the proliferation of social media websites such as MySpace, Facebook and Twitter has also been slow[422]. Based on the infrastructural limits and internet access, the first recipients of internet have been the youth in the urban centres[423]. In the initial years of social media websites, concepts of citizen journalism particularly with reference to 2009 Iran elections also influenced social media users in Pakistan[424]. An example of such a citizen engagement was witnessed during the lawyers' movement (also referred to as Movement for the restoration of Judiciary) in 2009[425]. Interestingly, two years prior to the lawyers' movement, when General Musharraf suspended the constitution in 2007 and dismissed the Supreme Court's Chief Justice Iftikhar Chaudhry, a number of flash mob demonstrations against the emergency clampdown took place all over Pakistan. Considering the low internet penetration, these flash mobs were organised via SMS campaigns as they had a wider reach than internet users (see Figure 7)[426].

Similarly, during the devastating floods of 2010-11, citizens engaged in social/ civic activism to organise aid and rescue missions[427]. Spearheaded in the major and unaffected cities of Pakistan, social media was used extensively to enlist volunteers, charity and supplies which were then distributed through verified social 'networks of networks' to ensure aid reached the deserving[428]. The roll-out of 3G and 4G versions of GSM that have brought data, speed, and omnipresence to mobile networks has also helped them usurp wired internet connections as the fastest and quickest way to access the Internet.

---

421   World Bank: 2013a

422   What are you doing on Facebook? (2011a, February 22), Daily Dawn. Retrieved 23.10.2012 from https://www.dawn.com/news/608010

423   Warraich, H. (2011, March 18). Pakistan's social media landscape. Foreign Policy. Retrieved 04.05.2014 from https://foreignpolicy.com/2011/03/18/pakistans-social-media-landscape/

424   Yusuf, H. and Schoemaker, E. (2013). The media of Pakistan: Fostering inclusion in a fragile democracy? (Policy Briefing #9). Retrieved 01.10.2013 from http://downloads.bbc.co.uk/mediaaction/pdf/bbc_media_action_pakistan_policy_briefing.pdf

425   Michaelsen, M. (2011). New Media vs. Old Politics. Retrieved 22.02.2012 from http://www.fes-asia.org/media/publication/2011_NewMediaVsOldPolitics_Pakistan_fesmediaAsiaSeries_Michaelsen.pdf

426   E-Resistance Blooms in Pakistan (2007, November 13), Spiegel Online. Retrieved 02.02.2009 from https://www.spiegel.de/international/business/cyber-demonstrations-e-resistance-blooms-in-pakistan-a-517023.html

427   Murthy, D. and Longwell, S. A. (2013). Twitter and Disasters. Information, Communication & Society, 16(6), 837-855

428   Kugelman, M. (2011, July 11). Pakistan's demographic dilemma. Foreign Policy. Retrieved 30.06.2013 from https://foreignpolicy.com/2011/07/11/pakistans-demographic-dilemma/

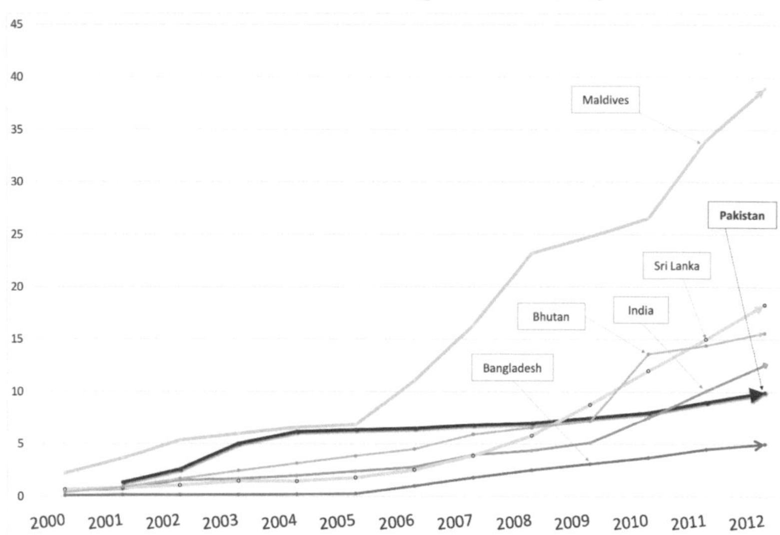

Source: Own Illustration based on World Development Indicators[429]

Steady use of social media by internet users in Pakistan has also encouraged the government to integrate these interactive tools for information dissemination[430]. Currently several ministries and departments are starting to use social media tools such as Facebook, Instagram, Twitter and YouTube as an alternative medium of publicity and engagement.

3.5. Concluding Remarks

This chapter has tried to focus on Pakistan's brief history, its social and economic development along with various developments in the field of information technology since the British colonial period through 1990s. Pakistan's major problem, as this chapter tries to argue, lies in its uneven development of the rural and urban land. The rural area of Pakistan, where the majority of Pakistanis resides, has been kept away from the infrastructural developments, where education and literacy standards are extremely low and where there is less penetration of the ICTs. Similarly, the bureaucracy of Pakistan which builds upon the administrative system left by the British does not seem to act as service provider. Its approach has so far been to control and administer the population. That is why this chapter puts emphasis on service delivery and public-sector

---

429   World Bank: 2013a
430   Strategic media cell (2015c, November 02), Daily Dawn. Retrieved 15.11.2015 from https://epaper.dawn.com/DetailImage.php?StoryImage=02_11_2015_005_005

100

reform which have been the major problems. Former military dictator Pervez Musharraf attempted to bring in a devolution plan after 2000. His tenure has coincided with not only the deregulation in telecommunication sector but also formation of several key departments and policies such as Electronic Government Directorate, NADRA, electronic transaction ordinance, eGovernment strategy and/or branchless banking.

However, we also saw that in order to approach the subject of eGovernment in developing countries as argued in chapter 2 requires an iterative approach at public-sector reform before the application of ICTs. The very confusion or at least the perceived confusion that lies in Pakistan's civil bureaucratic structure seems to further create entrenched elite oligopolies and exacerbate rural-urban divides. The unequal division of resources in the country has had a direct effect on the socio-economic growth and educational standards of Pakistan[431]. A number of developments in the field of information and communication technologies from telegraph to telephone, radio to television, cable television to mobile phones and internet to social media platforms depicted in this chapter provide an insight in understanding several scenarios. On the one hand, it helps in understanding the current communication technologies and their adoption level while on the other hand, it also helps us understand the persistent digital divide that exists in Pakistan. In addition, this chapter also tried to highlight the different ethnic and linguistic groups that are present in Pakistan. In order to approach the research queries raised in this study, the factors of digital divide, local languages, language of the technology and administrative structures are crucial to the role and development of eGovernment. Considering the huge gaps in rural-urban social and physical infrastructure, the question that this study would be dealing with is as to what kind of eGovernment system Pakistan follows for its extremely heterogeneous societal setup. Secondly, how do such eGovernments work or do they even work at all? What language does such eGovernment system in Pakistan use to communicate in? Can the mass media such as television, radio, mobile phones, digital communication media such as digital social media as presented in this study also be considered as tools and strategies for eGovernment? The following chapters give an account of current trends of ICTs and eGovernment development with the help of case studies and try to see the technological developments in terms of Pakistan's unequal geographical and human growth.

---

431 Siddiqui, S. (2012). Education, Inequalities, and Freedom: A Sociopolitical Critique. Islamabad: Narratives Publication

Chapter 4

Electronic Government Directorate (EGD)

In the previous chapter, we witnessed various developmental stages Pakistan has gone through since its independence in 1947. These included social development, governance and public administration infrastructure, progress in the print and mass media as well as the electronic media. We also noticed that most of the institutions, policies or administrative systems are derivatives of the civil organisational structure established during the British colonial rule. One of the important aspects that the previous chapter also highlighted was progress in the field of Information and Communication Technologies (ICTs) accelerated during the end of 1990s. The advancements in the ICTs have not only further improved but also enhanced during 2000s particularly in terms of investment, sophistication, outreach duly supported by the establishment of new institutional setups and public policy that reflects the changing nature of technology. It is the very reason that this study focuses on the developments made in the ICTs during the first 12 years of the new millennium and also try study their effect in the field of electronic public service delivery. This chapter focuses on the formation and functioning of a key institution namely Electronic Government Directorate (EGD) established by the Government of Pakistan to help take the digitisation drive forward and to introduce the electronic public service delivery mechanisms.

Until the end of 1990s, Pakistan's performance on teledensity chart as depicted in the previous chapter (see Figure 8) remained very low. From the beginning of 2000 the total number of fixed line telephone subscribers (per 100 inhabitants) started to increase. However, in comparison with other countries in the region such as India, Bangladesh or Nepal, the fixed-line telephone subscribers in Pakistan between 2000 through 2012 were recorded as less than 20 (per 100 inhabitants (see Figure 8). Among the SAARC countries, Pakistan is ranked on fourth (under 4 users) position with Sri Lanka being on the top (17 users) whereas Bangladesh on number seventh (under 2 users per 100 inhabitants)[432]. These numbers are important to remember because it is the fixed line telephone through which internet is provided particularly during late 1990s and beginning of 2000. Therefore, any electronic public service provided over internet would eventually and approximately reflect the fixed-lined telephone owners who can access it; escaping the majority of the population.

These numbers presented a very challenging task for any sort of electronic public service activity to successfully take place. Internet growth at the beginning of

---

432   World Bank: 2013a

2000 was gradually taking up and the number of subscribers was limited to 700,000 whereas the digital infrastructure was still evolving[433].

Figure 8: Pakistan and teledensity in South Asia

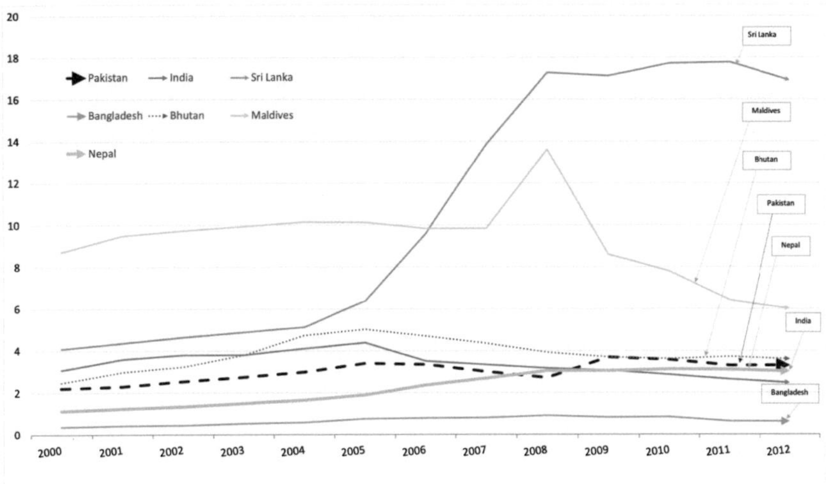

Source: Own illustration based on World Development Indicators[434]

Similarly, another important aspect for any electronic public service delivery that involves a financial transaction requires regulatory and legislative frameworks. Till the end of 1999 electronic banking laws did not exist and legal framework for electronic transactions or acts only commenced post 2000[435]. These facts are important to note and remember while discussing the case of eGovernment in Pakistan because not only did the institutions but also several key legislations for the provision of electronic public service delivery also took place within the first decade. In this regard, the first policy on developing understanding about information technology in Pakistan, setting the goals and target areas was promulgated by the Ministry of Science and Technology in August 2000[436]. The

---

433   Wolcott, P. and Goodman, S. E. (2000). The internet in Turkey and Pakistan: a comparative analysis.   Retrieved   04.05.2014   from https://www.researchgate.net/publication/264999624_The_Internet_in_Turkey_and_Pakistan_A _Comparative_Analysis

434   World Bank: 2013a

435   New act to regulate payment systems (2007c, June 16), Daily Dawn. Retrieved 04.05.2014 from https://www.dawn.com/news/251898; E-banking be made secure, says SBP chief (2003c, July 27), Daily Dawn. Retrieved 04.05.2014 from https://www.dawn.com/news/132128/karachi-e-banking-be-made-secure-says-sbp-chief

436   Ministry of Information Technology (Government of Pakistan). (2014). Pakistan IT Policy and Action   Plan   2000.   Retrieved   04.05.2014   from http://www.moit.gov.pk/gop/index.php?q=aHR0cDovLzE5Mi4xNjguNzAuMTM2L21vaXQvZ nJtRGV0YWlscy5hc3B4P2lkPTEwJmFtcDtvcHQ9cG9saWNpZXM%3D

103

following section reviews the salient features of this policy and explore factors how it became a starting point for the formation of different institutions and various other legislations.

4.1. IT Policy and Action Plan (2000)

IT Policy and Action Plan is the first official document by the Government of Pakistan that sets the stage for ten key IT policy strategies namely (i) Human Resource Development, (ii) Infrastructure Development, (iii) Software Industry Development, (iv) Hardware Industry Development, (v) Internet, (vi) Incentives, (vii) IT Promotion & Awareness, (viii) IT Usage, (ix) Legislation and (x) Regulations[437]. The guiding theme as per this policy was

"the Government shall be the facilitator and enabler to encourage the Private sector to drive the development in IT and Telecommunication"[438]

The major goal, therefore, of this policy is to provide the maximum opportunity to the private sector in Pakistan to lead the IT development initiatives particularly focusing on the telecommunications, development of databases, growth of local software industry whereas government role is seen more as a facilitator or arbitrator by providing for instance fiscal or regulatory incentives to the private sector[439]. The initial work on the preparation of the first IT policy was initiated during the Nawaz Sharif government under Vision 2010 programme[440]. Chairperson of the then Planning Commission of Pakistan, Ahsan Iqbal assigned National Database Organization (NDO) to prepare the first draft of the IT policy. Due to General Pervez Musharraf's military coup and subsequent elimination of Nawaz Sharif from his post of Prime Minister, the IT policy was never made functional[441]. As head of Ministry of Science and Technology (MoST) Government of Pakistan, during the military rule of General Musharraf, Prof. Dr. Atta-ur-Rehman took over the task of reformulating the IT policy draft[442]. In this regard, Prof. Rehman invited over 250 professionals from the IT community mostly from Pakistan as well as few expatriates who took part in consultative

---

437   ibid.

438   ibid., p. 4

439   Masood, J. and Malik, S. .pk. In Librero and Arinto (Eds.). Digital Review of Asia Pacific 2007–2008. (pp. 263 - 267). Retrieved 04.05.2014 from http://www.digital-review.org/uploads/files/pdf/2007-2008/intro.pdf

440   Prime Minister's Office (2013). National Agenda for Real Change: Manifesto (PML-N). Retrieved 20.04.2013 from https://pmo.gov.pk/documents/manifesto.pdf

441   Sustainable Development Policy Institute (2014). "Profile: Prof Ahsan Iqbal (Board of Governors, SDPI)". 02.06.2014, from https://www.sdpi.org/about_sdpi/bog_details82.html.

442   Atta-ur-Rahman (2002). "Science and Technology in Pakistan: The Way Forward". Retrieved 02.12.2008 from https://www.sciencemag.org/careers/2002/09/science-and-technology-pakistan-way-forward.

meetings and strategic dialogues for four months[443]. These meetings resulted in the formation of 11 working groups whose deliberations eventually became part of the submissions and contributions to the development of first IT policy.

The first IT policy for the Government of Pakistan thus opened the door for a number of initiatives to take place such as establishment of software technology parks, virtual learning centres, digital local content development, IT education and the development of eCommerce industry in Pakistan[444]. However, for any electronic commerce activity to take place, Pakistan needed legal foundations and legislative authority to conduct any internet-based transactions and the development of such legal frameworks started during 2000-01[445].

4.2. Electronic Transaction Ordinance (2002):

IT Policy and Action Plan served as a starting point and a strategic blueprint for developments in the IT sector at the multiple levels[446]. One of the important aspects of this policy was among others, its support for the legislative frameworks towards creating a conducive electronic commerce environment[447]. As a matter of fact, till 2002, there were no legal frameworks or laws that supported the electronic transactions of business activities via internet[448]. Resultantly, Pakistan did not have any (local) online shopping stores from private sectors, let alone online electronic public services. IT Policy and Action Plan of 2000 provided the required legal and favourable ground for the promotion of eCommerce in Pakistan. Government of Pakistan started developing the first draft of electronic transaction law in 2001. In this regard, Ministry of Science and Technology set up an Information Technology Law Forum (ITLF) that invited proposals from the lawyers' community with expertise in IT[449]. Over the course of one year ITLF deliberated over various aspects of electronic transaction pertaining to trade and

---

443 Ministry of Science and Technology (2000a). IT Policy and Action Plan: Including initial steps required for updating and Implementing a modular National IT Policy & Plan. Retrieved 23.06.2008 from
http://anf.gov.pk/library/acts/Pakistan%20IT%20Policy%20and%20Action%20Plan.pdf

444 Ministry of Planning Development and Reform (2004). Public Sector Development Projects 2002-03. Retrieved 03.02.2009 from https://www.pc.gov.pk/uploads/archives/PSDP2002-2003.pdf

445 Ministry of Finance (2001). Interim Poverty Reduction Strategy Paper (I-PRSP). Retrieved 28.07.2008 from http://www.finance.gov.pk/poverty/iprsp_2.pdf

446 Shaukat, M. (2009). Developments of Information Technology, Telecom and E-Commerce in Business Environment of Pakistan: An Analysis of Banking and Manufacturing Sectors. Pakistan Journal of Social Sciences, 29(2), 259-278

447 Certification authority formed (2004, September 24), Daily Dawn. Retrieved 04.05.2014 from https://www.dawn.com/news/371664

448 State Bank of Pakistan. (2003). Banking System Review. Retrieved 20.08.2009 from http://www.sbp.org.pk/publications/bsr/bkg_system_review(2003).pdf

449 United Nations Economic and Social Commission for Asia and the Pacific (2003). Initiatives for e-commerce capacity-building of small and medium enterprises : proceedings and papers presented at the Regional Consultative Meeting on Initiatives for E-Commerce Capacity-Building of Small and Medium Enterprises, Seoul, 13-15 November 2002. New York: United Nations

investment as well as online sales and provision of trading rules in Pakistan. The Electronic Transaction Ordinance (ETO) was formally approved by the federal cabinet in 2002[450].

The ETO of 2002 comprised of nine chapters and deals with the basic understanding and recognition of electronic documents, certification service providers, certification council, jurisdiction and possible offences[451]. In this context one of the first developments that took place in the field of transactions, digital encryption and security procedures was to establish a regulatory institution that oversees the electronic transactions in public and private sectors[452]. Under the section 18 of ETO, Government of Pakistan established Electronic Certification Accreditation Council (ECAC) which is an autonomous organisation and functions under Ministry of Information Technology & Telecom (MoIT), Government of Pakistan. ECAC's role is to provide accreditation to any organisation interested in providing electronic public services under its regulatory framework. Therefore, any agency intending to provide eServices for public would require an accreditation from ECAC. The ETO's chapter 8 also provides a protection and security of digital transactions. In order to protect the unauthorised access or illegal access (or even attempt thereof) to the information systems in Pakistan, ETO also laid down a clause for an offence under this ordinance[453].

Till this point, neither Government to Citizen (G2C) nor Government to Business (G2B) electronic public services had been possible in Pakistan. However, with ETO's initiation in 2002, the legal cover was provided for the electronic commerce and electronic public service by the statutes defined in ETO[454]. Both the IT Policy (2000) and ETO (2002) laid the licit foundations for digital communication, infrastructure development and the electronic transaction of documents and finance[455].

4.3. Formation of Electronic Government Directorate (2002)

One of the key aspects of the first IT Policy and Action Plan (2000) was to implement the deliberations and recommendations of the eleven working groups which helped developed the first IT policy[456]. In this regard, the working group

---

450   Legal cover soon for e-signatures, -documents (2002d, September 03), 23.08.2009 from https://www.dawn.com/news/55507/legal-cover-soon-for-e-signatures-documents

451   State Bank of Pakistan. (2002a). An Ordinance: Electronic Transaction Ordinance (ETO). Retrieved 23.08.2009 from http://www.sbp.org.pk/about/act/ETC202.pdf

452   The Gazzette of Pakistan. (2013). Pakistan Telecommunication (Re-organization) Act. Retrieved 03.04.2013 from https://www.ecac.org.pk/assets/front/files/TelecomAct.pdf

453   State Bank of Pakistan: 2002a

454   E-commerce and its future in Pakistan (2003d, April 21), Daily Dawn. Retrieved 26.07.2009 from https://www.dawn.com/news/96984

455   Digital signature certificates: SECP, NIFT sign agreement (2008b, October 31), Business Recorder. Retrieved 23.07.2009 from https://fp.brecorder.com/2008/10/20081031828456/

456   Ministry of Science and Technology (2000b). IT Policy and Action Plan. Retrieved 02.09.2008 from

of "IT in Government and Database" put forth a number of proposals among others, improving efficiency and quality of public services by inducting ICTs across different levels of the government[457]. The working group recognised the dearth of IT knowledge in the public sector and proposed the acquisition of new hardware and software applications to gradually move towards paperless office. Similarly, within the context of providing efficient public services via ICTs, it is the same IT policy paper that dedicates a special section on the recognition and need for eGovernment model for Pakistan. Nevertheless, considering the limitations of IT across the public sector including the infrastructural and financial constraints, the working group found the development of eGovernment model to be a humongous task. This is precisely because the eGovernment does not mean only digitisation/automation of office with hardware and software tools, instead it also requires, among others, reengineering or reformation of government departments and its processes[458].

Based on the "IT in Government" working group's recommendations, it was suggested that the Government shall make IT literacy mandatory for current and future employees at provincial and federal level for the eGovernment in Pakistan to work[459]. Similarly, the government departments (federal and provincial) shall dedicate a minimum of 2% of their budget on IT services and infrastructure. Similarly, the working group also proposed the establishment of IT Board/Department at the provincial level that can help plan, coordinate and implement the IT projects. In this regard, the provinces of Sindh and NWFP (now Khyber Pakhtunkhwa (KP)) established IT boards at the provincial level in 2002[460]. In KP province the department later became a directorate whereas in Sindh it received a status of a provincial ministry. In the province of Punjab, the Information Technology Board already existed since 1999 whereas in Balochistan, Department of Science and Information Technology was established at the provincial level[461].

At the federal level, the Government of Pakistan established Electronic Government Directorate (EGD) in October 2002 under the IT and Telecommunication Division[462]. EGD will be considered as the first official institution in Pakistan that was mandated to implement electronic government

---

https://moitt.gov.pk/moit/userfiles1/file/policies/Pakistan%20IT%20Policy%20%20Action%20Plan%202000.pdf

457   Ministry of Science and Technology: 2000a

458   Gupta, M. and Jana, D. (2003). E-government evaluation: A framework and case study. Government Information Quarterly, 20(4), 365-387

459   Ministry of Science and Technology: 2000a

460   Software park, IT board inaugurated (2002e, February 17), Daily Dawn. Retrieved 04.05.2014 from https://www.dawn.com/news/22096; Sindh IT board constituted (2002f, November 22), Daily Dawn. Retrieved 04.05.2014 from https://www.dawn.com/news/67931

461   Punjab Information Technology Board (2011). "History". Retrieved 15.10.2011 from http://pitb.gov.pk/?q=history.

462   E-governance will help in eradication of corruption: Ata (2002g, November 05), Daily Dawn. Retrieved 04.05.2014 from https://www.dawn.com/news/65117

projects. Minister for Science and Technology, Prof. Atta-ur-Rehman, who inaugurated the EGD viewed the directorate not just as an implementation body but also someone that develops a capacity to outsource IT projects to the private sector and equally monitors their work. EGD's core group was comprised of five technology specialists whereas it was led by a programme coordinator who is supported by a Director General (Projects). Alvi Abdul Rahim, a former Secretary from the IT Commission was assigned to be programme coordinator whereas Syed Raza Abbas Shah was appointed as the Director General (Projects)[463].

One of the first tasks that EGD undertook was the development and maintenance of Government of Pakistan's first official web portal (www.pakistan.gov.pk). The website provided basic but structured information about all the federal ministries and their attached departments (see Figure 9). Similarly, it also offered information about the objectives of government; how the web portal could help provide better government services and information electronically. One of the key aspects of the portal's vision (in its own words) was its design being citizen-centric that provides a gateway and complementary channel for citizens dealing with the government[464]. Analysis of the web portal can be found in the section below.

Apart from the development of Pakistan's first official portal, EGD's main mandate involved planning of and technical support for the eGovernment projects at federal, provincial and district government level[465]. For that matter, EGD's tasks also included the provision of software standards and infrastructural approach required in the field of eGovernment (such as eHealth)[466]. Between 2002 through 2005, EGD was involved in developing the government portal, survey of all the federal ministries and their divisions for establishing Local Area Networks and hardware required, establishment of technology laboratories at 11 different government academies to provide IT training to the government staff as well as

---

463   While conducting a field work in Pakistan, the author visited Electronic Government Directorate first in Summer 2008 and then in Spring 2009. The EGD was located in a multi-storey building in Taimoor Chamber in Islamabad's commercial district, Blue Area on Fazl-ul-Haq Road. During these meetings, author interacted with the project directors (IT) and as a part of my empirical data collection also conducted expert interviews with them. While author remains indebted to the EGD staff members continued cooperation and courteousness, one technical aspect that has remained a part of research observation is the location and condition of the EGD's premises. Even though EGD was tasked Pakistan's official eGovernment policy and planning, the physical infrastructure of the EGD, the manual/traditional way of handling documentation (in a paper-based environment) as well as the lack of alternate electric supply (electricity failure during between 2008-2011 remained a serious problem in Pakistan) presented a sorry condition of an institution responsible for digitising ministries and department. On both occasions, author remained unlucky to have an opportunity to meet Mr. Shah. Nevertheless, author remains grateful to his staff members who entertained the research enquiries.

464   Government of Pakistan (2006b). "Vision and Objectives: About Pakistan.Gov". Retrieved 01.11.2007 from http://www.pakistan.gov.pk/AboutPakGov.jsp#vis.

465   Rs452 million data centre project approved: Communication within depts (2004b, June 09), Daily Dawn. Retrieved 04.05.2014 from https://www.dawn.com/news/361312

466   Hospitals to be computerised (2006c, June 13), Daily Dawn. Retrieved 04.05.2014 from https://www.dawn.com/news/196753/hospitals-to-be-computerised

identification of areas of automation at the Ministry of Science and Technology that could improve the operations and efficiency within the ministry[467].

Figure 9: Pakistan's first official web portal

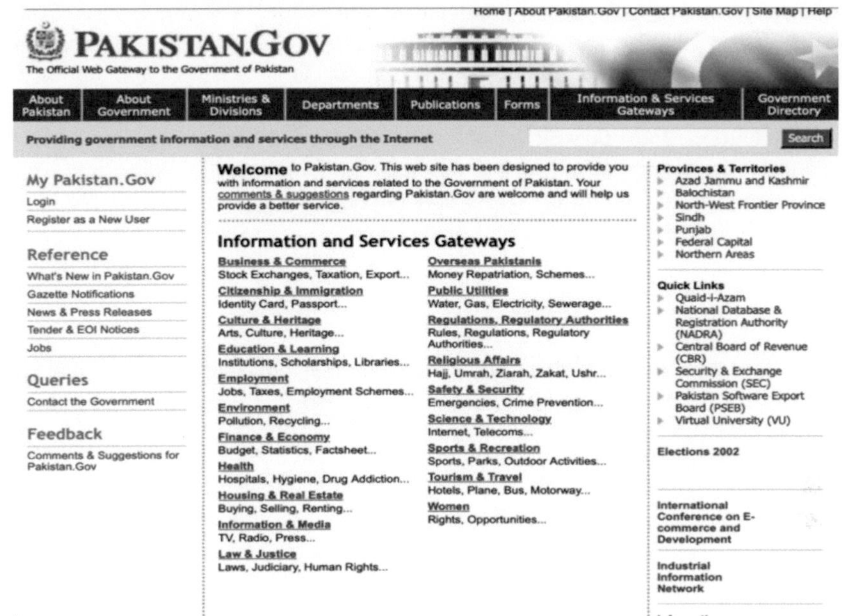

Source: Government of Pakistan[468]

Alongside, EGD also prepared a detailed 56-page documentation about the standards that helped EGD communicate and ensured quality approaches adopted in the design and implementation of the websites of government ministries/departments/divisions[469]. The idea behind Web Standards and Guidelines is to simultaneously engage with and involve officials and staff of different ministries to take ownership and responsibility of managing and maintaining various government websites.

---

467 Electronic Government Directorate (2009b). Projects Implemented by EGD. Retrieved 30.11.2010 from http://www.pakistan.gov.pk/e-government-directorate/projects/projects.jsp

468 Government of Pakistan (2006a). "Government of Pakistan's official webportal ". Retrieved 01.11.2007 from www.pakistan.gov.pk.

469 Electronic Government Directorate (2009a). "Web Standards and Guidelines". Retrieved 27.03.2009 from http://www.pakistan.gov.pk/e-government-directorate/standards/postimplement.jsp.

4.4. Pakistan's first official eGovernment Strategy

Along with the projects mentioned above, between 2002-2005 EGD had been working towards drafting Pakistan's first eGovernment strategy. In May 2005, the EGD introduced an official eGovernment strategy paper *"E-Government Strategy and 5-Year Plan for the Federal Government"*[470]. This paper discussed the background information about eGovernment in Pakistan, its underlying principles, eGovernment strategy for Pakistan, eServices for citizens, 5-Year implementation plan as well as benefits of the eGovernment programme for Pakistan. This document relies upon among others, National IT Policy and Action Plan (2000) as a foundation stone and encompasses Pakistan's technological position further within the eGovernment scenario.

4.4.1. Definition and Scope

The EGD's eGovernment strategy for Pakistan does not seem very different from the ones that are already floated in the academic literature or in practice. According to Pakistan's eGovernment strategy:

"E-Government is defined as the usage of Information and Communication Technologies (ICT) to support processes within the government as well as for the delivery of services to its consumers, including other organizations, citizens as well as businesses." (Electronic Government Directorate, p. 7)

Pakistan's eGovernment strategy paper takes an international understanding towards eGovernment where governments interact with business and citizens using ICTs. Similarly, it also divides the phases of eServices into four different categories, namely (i). informational, (ii). Interactive, (iii). Transactional and (iv). Collaborative (p. 8)

These phases are not dissimilar to the ones discussed in Chapter 2 however, there seems to be a confusion in the strategy paper's understanding of eServices and eGovernment. The paper interchangeably uses the terminologies of eGovernment and eServices to elaborate the four phases, whereas both eGovernment and eService are related but different concepts[471]. eGovernment is a much broader concept and eServices are just one aspect of it. The four phases that the strategy paper refers to as the phases of eServices seems to be the phases of eGovernment. Similarly, the strategy paper also does not exactly define as to what constitutes an

---

470   Electronic Government Directorate (2005a). E-Government Strategy and 5-Year Plan for the Federal Government. Retrieved 16.12.2010 from http://202.83.164.28:9080/egdsite05/downloads/E-Government%20Strategy%20and%205-Year%20Plan%20(19%5B1%5D.06.2005).pdf

471   Rust, R. (2001). The rise of e-service. Journal of Service Research, 3(4), 283-284

eService. While defining the phases of eService, the strategy paper continuously refers to the word 'online interaction' but what it does not elaborate is how this online interaction is going to work in real life[472]. Does this online interaction mean government to citizen interaction via website or will there be other mechanisms of online interaction? The strategy paper does not dwell on details about the online interaction aspect.

The strategy seems to have another problem while explaining the scope of eGovernment (also depicted as an illustration, reproduced in this study below, see Figure 10). The strategy paper defines the scope eGovernment in the context of federal government where government interacts at four different levels, that is, Government to Government (G2G), Government to Employee (G2E), Government to Citizen (G2C) and Government to Business (G2B)[473]. According to this paper's understanding of G2C, the G stands for federal government whereas C stands for citizens and it includes all citizens of Pakistan (p.8)[474]. From this elaboration, it appears that it would be the federal government who would provide eServices to all the citizens. As discussed in previous chapter, Pakistan is a federation with different provinces and each province with its own provincial government. If every province has its own provincial autonomy, it remains unclear from the illustration below as to how the federal government can be solely responsible for eServices for all citizens.

Figure 10: Electronic Government Directorate's eGovernment scenario

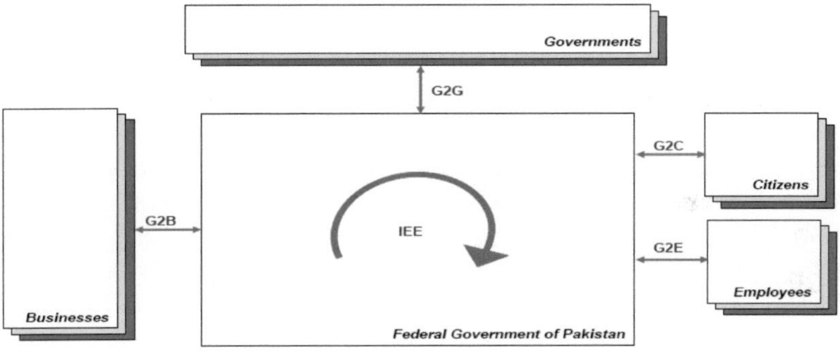

Source: EGD[475]

The salient features of Pakistan's first eGovernment strategy can be divided into five parts. It involves (i) deployment of the basic IT infrastructure to all government agencies and its promotion (ii) identification of applications that are common to many or all divisions. These applications include internal

---

472   Electronic Government Directorate: 2005a
473   ibid.
474   ibid.
475   ibid.

communication, human resources, budgeting, project management, document/file management, as well as collaboration to be implemented in the Ministry of IT. The idea behind these features while on one hand advocates the digital connectivity and IT infrastructure development at the ministerial level, on the other hand it backs the idea of developing standardised and most commonly used software applications that are applicable for all or majority of ministries/department. The idea of common application helps avoid the duplication of (amount of) work and in other words circumvents the reinvention of wheel.

4.4.2.Salient Features of Pakistan eGovernment Strategy

Similarly, the 5 years' action plan also emphasises (iii) identification of agency-specific applications to be implemented through the support of EGD[476]. An important aspect of the eGovernment strategy is also the (iv) standardisation with reference to software architecture, methodologies and best practices for the development of eGovernment application. EGD's approach towards standardisation appears to focus on sustainable development of eGovernment application by developing a framework that can work as a blueprint for future eGovernment application or eService development by any ministry or related department. As its fifth salient feature of the eGovernment strategy, (v) EGD backs the enabling environment that may provide EGD more autonomy to avoid delays.

Five salient features of Pakistan eGovernment strategy reflects, particularly with reference to standardisation and development of common application, that EGD recognises and acknowledges the adoption and acceptance of eGovernment at the national scale. However, there are a number of reality gaps which Pakistan's first eGovernment strategy fails to address. For instance, considering the fact that between 2000 through 2011, internet users in Pakistan have remained below 12 (per 100 users), the EGD's strategy paper stressed the web based interactions (see Figure 11). The strategy plan does not seem to refer to the technology adoption or diffusion of ICTs in Pakistan. Then, it refers to the question of "how and what information citizens can access through eGovernment portals" rather than enlisting services that could be availed through the portals. Countries leading in eGovernment initiatives such as USA, South Korea or Singapore (mentioned as best practices in the said strategy paper), have entirely different ICT growth, standards of teledensity or eReadiness indicators in comparison to Pakistan as also depictured in Figure 11. Similarly, the strategy paper does not clearly identify the main G2C services to be provided to citizens.

Nevertheless, the strategy paper tries to present a comprehensive five-year plan from 2005 through 2010 on how the above-mentioned salient features will be

---

476   ibid.

112

implemented. In July 2012, the EGD rolled out a second eGovernment strategy plan for the Federal Government, which apparently is the abridged version of the strategy plan approved in 2005[477]. Although, the 2012 strategy plan argues to be the revised version of the previous one, keeping in view the new realities and present condition, it however, does not vary much from the 2005 strategy. There has been an addition of new module on security of government information considering Pakistan's vulnerable security situation between 2007 and 2012. Post 9/11 and particularly within the Pakistan-Afghanistan scenario, Pakistan's security details and its assets had come under international scrutiny, the 2012 eGovernment strategy paper also reflects and acknowledges that Pakistan has a considerably weak IT infrastructure[478].

Since Pakistan's official portal is designed, prepared and maintained by the EGD, the revised eGovernment strategy plan suggests a secure storage of government and official information in a dedicated data centre at the EGD, which shall not be connected to the Internet. This in turn is actually a second acknowledgement that as of 2014, the EGD has not made any arrangements to keep this otherwise sensitive data in secure computers. Similarly, the strategy paper does not seem to provide any detail to which internationally recognised security standards are going to conform to (such as International Standard Organisation, ISO).

From 2002 to 2012, the EGD has existed as a government organisation with a legislative and financial mandate under the Ministry of Information Technology (MoIT), however, from 2012 onwards, the EGD has been merged with Pakistan Computer Bureau (PCB) which also functions under the MoIT. Although this merger could possibly result in better and joint efforts but the strategy paper admits that there is a lack of human resource not just in the ex-EGD but even the combined staff strength of the EGD and PCB cannot correspond to the eGovernment requirements (see Discussion for response to and reception of EGD projects).

---

477   Electronic Government Directorate (2012). Abridged Version: E-Government Strategy for the Federal Government. Retrieved 01.12.2013 from http://e-government.gov.pk/gop/index.php?q=aHR0cDovLzE5Mi4xNjguNzAuMTM2L2VnZHdlYi91c2VyZmlsZXMxL2ZpbGUvcGVyy9aW5hbEFicmlkZ2VkJTIwRUdvdmVybm1lbnRTdHJhd GVneSUyMFdlYjEucGRm

478   There have been several incidents when Pakistani government websites were subjected to hacking attempts which highlights the fragile security parameters in place for the protection of government websites. See for instance: The futility of Indo-Pak cyber wars (2011, July 28), 04.05.2014 from https://www.dawn.com/news/647571; Indian hackers hack www.multan.gov.pk (2012, May 22), The Nation. Retrieved 04.05.2014 from https://nation.com.pk/22-May-2012/indian-hackers-hack-www-multan-gov-pk

Source: Own illustration based on World Development Indicators[479]

eServices, the core of Pakistan's eGovernment strategy, sits on top of its strategic framework. Still there seems to be a confusion as to how it would like to define eServices, what would constitute an eService and more importantly which services would be considered as must-have eServices, as has been the case with other eGovernment initiatives around the world[480]. Similarly, the revised eGovernment strategy paper provides no references to the e-readiness indicators, benchmarking mechanisms, monitoring and evaluation mechanisms. These factors as discussed in the Chapter 2 have been given paramount importance for envisaging any eGovernment implementation in both developing and developed countries.

From the eight elements promulgated in 2005 eGovernment strategy, it appears that the concepts of strategic framework, human resource, enabling environment, basic infrastructure, common applications, agency-specific applications, eServices and standards have been merely reworded in 2012 strategy paper, with major change being merger of EGD with PCB. It may have been useful if after a seven-year of review, the revised strategy paper had been highlighted or had given importance to major legal acts related to eGovernment such as e-signature act, information security act, digital divide elimination, cyber infrastructure protection. In major eGovernment initiatives around the world, these legal acts have assumed a primary importance as hallmarks of the inception of eService delivery[481].

---

479   World Bank: 2013a

480   Ancarani, A. (2005). Towards quality e-service in the public sector: The evolution of web sites in the local public service sector. Managing Service Quality: An International Journal, 15(1), 6-23

481   Zhao, J. J. and Zhao, S. Y. (2010). Opportunities and threats: A security assessment of state e-government websites. Government Information Quarterly, 27(1), 49-56; Prasad, K. (2012). E-

114

The communication channels for the electronic public service delivery between years 2000 through 2010 have also gone through major transformation not just around the world but also within Pakistan (as we shall also later see in this study)[482]. The 2012 eGovernment strategy does not seem to emphasise or acknowledge the need or presence of multiple channels of communication for the eService delivery. While realising the drop in the Internet penetration and the rise in mobile phone usage, the eGovernment strategy paper does however stress the need for mobile service delivery platform through which government information could be shared not just via web portal but also through mobile phones.

One of the most problematic aspects of both 2005 and 2012 eGovernment strategy papers is their approach towards local and multilingual content. As discussed in previous chapter that every province in Pakistan has its own provincial language which even dominates the national language Urdu and official language (English). Both the strategy paper advocates for developing software applications that would translate the English language content on Pakistani government websites into Urdu. It is interesting that while on the one hand Pakistan's eGovernment strategy paper acknowledges and recognises the outreach and content base of Urdu, yet on the other hand it still chooses to speak of language translations instead of supporting the production of local content. While the work and debate on multilingualism on the internet as well as in the eServices has matured over the past ten years, the 2012 eGovernment strategy also seems to have ignored this aspect which otherwise could have been a decisive factor in setting up a vision for the electronic public service paradigm in Pakistan[483].

4.5. Analysis of Pakistan's official web portal

Pakistan's official web portal (www.pakistan.gov.pk) has been the responsibility of EGD. Since its inception in 2002, the portal has been revamped multiple times. This section tries to analyse Pakistan's official web portal in order to gauge its

---

governance policy for modernizing government through digital democracy in India. Journal of Information Policy, 2, 183-203

482 Contributions by Pieterson and Schellong about the developments in and growth of service delivery in channels for eGovernment implementation provide a very stimulating read. See: Meijer, A., Boersma, K. and Wagenaar, P., (Eds.) (2009). ICTs, citizens and governance: After the hype! Amsterdam: IOS Press; For analysis of channel choice also see Pieterson, W. (2010). Citizens and service channels: Channel choice and channel management implications. International Journal of Electronic Government Research, 6(2), 37-53

483 It is interesting to note that EGD's parent organisation Ministry of IT has been emphasising on development of local content. See for example: Ministry of Information Technology (2006). National ICT R&D Fund Policy Framework. Retrieved 04.05.2014 from https://moitt.gov.pk/SiteImage/Misc/files/National%20ICT%20R%26D%20Fund%20Policy%2 0Framework%2C%202006.pdf; Similarly, there also seems to be an ignorance on EGD's part about the private initiatives in the production of local language content taking place in Pakistan, see for instance: Mir, A. (2010). "Local Web Content: Mobile Operators Sign MOUs, Telenor has an Urdu Website". Retrieved 04.05.2014 from http://telecompk.net/2010/12/29/local-web-content-mobile-operators-sign-mous-telenor-has-an-urdu-website/.

practicality and functionality. In order to analyse the web content of the portals, this analysis draws Darrell West's method of content analysis of websites which is based on a number of features that the website offers.

The front page of Pakistan's official web portal offers six sections: (a) Home (b) About (c) Government (d) eServices (e) Business (f) Tourism. Most of the information provided in these sections is not hosted on the web portal. Every section refers to an external website of a corresponding governmental agency whose website link may not work or the page might be under construction. The web portal during 2007 provided information within one single website. Downloadable forms, which are the most initial phase of the web portal strategy, are missing. As of 2012, the portal referred to the corresponding webpage of Planning Commission of Pakistan where there is no indication of the location of the forms. Downloadable forms are not only helpful as they provide instructions, prerequisites and procedures about the public service(s) but they are one of the initial and important features in any eGovernment implementation[484]. In Pakistan's first web portal (see Figure 9) there was a special section on downloadable forms. This section not provided provision to download but to also know the corresponding agency/department/ministry responsible for the functioning of the public service. In every new version of Pakistan's official web portal this section has been taken offline.

Figure 12: Government of Pakistan's official web portal

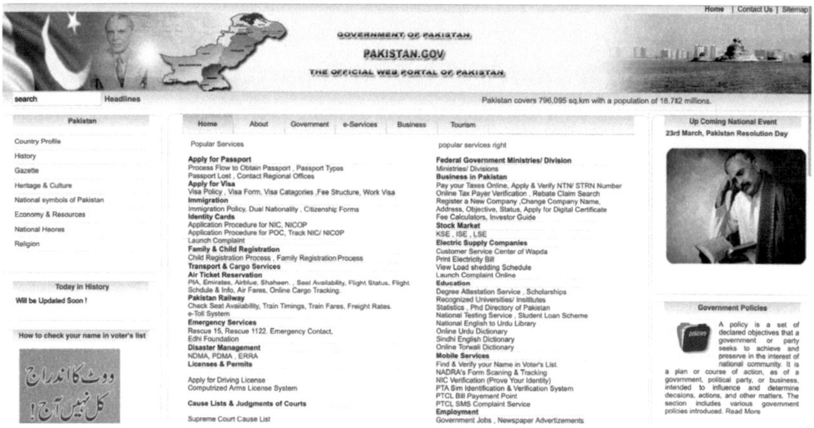

Source: www.pakistan.gov.pk

The section on eServices in the official web portal promises to provide information about Health, Education, Statistics, Passport, Visa, Immigration &

---

484 Shah, M. (2007). E-governance in India: Dream or reality? International Journal of Education and Development using Information and Communication Technology, 3(2), 125-137

Identity, Registration & Verification, Forms & Publications, Prices & Bills & Reservations, Banking & Insurance. It is interesting to note that none of the services provides any facility to interact with the government but were static, not updated and missing actionable content. If the government's web portal is analysed in the light of four-stage model (as presented in Chapter 2), eGovernment takes place in four parallel stages of catalogue, transaction, vertical integration and horizontal integration[485]. For most governments, this model has become a benchmark and starting point for deploying eGovernment applications. Given the services and functionality available on Pakistan's official web portal, it appears to correspond to the ideal of four-stage eGovernment model envisioned by Layne & Lee in different ways. As for the catalogue stage, which is the foremost stage in eGovernment, it appears that the official web portal faced difficulties in qualifying for the first stage. The web links are redirected to the concerned department and ministry; however, these links are not regularly updated. If the concerned ministry or department changes the link in their own website, that link will result in an error on the official web portal. In June 2008, the EGD launched a new content management system (CMS). At the operational level the web portal gave a number of errors, which seemed to reflect a lack of expertise in software handling. Till the end of 2008, the entire directory tree of the government portal was accessible, making it vulnerable. One of the most problematic issues with the website is that the entire information is provided in English language. Although the eGovernment strategy paper (as discussed above) has emphasised the use of a multilingual portal, the option of local language content remains unavailable. About 95% of the government portals provide all sorts of information in English language, restricting its use to only select English-speaking citizenry which is very small in number[486]. As indicated in Chapter 3, every province in Pakistan has a different language (which in itself has different dialects) whereas Urdu as a national language is understood by about 61%. In a country where more than 50% of the population cannot understand English, yet Pakistan's official web portal continues to offer content in English language[487].

4.5.1. Language and the colonial legacy

The problem of multilingual content can be traced back to Pakistan's colonial legacy which Chapter 3 has also briefly touched upon. Even though Pakistan's

---

485 Layne, K., et al.: 2001

486 For informed reading on English speaking elites in Pakistan see Rahman, T. (2005). Passports to privilege: The English-medium schools in Pakistan. Peace and Democracy in South Asia, 1(1), 24-44; Rahman, T. (2009). Language ideology, identity and the commodification of language in the call centers of Pakistan. Language in Society, 38(2), 233-258

487 Pinon, R. and Haydon, J. (2010). English Language Quantitative Indicators: Cameroon, Nigeria, Rwanda, Bangladesh and Pakistan. In A custom report compiled by Euromonitor International for the British Council.Retrieved 04.05.2014 from http://teachingenglish.britishcouncil.org.cn/sites/teacheng/files/Euromonitor%20Report%20A4. pdf

national language is Urdu, the official language is still English as it was in British India. Although English was supposed to continue only temporarily as an official language till the national language(s) could replace it, Pakistan's anglicised ruling elite's patronage of English has to date continued[488]. This ruling elite as we saw in chapter 3 exists in various forms, such as the bureaucracy who have moulded themselves in the traditions of the British[489]. Similarly, Pakistan's policy of giving way to parallel education standards of Urdu and English medium schooling has also created a divide between schooling systems[490]. At the parliamentary level, even though some of the sessions take place in Urdu, the documentation of parliamentary affairs as well as the judicial documentation is carried out in English[491]. The patterns of colonial legacy can also be found in other colonies of Britain in African and Asian countries. Literature on the government websites from Sub-Saharan Africa faces the same issue of language[492]. In a survey on the official websites of Sub Saharan Africa, Akakandelwa found that the majority of the web sites were written in English[493]. There were only a few that offered bilingual or multilingual approaches. Most web sites did not contain statistics, calendars of forthcoming events, and/or facilities for feedback which also seems to be the case with Pakistan's official web portal.

4.5.2. Mapping of federal and provincial government web portals

In order to analyse and to generate an index of the level of functionality of Pakistan's government web portals, this study employs Darrel West's method and extends it to the case of eGovernment portals chosen for this study[494]. According to the West's method, each website receives a score on the basis of a list of features that are required or strongly enhance the usability of eGovernment websites[495]. The scoring method checks for a list of 18 essential features and weighs them by a factor of 4 and awards additional points for extra features thereof. The formula applied to compute the web portal is as below:

eGovernment index for site i

---

488   Rahman, T. (1997b). The medium of instruction controversy in Pakistan. Journal of Multilingual and Multicultural Development, 18(2), 145-154

489   Mahboob, A. (2002). No English, no future! Language Policy in Pakistan. In Obeng and Hartford (Eds.), Political independence with linguistic servitude: The politics about languages in the developing world (pp. 15-39). New York: Nova Science Publishers

490   Rahman, T. (1997a). The Urdu—English Controversy in Pakistan. Modern Asian Studies, 31(1), 177-207

491   Akram, M. and Mahmood, A. (2007). The status and teaching of English in Pakistan. Language in India, 7(12), 1-7

492   Mutula, S. M.: 2005

493   Akakandelwa, A. (2011). An exploratory survey of the SADC e-government web sites. Library Review, 60(5), 421-431

494   West, D. M.: 2005

495   West, D. M.: 2007; West, D. M.: 2004

ei = 4fi + xi

fi=the number of features present on website i,$0 \leq fi \leq 18$

xi=the number of online executable services on website i, $0 \leq xi \leq 28$

Following a similar method for all government websites, the national eGovernment index can be computed by averaging the score. The essential 18 features include the provision of publications, databases, audio clips, video clips, local language access, not having ads, not having premium fees, not having user fees, disability access, having privacy policies, security policies, allowing digital signatures on transactions, an option to pay via credit cards, email contact information, areas to post comments, option for email updates, option for website personalisation, PDA (personal digital assistant) or mobile accessibility. Extra features such as live chat support, first-time user guide or price-comparison of food provisions also increase the score by one unit[496].

In this exercise of indexing the eGovernment websites of Pakistan, seven main websites comprising a national/federal level and six at the state/provincial level are chosen (see Table 4). These are the official websites of Government of Pakistan or their federated units and are their main points of contacts on the web. It must be noted that all the websites analysed were in English and did not have provisions for translation to a local language rather than a foreign language.

Table 4: List of Pakistan eGovernment Portals analysed

Nr.	Name	Status	Web Portal
1	Government	Federal Government	www.pakistan.gov.pk
2	Gilgit-Baltistan	Provincial Government	www.gilgitbaltistan.gov.pk
3	Khyber Pakhtunkhwa	Provincial Government	www.khyberpakhtunkhwa.gov.pk
4	Punjab	Provincial Government	www.punjab.gov.pk
5	Balochistan	Provincial Government	www.balochistan.gov.pk
6	Sindh	Provincial Government	www.sindh.gov.pk
7	Federally Administered Tribal Areas	Federal Entity	www.fata.gov.pk

Source: Own illustration

---

496   Rorissa's content analysis of African and Asian eGovernment websites is a useful study in understanding the range of web portal analysis approach based on West's method. See Rorissa, A., Gharawi, M. and Demissie, D. (2010). A tale of two continents: Contents of African and Asian e-government Websites. In 43rd Hawaii International Conference on System Sciences. (pp. 1-9). IEEE. Retrieved 04.05.2014 from https://ieeexplore.ieee.org/document/5428303

The Government of Pakistan website (www.pakistan.gov.pk) hosts a variety of services including updated gazettes on national laws, instructions to apply for and download forms for official applications, provisions to check voter list status, guides for business and tourism, paying taxes and registration of marriage and births, tenders issued by the public procurement authority, etc. However, most of these additional provisions were linked to the respective webpage of the constituent government agencies making this webpage into an online catalogue of sorts. A number of additional features offered on the website were outdated or non-functional web links and failed their purpose despite the intended facilitation. Amongst notable services that were functional included for instance, provisions to check daily updated prices from grain and farmer markets, timing of power shutdowns in different areas and checking up on the status of a complaint that had been lodged either online or offline.

The websites of the provinces and federated units focussed on updates from the provincial legislature and the government and provided a directory for jobs and tenders advertised by the state. With the exception of the Punjab government website (punjab.gov.pk) and to some extent the Khyber Pakhtunkhwa website (http://khyberpakhtunkhwa.gov.pk), the other contained merely updated static information and a number of web links did not work. The Sindh government website (sindh.gov.pk) is the only official website to provide a mail inbox to its registered users allowing them to introduce some customisation in their usage of the website. None of the websites charges a user fee or a premium fee for usage nor does any of them offer a complete transaction online. The most advanced options relate to the second stage of interaction of the four-stage model as they allow users to download forms and then submit them in physical offices.

Of the eighteen criteria in West's approach, five are most conspicuous by their absence in this study's sample. These are the privacy policy, existence of visible security policy, disability access, mobile device or PDA support and existence of a forum for users to post comments. Given that a few of them disperse information such as voter list status and car registration details, these should not be discounted due to the major informational components of the websites. This means that given a national identity card number, one could search through a voter list without any authentication or passport system required. Any further expansion of such a system has the tendency to severely compromise the citizen's privacy and also raises concerns for data protection[497].

National ICT policies have a large impact on adoption of technology especially where legal voids existed like digital signatures and credit card payment gateways. Through the Electronic Transactions Ordinance 2002 and Payment System and Electronic Fund Transfers 2007, both of these domains were covered

---

497 Otjacques, B., Hitzelberger, P. and Feltz, F. (2007). Interoperability of e-government information systems: Issues of identification and data sharing. Journal of management information systems, 23(4), 29-51

under them respectively, the adoption of this technology seems to be regressive[498]. Digital signature which is core to completing both non-financial and financial transfers consists of a process of authorisation and clearance about which the web portals do not provide any information[499].

On the user side, availability of content in local language is the first stumbling block and this is coupled with the accessibility options or the lack of them. Lack of local language content in the government web portals is not only an issue in Pakistan but several other developing countries also suffer from the same dilemma[500]. Given that most of the internet users in Pakistan lack a stable internet connection and use their phones as a means of communication, the mobile or PDA platform support can be an essential facet of all government websites[501]. Providing forums on these websites or a place to comment would also enhance user experience and tends to populate the website with personal usage or interaction stories that help enable e-participation of the citizens[502]. The trust factor in adoption of technology has been a stumbling block for eGovernment strategies and it needs to be fostered through an efficient and responsive web platform which is not only regularly updated but also follows up on complaints and processes that have been initiated through either offline or online channel[503].

Table 5 provides a detailed analysis of seven web portals of the Government of Pakistan chosen as a part of this research. The average score for the sample set is 30.8 out of the maximum 72. Similarly, the average score in additional features is 1.5 bringing the national average score to be 32.3 out of 100 (see Table 5). The province of Punjab has enhanced its online presence by involving private Internet companies that have designed a more colourful and image rich interface, however, this might be at a cost of the utility given slow internet speeds. The Khyber Pakhtunkhwa website has a detailed location finder which informs the user of the closest hospitals, family care centres, places of worship, and community animal husbandry services. The number of additional services also point to an attempt to go beyond the interaction level or expanding the scope of digitised services. However, the usability and basic features need to be addressed before future expansions.

It is also pertinent to note here the maturity level of the cataloguing stage activities in web portals. The catalogue stage ensures that the government provides

---

498   Daily Dawn: 2002h, September 12

499   Corradini, F., Paganelli, E. and Polzonetti, A. (2007). The e-Government digital credentials. International Journal of Electronic Governance, 1(1), 17-37

500   Felix Librero (Ed.) (2008). Digital Review of Asia Pacific 2007-2008. Retrieved 28.10.2010 from http://www.digital-review.org/uploads/files/pdf/2007-2008/intro.pdf

501   Ebrahim, Z. and Irani, Z. (2005). E-government adoption: architecture and barriers. Business Process Management Journal, 11(5), 589-611

502   Chun, S. and Cho, J.-S. (2012). E-participation and transparent policy decision making. Information Polity, 17(2), 129-145

503   Parent, M., Vandebeek, C. A. and Gemino, A. C. (2005). Building citizen trust through e-government. Government Information Quarterly, 22(4), 720-736

procedural directions accompanied by contact details in the web portal. However, not all the web portals chosen for this study provided email, telephone and contact details. This seems to be a missed opportunity and a big obstacle for adoption of these services not only by experienced citizens trying to shift online but also a lot of the younger people as one of the purposes of such web portal is to also increase and enhance the citizen engagement with the government[504]. eGovernment websites are not only avenues for better public service delivery but they can also serve as an important link to digitise information and manage and standardise records[505]. Wider publicization of these websites and incentivizing people to use them may drive higher turnover rates and adoption for eGovernment services which in turn strengthen government backing for their online projects[506].

504   Chen, Y.-C. and Dimitrova, D. V. (2006). Electronic government and online engagement: citizen interaction with government via web portals. International Journal of Electronic Government Research, 2(1), 54-76

505   Reddick, C. G. (2005). Citizen interaction with e-government: From the streets to servers? Government Information Quarterly, 22(1), 38-57

506   Chadwick, A. and Howard, P. N. (2010). Routledge handbook of Internet politics. London: Taylor & Francis

Table 5: eGovernment website index analysis of Pakistan

	fata.gov.pk	balochistan.gov.pk	gilgitbaltistan.gov.pk	punjab.gov.pk	sindh.gov.pk	khyberpakhtunkhwa.gov.pk	pakistan.gov.pk
Publications	1	1	1	1	1	1	1
Databases	1	1	1	1	1	1	1
Audio Clips	0	0	0	1	0	1	1
Video Clips	1	1	1	1	1	1	1
Local Language Access	0	0	0	0	0	0	0
Not Having Ads	1	1	1	1	1	1	1
Not Having Premium Fee	1	1	1	1	1	1	1
W3C Disability Access	0	0	0	0	0	0	0
Having Privacy Policies	0	0	0	0	0	0	0
No User Fee	1	1	1	1	1	1	1
Having Security Policies	0	0	0	0	0	0	0
Allowing Digital Signatures	0	0	0	0	0	0	0
Pay via Credit Cards	0	0	0	0	0	0	0
Email Contact	0	1	1	1	1	1	0
Areas to Post Comments	0	0	0	0	0	1	1
Option for Email Updates	0	0	0	0	1	0	0
Personalization of Website	0	0	0	0	1	0	0
Handheld Device Accessibility	0	0	0	0	0	0	0
Weightage by 4	24	28	28	32	36	36	32
Additional Services	0	0	0	3	0	2	6
Total	24	28	28	35	36	38	38

Source: Own Illustration based on Pakistan eGovernment portals and method adopted from West[507]

---

507 West, D. M.: 2007; West, D. M.: 2004; West, D. M.: 2005

4.6. EGD's Government to Citizen projects

In addition to the web portal, the EGD has also invested its resources in projects that are focused on G2C facilitation, though they are not offered as a part of web-based mechanisms. These projects are part of the citizens' needs related to health, police, citizen registration and/or digitisation of land record. Although EGD is mostly involved in eGovernment projects in the G2G and G2B categories, the following table lists some of the eGovernment projects EGD has attempted that can be categorised under G2C category.

Table 6: Projects undertaken by the EGD

eService	Brief Description	E-Enabled Public Services	Status	Operational Status
Tele Medicine for Rural Areas – Holy Family Hospital, Lahore & JPMC, Karachi	To provide people living in rural/remote areas of Pakistan, cost effective means for seeking consultation, advice and treatment from specialist doctors based in big urban centered hospitals.	Patients' Remote Consultations, Remote Health Check-ups etc.	Necessary equipment as per projects' scope are deployed since 2007. Consultation provided in past years to more than 12,000 patients in rural areas from major hospitals	All the developed modules are operating fully electronically from major hospitals.
E-services at Chief / Deputy Commissioner Office –ICT	Automation of the core citizen services in a District	Domicile Issuance System, Citizenship Certificate Management System, International Driving Permit System, Arms' License Management System, Route Permit for Islamabad Transportation Authority.	All the services are automated and deployed at ICT administration since Mar 2008. Applications for International Driving License, Domicile Certificate and route permits functional at ICT administration.	The organization is operating partially electronically. The modules mentioned in status are operating electronically while the processes related to remaining modules are manually performed.
E-services at Capital Development Authority (CDA)	Automation of the core services along with automating the	Web Based Property Management, Web Based Billing System (Water &	All the services are automated and deployed at CDA since June 2008. At present, only Complaint	The organization is operating partially electronically, only the Complaint

eService	Brief Description	E-Enabled Public Services	Status	Operational Status
	back-office processes of CDA Office.	Property tax), File Tracking System, Complaint Management System (Complaints received at Call Centre)	Management System (CMS) is fully functional. Citizens are able to lodge their complaints through telephone or website.	Management System module is operating electronically while the process related to remaining modules are manually performed.
E-enablement of Islamabad Police	To transform the traditional police stations of Pakistan into E-powered police stations. The Project scope includes development of citizen services and back end processes of Islamabad Police. Further, some of the applications including e-FIR includes URDU interface.	Automation of Driver/Learner's License Application process, Automation of Internal Registers e.g. FIR, Zimni, Challan etc.	All applications were developed and deployed. Driving License Application is fully functional since March 2007. Further, some internal registers e.g. Zimni, FIR, Challan etc. are also functional in 8 out of 24 Police Stations	The Zimini, FIR and Challan modules are operating electronically as well as manually, while the Driving License Application is operating fully electronically. One of the barriers in completely shifting to electronic mode vis-à-vis Zimni & FIR module is proper awareness and enforcement of electronic transaction ordinances within courts and other organizations.
Health Management Information System (HMIS) and Networking facilities at PIMS, CDA Hospital, Sheikh Zyed Medical Complex and JPMC (Jinnah Postgraduate	Installation of Networking facilities and deployment of Health Management Information System (HMIS) at following hospitals: PIMS, Islamabad CDA Hospital, Islamabad, Sheikh	Patient Registration, Out Door Patient Department (OPD) Management, In Door Patient Management	Applications are functional at PIMS, Islamabad and CDA Hospital, Islamabad since June, 2007 and December 2007 respectively. The deployment of HMIS is under process at Sheikh Zyed Medical	70% functions of the HMIS are operating electronically in PIMS, 30% of the HMIS is operating electronically in CDA Hospital, while SZMC, Lahore and JPMC Karachi are operating

eService	Brief Description	E-Enabled Public Services	Status	Operational Status
Medical Centre), Karachi	Zyed, Lahore, JPMC, Karachi The HMIS under development at SZMC includes the Urdu Interfaces.		Complex, Lahore. The deployment process at JPMC, Karachi is yet to be initiated.	manually as the system is under development stages in these hospitals.
E-Services for Ministry of Health, Islamabad	To provide the basic ICT Infrastructure to Ministry of Health and automation of Business Processes / Agency Specific Applications for Ministry of Health	Free Medical Treatment Request System, Issuance of NOC System, Statistical System for Health-Related Facilities, Drug Manufacturing License Management System, Drug Registration, Import /Export System	Under Development	Currently Manual
Land Revenue Records Management System in rural areas of ICT	To computerize the record which is maintained by the revenue officers and various related departments all across the Islamabad Capital Territory (ICT) in the form of registers. The main theme is to make all the material available and easily accessible as a single electronic database. The application includes the interfaces in English and Urdu	Access to and Land records	Under Development	Currently Manual
Economic Development mapping in 5 districts of	To create a GIS-based data infrastructure, Economic Development Map	GIS-based economic development map of Pakistan, Decision Support	Under Development	Currently Manual

126

eService	Brief Description	E-Enabled Public Services	Status	Operational Status
Pakistan (Pilot Project)	(EDM), both at Provincial and Federal level thereby creating economic development map for the country covering following sectors: Education Health Roads Population Welfare	System for better and effective planning and uniform distribution of development resources		
On-line Tracking system for cargo handling, Freight wagons and Locomotives	To establish an ICT-based solution to track the locomotives and freight Wagons so as to enable Pakistan Railways to efficiently manage its cargo handling business	Online Tracking of Locomotives and Cargo Wagons	Under Development	Currently Manual

Source: Electronic Government Directorate

The data for Table 6 was collected from the EGD during the field work conducted in 2008/2009. Unfortunately, individual monitoring reports were either not available or not shared by the EGD. However, the review of these nine projects highlights some important aspects for the growth and development of eGovernment in Pakistan.

The nine projects presented here point at one of the crucial aspects that they are all focused on automation or computerisation of manual data to a digital form. One of the first core concerns of any eGovernment application or electronic public service is the availability of physical data into a digital form. It is only after the readiness of digital data that the electronic public services can be conceived and initiated[508]. Once the data is available, only then digital communication delivery channels can be thought of. That is, which digital mechanism(s) will be used to deliver the data, will it be via internet, wireless or a simple phone line[509]. Similarly, another key concern for any eGovernment application/software is also the product or a tangible gadget with which the citizen will receive the eServices. It can be a web-based technology accessible via web browser with a computer

508   Zhou, Z. and Gao, F. (2007). E-government and Knowledge Management. International Journal of Computer Science and Network Security, 7(6), 285-289

509   Kernaghan, K. (2005). Moving towards the virtual state: integrating services and service channels for citizen-centred delivery. International Review of Administrative Sciences, 71(1), 119-131

connection, it can be a mobile phone (either smart or feature phone) or it can employ any other front-end technology such as kiosk[510]. However, the basic step is the availability of digital data around which the services can be built. The list of projects in Table 6, however, reveal that despite the major revolution in computer industry over a number of decades, the Government of Pakistan's institutions, departments and ministries did not maintain any digital records. For instance, project 6 – 9 have been entirely handled manually and need to be digitised from scratch. Similarly, rest of the project 1 – 5 are also only partially digitised.

Another factor, that is also linked to the aspect of digitisation is the eGovernment category, that is, either the project is related to eLearning, eVoting, eDemocracy or eAdministration. Most of the projects in Table 6 are of administrative nature in which EGD has tried to set up and introduce electronic mechanisms for the government departments to perform administrative tasks mostly via computers. Similarly, based on the type of eGovernment, only one out of nine projects fall under G2C category whereas rest of the projects are either G2G (6 out 9 projects) and the remaining two belong to the G2B category.

While reviewing the health management projects conducted by the EGD for hospitals, even though the digitisation of patient data would be very useful for the hospital and eventually for the citizens, however, in a typical eGovernment scenario there is basically no direct contact between the citizen and any (communication) technology (or device). It is predominantly the hospital staff who is utilizing the health management system and perhaps a mean for epidemiological measure or statistics for recourse to public health. Similarly, EGD itself also did not categorise or characterize these projects according to their target groups. The matrix or a table based on the projects listed in Table 6 through which one could get an idea about EGD projects' category, type, outreach, scope and citizen-technology interaction is constructed for this study.

Table 7 re-presents the Table 6 by introducing six different classifications based on the EGD projects. These include (1) institution type, (2) eGovernment category, (3) type, (4) scope, (5) Citizen-Technology interaction and (6) coverage. The main motivation behind this classification is to separate G2G and G2B from the G2C projects. Similarly, while classifications of category and type have been elaborated above, Citizen Technology Interaction (CTI) has an important characteristic. CTI here means if the citizen directly gets in touch with the technological tool or will the technology assist her/him to seek information/relevant data. For instance, in case of digitisation of hospital data, although citizen can receive benefits of such a digitisation. Doctors can better help the patients if there is a digital maintenance of medical tests, previous medical

---

510   Arch, A. and Hardy, B. (2005). E-Government: Accessible to All. Future Challenges for E-government. (pp. 54-74). Retrieved 04.05.2014 from https://www.finance.gov.au/sites/default/files/Future-Challenges-for-Egovernment-Volume02.pdf

history or diagnosis-related data. However, the patient her/himself does not interact with the technology itself.

The classification of scope is an additional but associated category related to Type. Scope helps in clarifying the aspect of impact, that is, if the project has a direct or immediate impact on citizens. Similarly, the last aspect of coverage helps in explaining the geographical reach of the projects.

Within the eGovernment category eAdministration, the project of digitisation of eServices at the Capital Development Authority (CDA) is the only project out of the given nine projects (see project number 3 in Table 7) whose type and scope are also G2C. Capital Development Authority (CDA) is a government organisation responsible for the management and development of Pakistan's capital Islamabad. The eServices developed by the EGD for the CDA include the property management system, utility bills for water and property tax and complaints management system[511]. All these systems are accessible for the public via CDA's web portal, which requires a web browser and a computer with internet connection. Therefore, EGD's project for CDA is only G2C service where citizen can personally interact with the technology to communicate with the government or in this case government entity (municipality to be precise) through electronic means.

Table 7: Characterisation of eGovernment projects initiated by the EGD

Nr	Institution Type	eGovernment Category	Type	Scope	Citizen-Technology Interaction	Coverage
1	Hospital	eHealth	G2B	G2C	No	Rural Districts
2	Government	eAdministration	G2G	G2C	No	Urban City
3	Government	eAdministration	G2C	G2C	Yes	Urban City
4	Government	eAdministration	G2G	G2C	No	Urban City
5	Hospital	eHealth	G2B	G2C	No	Urban Cities
6	Government	eAdministration	G2G	G2G	No	Urban City
7	Government	eAdministration	G2G	G2C	No	Urban City
8	Government	eAdministration	G2G	G2G	No	National
9	Government	eAdministration	G2G	G2B	No	National

Source: Own illustration based on Table 6

---

511 Capital Development Authority (2008). "Complaint & Enquiry Management System". Retrieved 28.08.2008 from www.cda.gov.pk/cda-latest/files/file.asp?var=cems.

Another aspect in Table 7 is that of coverage which highlights EGD's choice of projects, project partners, strategic preferences and eGovernment strategy. That is, how inclusive are these projects for a geographically and demographically diverse country such as Pakistan where there are stark differences between the rural and urban developments. The aspect of coverage gains more importance within the rural and urban debate because it would (re)define the preference of communication channels and choice of gadget in citizen-technology interaction. Most of the projects listed in Table 6 are focused either in Islamabad or in other urban cities such as Karachi or Lahore. Except the first project on telemedicine which aims to provide medical coverage to rural areas, rest of the projects have been implemented in the bigger cities. EGD with its own presence and jurisdiction limited to Islamabad along with the choice of projects explains EGD's own scope and limitation as an institution. There does not appear to be an equal balance and distribution in providing eServices for example to Pakistan's provincial capitals, bigger and urban centres, medium sized towns and villages.

4.7. Concluding Remarks

This chapter has attempted to explore the developments towards electronic public service delivery and initiation of eGovernment in Pakistan starting 2000. It has tried to focus on the major policies in particular IT Policy and Action Plan (2000) that became the backbone for all major initiatives such as development of IT departments and boards at the provincial level, Electronic Transaction Ordinance (2002) and subsequent formation of Electronic Certification Accreditation Council. An important aspect this chapter highlighted is the interest in and growth of eGovernment by the federal government particularly led by the Federal Minister of Science and Technology, Prof. Dr. Atta-ur-Rahman. He has been the driving force behind initiating several policies and institution building. His efforts resulted in, as this chapter has tried to underscore, the establishment of Electronic Government Directorate under the Ministry of Information Technology. EGD will be considered as the first official institution for policy planning and implementation of eGovernment and digitisation programmes, projects and framework in Pakistan. Similarly, we also saw that the first official IT policy was the result of a collaborative effort that brought together the IT community from public and private sector. These concerted and mutual efforts have led Pakistan to officially delve in the field of eGovernment.

However, during the development of first IT policy of Pakistan, particularly with reference to the Working Groups and individuals involved, it seemed that the Information Technology discourse was predominantly the level playing field for the IT experts. There seemed to be a strong need of participation by other civil society actors and interest groups such as advocacy groups and academia from humanities, political and social science background. It is the civil society along with the social scientists who are relatively more connected with a broad spectrum of public, who can correlate with the emerging behavioural and cultural changes in society and whose contribution could help identify or fill design-reality gaps either in policy making or in a technological development. This particular aspect can be strongly felt in the first IT Policy (2000) as well as in the eGovernment strategy paper (2005) with reference to the idea of 'online interaction' and the way Government of Pakistan plans to interact with its citizens considering an overall low teledensity (see Figure 8). Nonetheless, the development of institutions and policies provided basic start for government to at least proceed with the eGovernment discourse. The initial work on developing prototypes for agency-specific applications and standardisation of software architecture across Pakistani ministries as proposed by the EGD showed an understanding towards developing single and customisable Document Management Systems instead of involving multiple vendors. It is unfortunate that EGD's own performance and continued irreverence for its mandate at the hands of bureaucracy led to its decline (EGD's decline is further elaborated in the Discussion section). The following chapter focusses on the development of another government institution that not only enjoyed more autonomy and bureaucratic support but also exploited the gaps left by the EGD particularly with reference to the use and exploitation of communication patterns and channels.

Chapter 5

NADRA's Kiosk and eSahulat Programme

One of the important takeaway from Chapter 4 is Pakistan's concerted efforts in pursuing the field of eGovernment and taking institutional and legislative steps towards developing electronic public service delivery. In this context, we note that between 2000 through 2005 the platforms provided by several initiatives such as first IT Policy (2000), Electronic Transaction Ordinance (2002), Electronic Government Directorate (2002) and eGovernment Strategy (2005), led to concrete steps towards conceiving eGovernment applications. In the previous chapter, we also note that EGD mostly focussed on projects related to computerisation/automation and developing prototype for eGovernment platform for Pakistan. In due process, it did not develop any G2C service that could facilitate citizens for instance, submitting online applications, paying utility bills or applying for passport and driver licence at the national level.

This chapter focusses on the eGovernment initiatives undertaken by another government entity namely National Database Registration Authority (NADRA). It will highlight the gaps left by EGD and communication patterns that were employed by NADRA to initiate their own set of eGovernment applications. While the previous chapters have pointed out the importance of language, culture, local capacities and expansion or diffusion of the new ICTs, this chapter will particularly focus on two eServices developed by NADRA namely kiosk machine and eSahulat. Both these services have not only become a benchmark for other eService providers but kiosk and especially eSahulat continue to contribute to the socio-economic sector as well. This chapter provides details about these projects, their functionality, operations, their usage pattern as well as their individual analysis based on the data received from NADRA as part of this research. The main objective of analysing kiosk and eSahulat projects is to depict citizen usage patterns and try to contextualise the electronic public service delivery scenario created by NADRA in Pakistan.

5.1. Origins and Formation of National Database Registration Authority

Ministry of Interior in Pakistan performs several executive and policy functions apart from ensuring internal security. Nationality, citizenship and naturalisation, immigration, passports, regulation of entry and exit of foreigners fall under the jurisdiction of Ministry of Interior. Its other responsibilities also include policy coordination and higher training of civil defence, anti-smuggling measures and enforcement of anti-corruption laws, control and administration of Federal

Investigation Agency (FIA)[512]. NADRA was established in 2000 to develop and issue Computerised National Identity Cards (CNIC) to Pakistani citizens and it functions under the Ministry of Interior[513]. NADRA's expedition in G2C electronic public services has only been possible due to its proficiency in producing and later maintaining the computerised citizen database. Prior to 2000, a centralised database of citizen did not exist. A citizen database did exist but in the form of several manual registers under provision of provincial departments such as Department of Local Bodies and Rural Development headed by the Civil Registration Organization Ministry of Interior[514].

Till 1973, there was no formal statistical database of citizens maintained by the Government of Pakistan. After the partition of India in 1947, problem of governance and administration as pointed out in Chapter 3 also impacted the citizen registration system. Registration system for birth and death that was practiced in pre-1947 during the British colonial period was carried over in young state of Pakistan. In the urban cities head of the household or the medical staff representative was responsible to register birth or death in their respective municipal committees[515]. In the rural areas, a different procedure was followed. The village head kept two registers (one for recording birth and other for deaths) which were regularly provided to local police station. Police stations' clerical staff recorded these entries in their own registers and copies of these would be submitted to District Health Officers. These data would be forwarded to Ministry of Health for compilation and preparation of reports. Under Basic Democracies Order of 1959 and Municipal Ordinance of 1960, registration of vital events was made mandatory to be registered by heads of household in local union councils or municipal committees[516].

However, these civil registrations did not lead to any issuance of identification cards. Between 1950 through 1971 Personal Identity System remained a subject of discussion that was supposed to register Pakistani citizens as well as the migrants coming from India to live in Pakistan. In 1971, East Pakistan broke away from its western wing to form an independent country Bangladesh that brought a

---

512   Ministry of Interior (Government of Pakistan) (2014). "Rules of Business". Retrieved 04.05.2015 from https://www.interior.gov.pk/index.php/business-rules.

513   National Database Registration Authority (2014b). "About Ministry of Interior". Retrieved 04.05.2014 from http://nadra.gov.pk/index.php/about-us/ministry-of-interior/about-moi.

514   International Institute for Vital Registration and Statistics. (1995). Organization of National Civil Registration and Vital Statistics Systems: An Update. Retrieved 04.05.2014 from https://www.cdc.gov/nchs/data/isp/063_organization_of_the_national_civil_registration_and_vi tal_stat_systm_an_update.pdf

515   Alvi, A. M. (1993). National Registration System in Pakistan. In East and Sourth Asian Workshop on Strategies for Accelerating the Improvement of Civil Registration and Vital Statistics Systems. 29 November - 3 December 1993, Beijing, China. (pp. 1-19). Retrieved 04.05.2014 from https://unstats.un.org/unsd/demographic/meetings/wshops/1993_China_CRVS/docs/1993_Doc. 11_Pakistan.pdf

516   See Section 46 on registration of birth, death and marriages in Gazette of Pakistan (1960). Ordinance No. X of 1960: Municipal Administration Ordinance. Retrieved 04.05.2014 from https://cmsdata.iucn.org/downloads/municipal_administration_ordinance_1960.pdf

number of national security concerns for the remaining (West/current) Pakistan. The idea of unique identification of Pakistani citizens became a priority in the aftermath of fall of Dhaka. It was in 1972 that the then government decided to revamp the system of citizen registration[517]. The then Prime Minister Zulfiqar Ali Bhutto (1928-1979) approved the formation of a separate department (to maintain statistical data of the citizens) in the meeting of federal cabinet on 20th December 1972[518]. In this regard, National Registration System was established in the form of National Registration Act (NRA) in 1973. Under NRA, Directorate Generate of Registration (DGR) as an organisation was formed to register every citizen and issue national identity cards. NRA's clause 4 envisioned the development of Population Register for Pakistan by administrative units. Such a register would help maintain all vital events also including occupational and residential details[519].

With announcement of Article 30 of the Second Amendment of the Constitution of Pakistan, government started to record the identification and maintenance of statistical database of Pakistani citizens[520]. The NRA also replaced the previously used Personal Identity System of Pakistan that had been in discussion till 1971[521]. According to NRA, identity card details based on the statistical database enabled the government to record citizen's photograph, manual thumb impression (ink based), address and demographic data on government registers[522]. The population register continued to grow and government also employed computers to record citizen data. However, this data was not electronically linked and neither was it mandated to develop computerised identity cards. In 1993, the Director General of Registration at Ministry of Interior authorised a pilot project under the name of "Automation of Registration Scheme" for twin cities of Rawalpindi/Islamabad to computerise the citizen database along with registration of births, deaths and change of marital status[523]. The project was extended to rest of the country over the course of five years. However, due to lack of interconnectivity among the databases, the problem of, among others, citizen identification and verification remained a challenge for Directorate General of Registration.

---

517   Malik, M. T. (2010). National Identity Management in Pakistan. Retrieved 04.05.2014 from http://siteresources.worldbank.org/INTEDEVELOPMENT/Resources/Malik-1.pdf

518   From an idea to reality: Pakistan's smart card (2012, November 11), The News. Retrieved 25.01.2013 from https://www.thenews.com.pk/archive/print/395365-from-an-idea-to-reality-pakistan's-smart-card

519   Alvi, A. M.: 1993

520   Gelb, A. (2014). Technology in the Service of Development: The NADRA Story. Retrieved 28.12.2014 from https://www.cgdev.org/sites/default/files/CGD-Essay-Malik_NADRA-Story_0.pdf

521   Basit, M. A. (2012). Pakistan Citizenship & NADRA Laws. Lahore: Federal Law House

522   As per the NRA, women were exempted to have the photographs in their identity cards based on strictures of veil (or purdah). Dhagamwar, V. (1995). Elections in Pakistan, 1993. Indian Journal of Gender Studies, 2(1), 115-120

523   Federal Ombudsman Secretariat (2015). Report on the functioning of National Database Registration Authority. Retrieved 15.07.2015 from http://www.mohtasib.gov.pk/images/pdfs/NADRA.pdf

134

5.1.1. 1998 Census and formation of National Database Organisation

In 1997, Nawaz Sharif government was preparing for the fifth national census to be held in 1998[524]. In order to develop a comprehensive citizen database, record social and fiscal indicators as well as to document the economic parameters as an added exercise within the national census, a special organisation was set up to assist with these tasks. In this regard, National Database Organisation (NDO) was established as an attached department under Ministry of Interior[525]. The primary role of NDO was to record and handle the data that was collected through National Data Forms (NDF) as a part of fifth national census conducted in March 1998[526]. On 12th October 1998, the then General Pervez Musharraf overthrew the Nawaz Sharif government and assumed a role of the chief executive of Pakistan[527]. NDO continued to work under Musharraf's regime as well and completed the scanning and data entry of over 64 million records collected through NDF and eventually NDO succeeded in preparing a computerised citizen's database[528]. It was during Musharraf's military rule that it was decided that the NDO and Directorate General of Registration will be merged to establish a new public-sector organisation that will issue the CNICs[529]. On March 10, 2000, Musharraf passed a new ordinance for the merger creating the National Database Registration Authority (NADRA) with a primary goal of developing a scalable and digital National Data Warehouse (NDH) that could support numerous database management systems at national level[530].

The secondary goals behind establishing NADRA included, among others to develop CNIC for Pakistani citizens with each having a unique identification number as well as to digitise other civil registration matters related to death and

---

524 Goodwin, P. (1998). Pakistan prepares to redraw population map. Retrieved 04.05.2014 from http://news.bbc.co.uk/2/hi/despatches/61407.stm

525 Chaudhry, A. and Ahmad, I. (2011). Examining the relationship of work-life conflict and employee performance (a case from NADRA Pakistan). International journal of business and management, 6(10), 170-177

526 Waiting for the NIC: Dateline Islamabad (2002c, February 12), Daily Dawn. Retrieved 04.05.2014 from https://www.dawn.com/news/1062686/dawn-features-february-12-2002

527 Coup in Pakistan: The Overview; Pakistan Army seizes power hours after Prime Minister Dismisses his Military Chief (1999, October 13), The New York Times. Retrieved 09.10.2010 from https://www.nytimes.com/1999/10/13/world/coup-pakistan-overview-pakistan-army-seizes-power-hours-after-prime-minister.html

528 National Database Registration Authority (2006). "Objectives Achieved by NADRA". Retrieved 09.12.2006 from http://www.nadra.gov.pk/objectives.html.

529 National Database Registration Authority (2008a). "History of NADRA". Retrieved 18.08.2008 from http://www.nadra.gov.pk/site/351/default.aspx.

530 Hakeem, A. A. (2009). Smart National Identity Card in Pakistan. Presentation Slides. Retrieved 04.05.2014 from http://siteresources.worldbank.org/EXTSAFETYNETSANDTRANSFERS/Resources/Smart-National-ID-cards-in-Pakistan.pdf

birth[531]. In doing so, NADRA had set its aim in establishing databases at national, provincial and district levels that can have an interface with each other (or in other terms, they are able to digitally exchange information). In this context, NADRA emerged as a public-sector entity with an independent corporate body with requisite autonomy to operate independently with an aim to establish a multipurpose, computerised database and to ensure the security and protection of data and information[532].

NADRA's administrative structure is comprised of a headquarter in Islamabad, five provincial headquarters in Peshawar, Islamabad, Lahore, Karachi, Quetta and three regional headquarters in Sukkur, Multan and Sargodha[533]. NADRA's headquarter in Islamabad is located in the protected area, the so-called Red Zone where most of the government buildings such as Parliament, Senate, Supreme Court, and the foreign embassies in Diplomatic Enclave are also situated[534]. Coincidentally, NADRA is housed in State Bank Building which happens to be a former Parliament Hall where the 1973 constitution was approved, the same constitution that led to the formation of first statistical database of citizens[535].

5.1.2. NADRA's digital capacity and infrastructure

NADRA's digital infrastructure and its underlying database management system is locally produced, that is, it does not seek any third party or outsourcing services from a software firm or consultancy[536]. NADRA develops its own software(s) under its technology division headed by Chief Technology Officer[537]. NADRA's technology division started indigenous development of a state-of–the-art centralised data warehouse and associated network infrastructure to produce secure CNICs[538]. The technology division is complemented by the information security wing, networks and communications department and a data warehouse. All the technology divisions are further subdivided into software engineering, and

531   National Database Registration Authority (2008e). "Goals and Achievements ". Retrieved 10.09.2009 from http://www.nadra.gov.pk/site/353/default.aspx.

532   Gazette of Pakistan (2009). Ordinance No. LIX of 2009. Retrieved 20.11.2011 from http://www.na.gov.pk/uploads/documents/1302654535_961.pdf

533   National Database Registration Authority (2007). "History of NADRA". Retrieved 17.08.2008 from http://nadra.gov.pk/site/351/default.aspx.

534   European Country of Origin Information Network (2013). Bericht zur Fact Finding Mission Pakistan. Retrieved 03.06.2014 from https://www.ecoi.net/en/file/local/1068808/1729_1374674206_ffm-bericht-pakistan-2013-06.pdf

535   National Database Registration Authority: 2007

536   Sardar, A. (2007). Information System Reforms for Improving Governance: Case Study NADRA. Presentation Slides. Retrieved 04.05.2014 from http://siteresources.worldbank.org/PSGLP/Resources/InformationSystemReformsforImproving Governance.pdf

537   Interview with Chief Technology Officer, Mr. Usman Mobin. Islamabad, 09.03.2009

538   Nadra to compile data of 140m people (2002b, August 11), Daily Dawn. Retrieved 04.05.2014 from https://www.dawn.com/news/52097

also research and development departments. NADRA's Data Warehouse originally planned to digitise the 1998 census data and hosted up to 4.2 Terabytes of data by 2002 that was able to accommodate an extension of 100 Terabytes and more[539].

From 2001 to 2003 NADRA was able to produce about 31 million CNICs that were centrally created and maintained at NADRA's headquarter in Islamabad whereas the data was gathered through its site offices all over the country[540]. Over the number of years, NADRA's digital capacity has sharply increased and their hardware comprises all sorts of sophisticated technologies such as mini frame computers, biometric sensors and detectors, signature and form scanners, thumb digitiser and other networking technologies such as satellite-link, fibre optic connection and webservers to communicate with regional headquarters[541]. One of the unique features of NADRA's technological innovation is its emphasis on producing databases that are able to store, read, process and search data in local languages such as Urdu and Sindhi[542]. Since its inception in 2000 NADRA's data warehouse has produced a 91.3 million CNICs[543]. It is continuously being used to run various day-to-day operations in transaction processing, business intelligence and decision-support applications.

5.1.3. NADRA's foray into eGovernment services and implementation

NADRA's CNIC can be considered as one of the first eGovernment applications that Pakistan initiated in 2000. Even though in the initial phase (2002 through 2004), the CNIC did not provide any G2C services, nevertheless CNIC was a first electronic/digitised document with the help of which government was able to verify a person anywhere in the country[544]. This is precisely because unlike the old and manual method of statistical database, NADRA's databases are

---

539   National Database Registration Authority (2010). NADRA's Services Brochure. Retrieved 29.01.2010 from http://www.nadra.gov.pk/downloads/solutions/brochure.pdf

540   No extension in Dec 31 deadline for ID cards (2003g, November 15), Daily Dawn. Retrieved 04.05.2014 from https://www.dawn.com/news/125001

541   During a field/research visit to the NADRA data warehouse department, which houses the most sensitive citizen data, one is reminded of science fiction films whose instances can be found in Spielberg's Minority Report or in Mission Impossible. The data warehouse is the most guarded division within the NADRA headquarter in Islamabad. It requires several manual and digital security checks in order to enter the pool of digital processing machines maintaining the citizen database. See, for instance: Nadra to introduce new verification facility (2003f, July 28), Daily Dawn. Retrieved 04.05.2014 from https://www.dawn.com/news/132294

542   Daily Dawn: 2002b, August 11

543   Voter list keeps 7m CNIC holders out (2012a, September 04), The Express Tribune. Retrieved 04.05.2014 from https://tribune.com.pk/story/431030/voter-list-keeps-7m-cnic-holders-out/

544   In 2003 NADRA introduced digital verification mechanisms namely "verisys" and "biosys" aimed at online citizen verification for the financial institutions and banks. See Nadra's hi-tech system opens today (2003a, July 29), Daily Dawn. Retrieved 04.05.2014 from https://www.dawn.com/news/132510

electronically interlinked and every citizen has a unique identity card number. At present, the CNIC is being used as one of the major identification mechanisms in order to conduct regular activities such as buying or selling properties, opening a bank account or purchasing a mobile SIM card[545]. Accumulation of entire citizen database by a single public-sector entity and the development of CNIC provided a platform to NADRA to further extend its expertise within the domain of civil registry applications (as also mandated in NADRA Ordinance)[546]. In this context, NADRA started working on developing detailed civil registry modules[547]. These included on one hand computerised birth, death, marriage and divorce certificate and on the other hand Child Registration Certificate (CRC) and Family Registration Certificate (FRC). The idea behind computerisation of civil registry certificates lies in the fact that up to end of 1990s such certificates only existed in a manual form and more importantly there was no check and balance as to how many times such certificates were issued, with counterfeits being common[548].

Similarly, it was also difficult to prove the authenticity of these certificates as historically such certificates could also be used for identity thefts and fraudulent cases (as also pointed out in Chapter 3)[549]. Therefore, based on the existent citizen database, it became possible for NADRA to develop such a system which would later be provided to provincial and local governments that would issue such computerised vital civil registry certificates through union councils[550]. Another important identity document within the context of civil registry system is also the FRC[551]. Based on the family lineage, FRC provides the details of any Pakistani citizen in the form of computer generated document that contains basic details of her/his siblings and parents along with their photos. Similarly, second type of FRC can also be issued that contains person's spousal and children details. Such a system has been built to help prevent among others, cases of frauds that have been prevalent in Pakistan particularly in cases of inheritance, business and property deeds[552].

---

545   Get your expired IDs renewed, NADRA asks citizens (2011, June 15), Pakistan Today. Retrieved 14.08.2013 from https://www.pakistantoday.com.pk/2011/06/15/get-your-expired-ids-renewed-nadra-asks-citizens/

546   Chapter 4 (National Data Warehouse), 7.b (ii).

547   Nadra dispersing valuable data to government departments (2007c, April 30), Business Recorder. Retrieved 04.05.2014 from https://fp.brecorder.com/2007/04/20070430558432/

548   The importance of birth registration (2014, October 21), Daily Times. Retrieved 04.05.2015 from https://dailytimes.com.pk/102786/the-importance-of-birth-registration/

549   Gelb, A.: 2014

550   Though NADRA has tried to bring standardisation in the civil registry system, however, the cost of such certificates has often been criticised. See, for instance Computerisation comes at a price: Birth, death certificates (2007a, August 11), Daily Dawn. Retrieved 04.05.2014 from https://www.dawn.com/news/260696

551   National Database Registration Authority (2012b). "Family Registration Certificate". Retrieved 21.04.2012 from http://www.nadra.gov.pk/index.php?option=com_content&view=article&id=11&Itemid=14.

552   eGovernment: SECP's eServices greatly help business entities (2015, January 11), Business Recorder. Retrieved 04.05.2015 from https://fp.brecorder.com/2015/01/201501111141316/;

Continuing with its digitisation operations, NADRA's capacity developed over time and it started working towards computerisation of manual passport (system). In this regard, Planning Commission of Pakistan already started floating the idea of developing a biometric and Machine-Readable Passport (MRP) system in 2001[553]. When the tender notice for the MRP project was announced by the Immigration and Passport Department at the Ministry of Interior, NADRA also participated in the bidding along with the foreign bidders[554]. NADRA won the bid and started working towards developing MRP. It formally handed over the MRP project to Immigration and Passport department on 31st December 2004[555]. The MRP is equipped with fingerprint and facial recognition features along with multiple digital verification and encryption systems[556]. NADRA also extended its capacities to develop a border control system to help immigration department digitally patrol the national territories of Pakistan, particularly to prevent human trafficking and illegal immigration[557]. In this regard, it started developing the Automated Border Control (ABC) system that could be installed on every entry and exit ports of Pakistan. Under the backdrop of 9/11 incidents, a system called Personal Identification Secure Comparison and Evaluation System was initially developed in joint collaboration of the Federal Investigation Agency (FIA) and the Intelligence Bureau with an assistance from the United States[558].

Data Democratisation – ensuring Transparency in Disaster Relief (2010b, October 31), Business Recorder. Retrieved 04.05.2014 from https://fp.brecorder.com/2010/10/201010311118592/

553   Immigration identity system at airport (2001, December 09), Daily Dawn. Retrieved 11.05.2014 from https://www.dawn.com/news/9830/karachi-immigration-identity-system-at-airport

554   MRP facility handed over to passport dept (2004e, December 28), Daily Dawn. Retrieved 14.05.2014 from https://www.dawn.com/news/378252; New passports from September (2004a, March 31), Daily Dawn. Retrieved 13.05.2014 from https://www.dawn.com/news/355229

555   Immigration takes over passport project (2005b, January 01), Daily Dawn. Retrieved 04.05.2014 from https://www.dawn.com/news/378681/immigration-takes-over-passport-project; Machine readable passports in six months (2004c, January 11), Daily Dawn. Retrieved 14.05.2014 from https://www.dawn.com/news/390878

556   National Database Registration Authority: 2010; Old passports will be replaced by June 2006 (2004d, October 27, 2004), Daily Dawn. Retrieved 14.05.2014 from https://www.dawn.com/news/398323; Issuing of manual passports to stop by June 30 (2005d, May 07), Daily Dawn. Retrieved 04.05.2014 from https://www.dawn.com/news/138212

557   Karachi: Nadra terms MRP system perfect (2006a, April 2008), Daily Dawn. Retrieved 14.05.2014 from https://www.dawn.com/news/189724/karachi-nadra-terms-mrp-system-perfect; Process to issue passport, visa being revamped (2006b, October 20), Daily Dawn. Retrieved 04.05.2014 from https://www.dawn.com/news/215619/process-to-issue-passport-visa-being-revamped

558   National Assembly of Pakistan. (2009). Questions for Oral Answers and their Replies: National Assembly Secretariat (11th Session). Retrieved 04.05.2014 from http://www.na.gov.pk/uploads/documents/questions/1303078749_855.pdf

Figure 13: Organisational Structure of NADRA Headquarter

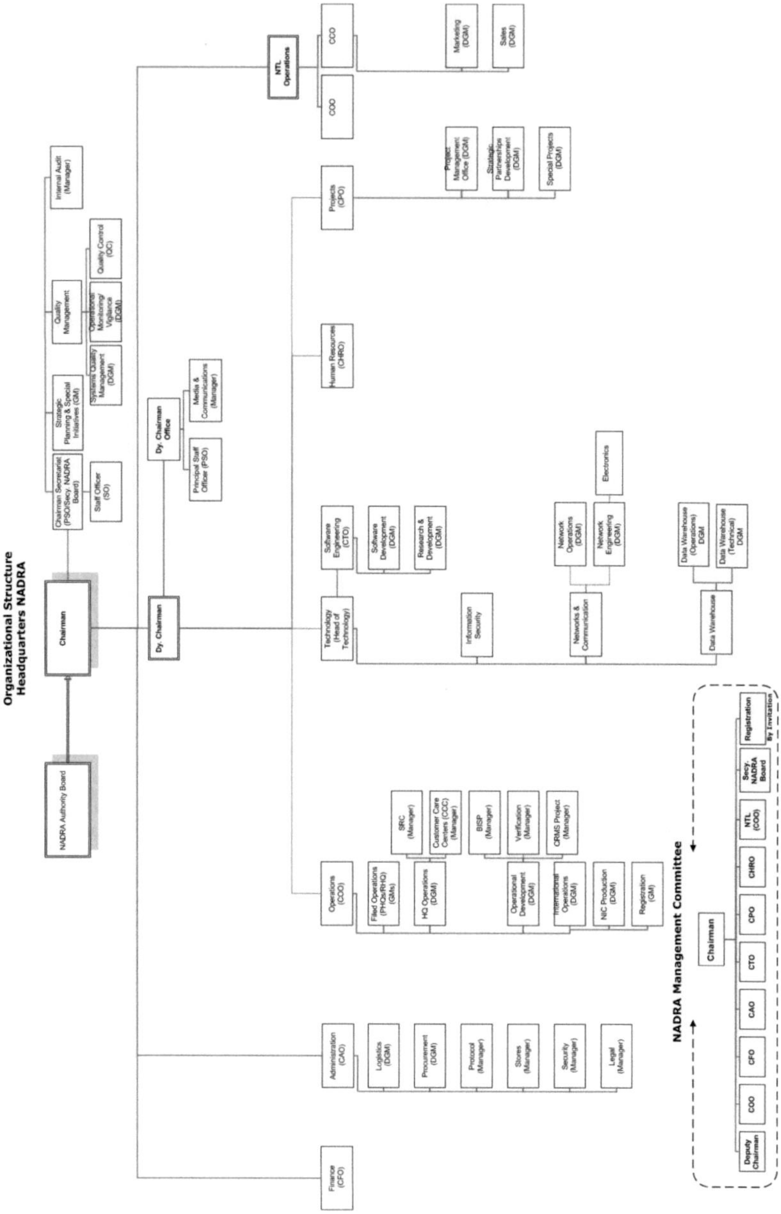

Source: NADRA [559]

---

559   National Database Registration Authority (2009b). NADRA's organisational structure. Retrieved
       09.06.2009 from http://nadra.gov.pk/images/organogram.jpg

At a later phase, FIA opted for a locally produced ABC developed by NADRA[560]. On the other hand, NADRA has been working on further enhancing its CNIC project, for instance to include more biometric and machine-readable features. Since 2012, NADRA is gradually replacing the CNIC with the Smart National Identity Card (SNIC) which is a chip based identity card incorporating 36 security features in the physical design of the SNIC[561]. The motivation behind the SNIC is to use it as a main identity instrument, for instance as a driving license, school card, and hospital ID. In addition to that, NADRA also collaborated with the State Life Insurance Company to offer death insurance worth Rs. 100,000 life insurance to every new SNIC holder[562].

Table 8: eGovernment Applications developed by NADRA since 2000

Identity Card Projects	Civil Registry System	Border and Aviation
Computerised National Identity Card (CNIC)	Child Registration Certificate (CRC)	Machine Readable Passport (MRP)
National Identity Card for Overseas Pakistanis (NICOP)	Family Registration Certificate (FRC)	Automated Border Control (ABC)
Pakistan Origin Card (POC)	Birth Certificate	
e-Driver's License	Death Certificate	
Smart National Identity Card (SNIC)	Marriage Certificate	
	Divorce Certificate	

Source: Own illustration based on NADRA's Service Brochure[563]

5.1.4. NADRA's pursuance of electronic public service delivery

As a nascent public-sector organisation, NADRA remained involved in Pakistan's most crucial, data-sensitive civil registry digitisation projects since 2000. Despite being subjected to continued criticism in the media, NADRA continued to consolidate its position with every other project it undertook after the initiation of CNIC[564]. The data available to NADRA and given the digitisation interface with

560 Deutscher Bundestag (German Federal Parliament) (2014). Deutscher Bundestag (18. Wahlperiode): Antwort der Bundesregierung; Drucksache 18/1271. Retrieved 14.05.2014 from http://dipbt.bundestag.de/doc/btd/18/012/1801271.pdf; Pakistan to replace 'insecure' US border watch software (2011a, June 08), The Express Tribune. Retrieved 04.05.2014 from https://tribune.com.pk/story/184568/pakistan-to-replace-insecure-us-border-watch-software/

561 NADRA issues 200,000 SNIC (2012a, December 17), Business Recorder. Retrieved 04.05.2014 from https://www.brecorder.com/2012/12/17/96294/

562 Accidental deaths: Nadra, State Life join hands to give insurance cover to people (2013a, January 24), Daily Dawn. Retrieved 04.05.2014 from https://www.dawn.com/news/781098

563 National Database Registration Authority: 2010

564 Nadra refuses to own its mistakes in CNICs (2003b, January 09), Daily Dawn. Retrieved 04.05.2014 from https://www.dawn.com/news/76685; In the first five year of its inception NADRA also worked towards computerisation of Pakistan's geography and in this regard, it also

other public and private entities mandated in NADRA's ordinance gave it a unique opportunity to further develop citizen-centric projects[565]. At the same time, development of CNIC and issuance of millions of computerised identity cards was also not an easy project for NADRA[566]. However, the exercise of conceiving CNIC, data modelling, data acquisition, technology required, formation of regional headquarters and expansion of NADRA's own digital network across Pakistan provided a unique opportunity. While registering every citizen, NADRA's staff personally interacted with millions of Pakistani citizens from each and every province, from bigger to smaller districts as well as remote towns. In order to register citizens in hilly or remote villages, NADRA sent its mobile multi-biometric registration centres in the form of digitally equipped vehicles to every possible nook and cranny of Pakistan[567]. In addition to that NADRA also collaborated with local NGOs to ensure the registration of more than 50% of Pakistan's rural population (for statistics on Pakistan's urban/rural divide, see chapter 3)[568]. These exercises provided NADRA numerous learning opportunities, among others, of experiencing the cultural and linguistic diversity across various provinces, varying literacy levels, professional and personal attitudes, digital aptitudes, communication patterns and above all, the digital divide in Pakistan[569].

Given the socio-economic disparities and limited tele-density in Pakistan and NADRA's own experience with digitisation of NDF and CNIC projects, it realised that web/computer-based model of electronic public service delivery

developed Geographical Information System as a part of its national spatial database. See for instance: Raza, F., Almas, M. and Ahmed, K. (2005). Land records information management system. In 25th Annual ESRI International User Conference, San Diego, California. Retrieved 04.05.2014 from http://proceedings.esri.com/library/userconf/proc05/papers/pap1279.pdf

565   Islamabad vehicle registration system to be computerised (2004a, May 02), Business Recorder. Retrieved 04.05.2014 from https://fp.brecorder.com/2004/05/20040502122986/

566   Haq, K. (2009). Technology and human development in South Asia. Oxford: Oxford Univ. Press

567   Nadra to set up more centres in Hyderabad (2003, December 20), Daily Dawn. Retrieved 03.05.2014 from https://www.dawn.com/news/130311

568   One of these collaborations is with women's right NGO PODA (Potohar Organisation for Development Advocacy) which assisted NADRA in facilitating the identity registration/update drive in rural areas of Pakistan. PODA's representatives helped prepare and testify the necessary documentation for the CNICs. See for instance: Potohar Organisation for Development Advocacy (2008). "Facilitating Rural Citizens in CNIC & Voter Registration Process". Retrieved 04.05.2014 from http://www.poda.org.pk/projects/15/?title=Facilitating-Rural-Citizens-in-CNIC-%26-Voter-Registration-Process.

569   In this context Pakistan's Ministry of Social Welfare report provides a good overview of NADRA's various lessons that NADRA continued to learn, also partly because of its collaborations with public-sector national and coordination with few international organisations such as UNICEF. See Ministry of Social Welfare and Special Education (2007). Pakistan's Consolidated Third and Fourth Periodic Report to the UN Committee on the Rights of the Child on the implementation of the Convention on the Rights of the Child. Retrieved 04.05.2014 from https://www2.ohchr.org/english/bodies/crc/docs/AdvanceVersions/CRC.C.PAK.4.pdf; NADRA's ex-Chairman Mr. Saleem Moin is often credited with his role to introduce Mobile Registration Van (as also discussed above) which took NADRA to Pakistan's rural heartlands. Chairman Nadra resigns (2008a, August 08), The Nation. Retrieved 04.05.2014 from https://nation.com.pk/08-Aug-2008/chairman-nadra-resigns

mechanisms may not receive a wide acceptance from the Pakistani public[570]. For a diverse country like Pakistan, web/computer based digital communication was relatively new. Considering the total number of internet users vis-à-vis the absence of eServices on government web portal developed by the EGD (see Chapter 4) also served as a testimony that for G2C services to progress in Pakistan, it may require a reliance on digital communication patterns that can actually complement the existing communication channels[571].

Figure 14: NADRA's mobile registration vehicle for rural areas

Source: NADRA[572]

Such channels of communication that are culturally inclusive and more importantly have less dependency on wired, fixed, costly and sophisticated front-end gadgetry represent the expectations of the public[573]. It is when NADRA

---

570   Herani, G. M. (2007). Computerized National Identify Card, NADRA KIOSKs and its prospects. Indian Journal of Science and Technology, 1(2), 1-5

571   There has been a continuous rise in the pursuance of alternative communication channels for digital public service delivery in the first decade of new millennium. The emphasis on discovery of such inclusive communication channels has been regularly discussed at various international fora also including the WSIS 2003 and 2005 (also discussed in Chapter 1 and 2). See for instance: International Telecommunication Union. (2013c). Contribution by Ethiopia: E-Government Strategy. Paper Presented at the Ministerial Roundtable at the WSIS Forum. Retrieved 04.05.2014 from https://unctad.org/meetings/en/Presentation/CSTD_2013_WSIS_Ethiopia_E-Gov_Strategy.pdf

572   National Database Registration Authority: 2010

573   The handbook put together by the International Finance Group provides an in-depth coverage on scope and innovation undertaken in alternative delivery channels and public services. See: International Finance Corporation. (2014). Alternative Delivery Channels and Technology. Retrieved 04.05.2015 from

started to think of alternatives to the web/computer-based eServices. The promulgation of Electronic Transaction Ordinance in 2002 (see Chapter 4) had already provided an opportunity to conduct any digital transaction since it was legally approved by the State Bank of Pakistan. Therefore, the field for electronic commerce had the essential requisites for its uptake. Post 2004, NADRA started to deliberate over developing its own eCommerce platform – a platform where citizens can authenticate themselves with their CNICs and perform various tasks primarily utility bill payments. These deliberations also included to develop a system that is compatible with the cash-based Pakistani society and that does not require detailed technical know-how. This is when NADRA started to consider developing a machine that is able to communicate in multiple languages and that is rich with graphics thereby reducing the user input to avoid confusions, and hence the idea of kiosk machine was born[574].

5.1.5. Kiosk: towards an understanding and use

Cambridge dictionary defines kiosk as "a small building where things such as sweets, drinks, or newspapers are sold through an open window"[575]. In a more contemporary and ICT scenario, kiosk refers to either a small single or multipurpose machine that enables the delivery of information or services[576]. Depending upon the ownership of kiosk, the information could be governmental or for private purposes. In recent years, kiosk machines have emerged as an alternative communication and service delivery channels in comparison to the web/internet[577]. The kiosk machines come in many variations. They can provide either unassisted service or as in most cases assisted.

Kiosk is also named as telecentre or knowledge centre, which has a wider scope. In ICT4D discourse, a kiosk is favoured in the rural areas where the information dissemination and service delivery has been in limited supply. The use of kiosk is to bridge this gap between the citizens and government[578]. For instance, in South Indian state of Tamil Nadu a small village called Kovalam lies in the coastal belt

https://openknowledge.worldbank.org/bitstream/handle/10986/25980/111344-WP-ADC-Handbook-2014-PUBLIC.pdf?sequence=1&isAllowed=y

574   National Database Registration Authority (2012c). "NADRA Kiosk Brochure". Retrieved 06.09.2012 from http://www.nadra.gov.pk/downloads/solutions/other-kiosk.pdf.

575   Kiosk. [Def. 2]. Cambridge Dictionary Online. In Cambridge Dictionary. Retrieved from https://dictionary.cambridge.org/dictionary/english/kiosk?q=kiosk

576   Bowonder, B. and Boddu, G. (2005). Internet kiosks for rural communities: using ICT platforms for reducing digital divide. International Journal of Services Technology and Management, 6(3), 356-378

577   International Telecommunication Union. (2011b). ICT success stories: Innovation and sustainability. Retrieved 04.05.2014 from https://www.itu.int/net/itunews/issues/2011/04/pdf/201104_32.pdf

578   Badshah, A., Khan, S., Garrido, M., Information, U. N. and Communication Technologies Task, F. (2005). Connected for development : information kiosks and sustainability. New York: UN Information and Communication Technologies Task Force

where the principal livelihood of villagers is fishing[579]. In 2005, Village Knowledge Centre (VKC) in Kovalam was established with an aim to provide basic and higher education, social and political awareness, and basic civic information. Using the weather reports from the VKC the fishermen could decide where and when to go fishing. Similarly, another initiative in Indian state of Karnataka uses kiosk centres to provide information about land records. The state of Karnataka computerised 20 million records of land ownership and the kiosk provides access to these. 6.7 million farmers use kiosk to seek a copy of their land record to perform tasks such as obtaining a bank loan[580]. The kiosk centres have helped to reduce the delay or bribes that the farmers face from middlemen such as village accountant[581].

5.2. NADRA's Kiosk

Until 2005, the concept of kiosk or knowledge centre in Pakistan was unheard of. There were initiatives such as Universal Service Fund that catered to bridge the rural-urban digital divide582. However, a multipurpose standalone machine that could work as an information centre through which financial transactions could be performed were not introduced. In terms of familiarity with kiosk, there were private initiatives in the form of Internet cafes and/or public call offices that served as a source of digital communication and information seeking purposes583. However, such cafes or PCOs neither had the capacity to provide the electronic public service delivery nor the mandate. With the kiosk machine, NADRA developed for the first time a multipurpose ATM styled machine that would function as an eCommerce platform. The basic idea behind kiosk was to develop a digital platform that could be used to perform electronic transactions using CNIC584.

For any electronic transaction in Pakistan, banking channel is a primary mode of performing monetary activities585. However, a distinguishing feature of NADRA's kiosk was that it reduced the need of a bank account. It primarily catered rural and urban poor who either could not afford an account. In this

579 Govindaraju, P. and Mabel, M. (2010). The status of information and communication technology in a coastal village: A case study. International Journal of Education and Development using Information and Communication Technology, 6(1), 128-135
580 Chawla, R. and Bhatnagar, S. (2004). Online Delivery of Land Titles to Rural Farmers in Karnataka, India. Retrieved 04.05.2014 from http://unpan1.un.org/intradoc/groups/public/documents/un-dpadm/unpan042925.pdf
581 Bhatnagar, S.: 2009
582 See Government of Pakistan's Cabinet Division's report on teledensity and USF in: Cabinet Division (2006). Year Book of Cabinet Division – 2005-06. Retrieved 04.05.2014 from http://cabinet.gov.pk/cabinet/userfiles1/file/Publications/year-book-2005-06.pdf
583 Batool, S. H., et al.: 2010
584 National Database Registration Authority: 2012c
585 See Business and Functions of Bank (Chapter 4) in: State Bank of Pakistan (1956). The State Bank of Pakistan Act, 1956. Retrieved 04.05.2014 from http://www.sbp.org.pk/about/act/SBP-Act.pdf

context CNIC functions as an authenticator. The software for the kiosk was conceived, designed and developed by NADRA in Pakistan[586]. In terms of design, CNIC sits at the top of kiosk's prototype as it is the primary means of citizen authentication and verification. Similarly, in order to increase the security around electronic transaction, NADRA introduced a security layer by adding a fingerprint as an additional verification measure[587]. In terms of a user interface, NADRA preferred a touch screen technology with coloured graphic display to increase the user friendliness around transaction procedure[588]. For every monetary transaction, the kiosk's functions as cash acceptor. Initially Nadra's eCommerce platform via kiosk focussed on the electronic payments of utility bills and citizen verification mechanisms. At a later stage, the purchase of mobile phone calling cards and payment of post-paid bill was also added to the kiosk[589].

Figure 15: NADRA's eCommerce Kiosk

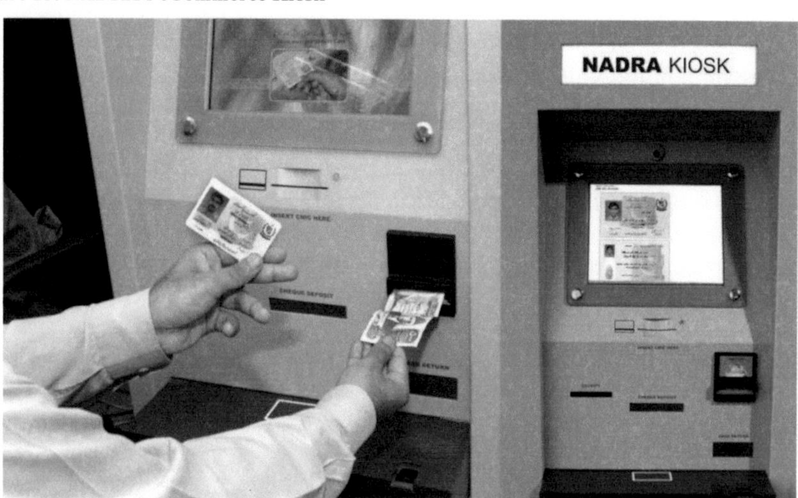

Source: NADRA[590]

5.2.1. Kiosk: application context

The idea behind utility bill payment and citizen verification is manifold and it relates to the structural problems at the administrative as well as governance level.

---

586   MoU for online payment of utility bills signed (2005b, August 10), Business Recorder. Retrieved 04.05.2014 from https://fp.brecorder.com/2005/08/20050810310524/
587   Nadra signs MoU for payment of bills (2005c, August 10), Daily Dawn. Retrieved 04.05.2014 from https://www.dawn.com/news/151636
588   National Database Registration Authority: 2012c
589   Nadra and PTCL sign contract (2006, March 03), Business Recorder. Retrieved 04.05.2014 from https://fp.brecorder.com/2006/03/20060303393678/
590   National Database Registration Authority: 2010

146

Prior to 2005, the only way utility bills could be paid was through the banks[591]. Banks operate under the guidelines of the State Bank of Pakistan, and are the only entity that are allowed to perform any financial transactions. With Pakistan's exponentially rising population, a bank's mandate is not only to work as a credit entity but they also accept bill payments. As pointed in Chapter 4 and 5, the initial absence and slow growth of online banking in Pakistan's predominantly cash-based society meant that the utility bills (such as electricity, gas, water and telephone) had to be manually and personally paid into the banks. The manual process of bill payment in Pakistan is to wait in long queues, in or outside banks that could take anywhere from several minutes to hours. Similarly, due to distribution discrepancies, the bills are often delivered almost towards the last date of payment that result into extra fine[592].

NADRA has tried to tap the potential of its huge citizen database and use the capacity provided by the market to extend its services into new avenues. In this context, NADRA initiated large scale collaboration agreements with major utility companies of Pakistan such as Sui Southern Gas Company Karachi, Water and Power Development Authority (WAPDA), Karachi Electric Supply Company (KESC), Sui Northern Gas Pipeline Ltd. (SNGPL) and Pakistan Telecommunications Company Ltd. (PTCL) in 2005[593]. Based on the memorandum of understanding between NADRA and the utility companies, it was agreed to electronically share the access to their consumer database with NADRA. In this way NADRA was able to digitally access the primary information about the citizen from utility companies' database that includes name of the utility service, consumer identification number, consumer name and monthly cost of the utility bill. NADRA's software was developed in a way that it maintained a kiosk-only profile of every citizen. With the help of CNIC database, NADRA maintains a profile for every citizen in its database, but for the purpose of eCommerce platform, NADRA's software is tailored to develop and maintain a kiosk-only profile for the citizens participating in the kiosk. In this way, NADRA was able to map the details of the utility bill(s) on to the corresponding citizen's profile whereas the citizen details came from the NADRA database. The profiling and its importance is explained further in the following section.

The kiosk machine project was formally inaugurated by the then President General Musharraf in October 2006[594]. Initially NADRA installed 36 machines in

591   Nadra advises gas consumers to get registered (2005d, August 30), Business Recorder. Retrieved 04.05.2014 from https://fp.brecorder.com/2005/08/20050830319049/

592   Tax imposed on surcharge: Late payment of power bills (2004f, September 29), Daily Dawn. Retrieved 04.05.2014 from https://www.dawn.com/news/372024

593   Nadra likely to sign MoUs with utility companies (2005e, April 16), Business Recorder. Retrieved 04.05.2014 from https://fp.brecorder.com/2005/04/20050416233198

594   National Database Registration Authority (2005). "President of Pakistan General Pervez Musharraf inaugurated the "NADRA Kiosk"". Retrieved 28.08.2008 from http://www.nadra.gov.pk/DesktopModules/top/topmore.aspx?tabID=0&ItemID=39&bID=0&M id=2925.

Karachi and Islamabad in the first phase whereas in 2006 it aimed at expanding its network to other cities such as Lahore, Rawalpindi, Faisalabad, Gujranwala, Multan, Peshawar, Hyderabad, Quetta and Sukkur[595]. Thus, by establishing such bill payment kiosks, not only relieved banks from an additional burden of encashing the utility bills but for the first time, citizens were able to use the electronic public service kiosk with the help of their CNIC without any need of web portal or computer. Kiosk introduced the concept of bill payment that works round the clock unlike the dependency of dealing with business hours limitations of banks.

Table 9: NADRA's Kiosk machine in brief

Purpose	User Interface	Features
eCommerce Platform	Bilingual (Urdu / English)	Standalone ATM-Styled machine
Payment of Utility Bills	Touch Screen	Cash Acceptor only
Citizen Verification	Rich Graphic Display	Bank-less Transactions

Source: NADRA[596]

Similarly, another service that comes on kiosk platform is the citizen verification mechanism. The basic idea behind this eService is to provide a citizen digital means of verifying the particulars of a fellow citizen before getting into any transaction with the same. Due to reliance on the CNICs as means to establish identity, there have been cases of fake identity documents used for business or domestic transaction such as inheritance[597]. There have also been several cases of identity fraud within the context of sale and purchase of real estate or vehicles[598]. Similarly, this service can also be helpful for the employers who can verify the genuineness of the potential employee at the time of hiring[599]. Citizen verification service is designed in the light of existing and manual process of verification that involves a visit to the notary public where the judicial representative would notarise the applicant on a stamp paper. Similarly, in a paper-based manual identity card before the arrival of CNIC, there was no electronic testing mechanism that could verify another person's identity as a real cardholder. NADRA in this regard relied on its citizen database and made the verification possible by offering this service via kiosk machine.

---

595   Nadra bill machines start functioning next week (2005a, November 28), Daily Dawn. Retrieved 04.05.2014 from https://www.dawn.com/news/167475

596   National Database Registration Authority: 2010

597   Holden, L. and Chaudhary, A. (2013). Daughters' inheritance, legal pluralism, and governance in Pakistan. The Journal of Legal Pluralism and Unofficial Law, 45(1), 104-123

598   When land grabbers rule (2014c, January 28), The Express Tribune. Retrieved 04.05.2014 from https://tribune.com.pk/story/664487/when-land-grabbers-rule/

599   National Database Registration Authority (2013b). "e-Commerce Platform". Retrieved 13.01.2013 from http://www.nadra.gov.pk/index.php?option=com_content&view=article&id=51&Itemid=102.

5.2.2. Kiosk at work

The kiosks are ATM-styled machines that are either available in the form of standalone machine or placed within a wall. The kiosk machine requires the CNIC to function and operate and offers a Graphical User Interface in the form of touch screen computer menu in either the Urdu or English language. In order to provide a secure mechanism, kiosk machines require a CNIC card and digital fingerprint scan of the person conducting a transaction. The functioning of kiosk machines is however, not very straight forward. The first-time visitors of the kiosk machine need to register or enrol themselves with the kiosk machine. Once enrolled or registered, kiosk's software creates an online/virtual account against the corresponding CNIC/person, NADRA refers to it as a virtual wallet[600]. Creating a kiosk account is a one-time exercise and after that this account can be accessed from any kiosk machine independent of the geographic location.

Once the kiosk account has been generated the person needs to navigate to the menu of bill payment. For this purpose, having the (copy of) utility bill(s) is important, at least for the first time to use the kiosk, to pay that particular utility bill. Within the bill payment menu on kiosk machine, person can enter the consumer identification number mentioned in the utility bill(s). As a result of user input, the kiosk machine shows the basic details related to the utility bill such as person's photograph along with the consumer number and name, utility company and due price to be paid[601]. Once the utility bill is mapped on to the wallet, the kiosk's software saves the bill details for future transaction thereby eliminating the need to carry the bill in the next visit. In this way, kiosk user can map other utility bills as well. Kiosk allows citizens to add and remove as many bills to their virtual wallet. This means that NADRA allows the citizens to pay bills on other's behalf. In addition to that, NADRA's kiosk software also allows account delegation. Within delegation scenario, any person can nominate other person(s) such as family members, friends or domestic helpers to pay bill on her/his behalf[602]. However, in this case the delegated person needs to know the CNIC number of the individual (the nominee). The procedure for creating a virtual wallet and the mapping of the bills remains the same for the nominated person as well.

Once the utility bills are mapped on to citizen's virtual wallet, the kiosk user is able to pay the bills via kiosk machine. Here, it is important to note that the kiosk user does not require any ATM, visa or debit card, instead the entire transaction

---

600   National Database Registration Authority: 2010

601   Nadra signs accord for KESC bills payment (2005a, September 09), Business Recorder. Retrieved 04.05.2014 from https://fp.brecorder.com/2005/09/20050909323033

602   Nadra bill payment machines to start work from next week (2005c, November 28), Business Recorder. Retrieved 04.05.2014 from https://fp.brecorder.com/2005/11/20051128358037

takes place with the help of CNIC and cash amount[603]. For that matter, kiosk user needs to enter the cash amount in the kiosk machine which gets registered with user's virtual wallet/account. Since kiosk machine functions on the model of cash acceptance therefore any additional money added to the virtual wallet are not returned to the user. If the kiosk user has inserted more than the amount required to pay the utility bill(s), the remaining amount will be added to the user's virtual wallet that can be used in future. Once the bill is paid through virtual wallet, the kiosk machine generates a transaction receipt on paper for the user. The following section dwells upon the key findings that were deduced through the data collected during an empirical study phase at the NADRA headquarter in Islamabad.

Figure 16: A view of kiosk machine in Lahore

Source: Author (a franchised kiosk machine in Lahore's Sanda area)

5.2.3. Kiosk typology

Between 2005 through 2007, NADRA launched three different variations of its eCommerce platform via kiosk machines. Initially in 2005 NADRA introduced

---

603 'Smart cards to greatly benefit private sector,' DG Operations, Nadra Technologies Limited (2014b, June 27), Business Recorder. Retrieved 04.05.2015 from https://fp.brecorder.com/2014/06/201406271197131/

only (i) standalone kiosk machine which was located in NADRA premises. Even though the basic idea of kiosk machine was to provide unassisted electronic public service however, the registration/enrolment with the kiosk machine to create a virtual account was relatively cumbersome for public. In this regard, NADRA assigned its staff members at its premises to help people register with the kiosk. By end of 2005, more than 0.2 million people registered themselves with the kiosks[604]. After observing the initial response, the amount of transactions generated by the kiosk machines and the potential it carried as an eCommerce platform, NADRA decided to launch the second variation of kiosk (ii) in the form of computer software which could be provided to its staff members at NADRA's offices. In this way people could actually pay their utility bills over the counter without having to interact with the kiosk machine[605].

NADRA found its eCommerce platform via kiosk machine as a potential business opportunity that was not only an innovative electronic public service delivery channel but it also presented a good prospect for (digital) entrepreneurship. In 2007, NADRA decided to initiate the third variation of its eCommerce platform by (iii) franchising the kiosk machines for general public with a purpose to spread the kiosk machines at the grass root level[606]. In this regard, NADRA's franchise model provided a self-employment opportunity to small businesses, particularly youth who could operate the franchised kiosks within their premises and also earn a small value for running the kiosk. NADRA's idea behind franchising was based on commercialising its kiosk machines in order to take the kiosks outside NADRA's premises and in the commercial centres where franchisee of the machine would look after the day to day operations and maintenance of the machine[607].

NADRA's incentive to the franchisee was the percentage share they could earn with every transaction. More the transactions, higher the revenue share for the franchisee. NADRA's proposed revenue sharing structure provided the franchisee an opportunity to earn Rs. 3/ with every utility bill transaction, whereas for every online verification they could earn Rs. 5/- per transaction and for purchase of pre-paid mobile phone calling card, franchisee could earn 1% of the original value[608]. In order to maximise revenue for its franchisees, NADRA invited applications

604   Daily Dawn: 2005a, November 28

605   National database Registration Authority (2009d). "Introduction to Kiosk: Frequently Asked Questions ". Retrieved 31.08.2009 from http://www.nadra.gov.pk/site/396/default.aspx#Frequently%20asked%20questions.

606   National Database Registration Authority (2008d). "NADRA has franchised the kiosk machines". Retrieved 09.03.2009 from http://www.nadra.gov.pk/DesktopModules/top/topmore.aspx?tabID=0&ItemID=46&bID=0&Mid=3026.

607   Nadra franchises kiosks (2007e, March 10), Daily Dawn. Retrieved 13.12.2009 from https://www.dawn.com/news/236682

608   National Database Registration Authority (2008b). "Kiosk: Franchise Model (Terms and Conditions)". Retrieved 09.03.2009 from http://www.nadra.gov.pk/Franchise_Model_terms_and_conditions.html.

from kiosk machines from the prospective candidates from particularly those areas where banks were not available. Similarly, these areas as per NADRA's franchise conditions should be densely populated which should serve at around three to four thousand housing units. Additionally, franchisees were asked to do their own marketing for the proposed kiosk sites[609].

Figure 17: Development of NADRA's Kiosk machine (2005 through 2007)

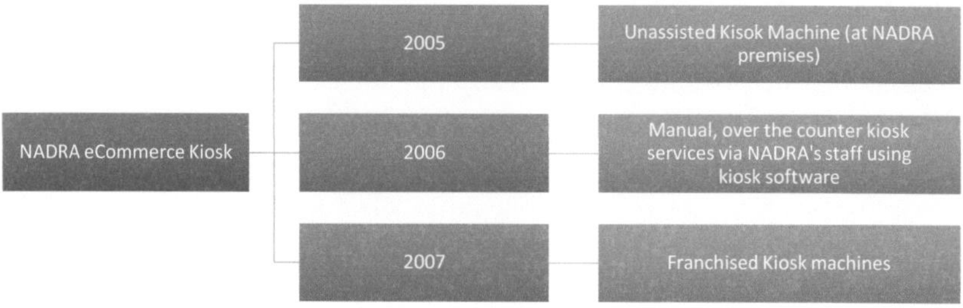

Source: Own illustration

5.2.4. Data collected from kiosk machines

From 2005 to 2007 NADRA's kiosk machines came in three different formats and these include (i) unassisted kiosk machines which were released in 2005 (ii) use of kiosk's software by NADRA's staff via computers started in 2006 and (iii) franchised kiosk machines outside the NADRA's premises. Most of these machines served the purpose of making public service easier as they were of the unassisted kind. However, visits to random kiosk machines revealed that though there was a curiosity in using the kiosk people faces problems in understanding the touch screen system, to register with the machine and to deposit money in the machine. As a part of this research endeavour, an analysis of the kiosk network was possible due to the data access granted by NADRA.

The primary focus was to identify the number of cities where kiosks machines had been installed, how many transactions had taken place at the total frequency of kiosk usage. By correlating this to the city and the target population around the kiosk, it could also provide information about which cities were more adaptive to kiosk machines. The amount of money that the kiosk collected was also enlisted. Finally, the statistics for the gender split in usage could also help estimate and project the gender wise technology adoption.

---

609   ibid.

152

5.2.5.Results

The focus raised above is more of a quantitative nature whose major purpose is to establish the usage patterns. The results of kiosk machines from its inception in 2005 through March 2009 revealed varied patterns of use. The total frequency of use in every city where kiosks were installed helped in understanding the people's reliance on technology. In addition, the NADRA also shared the results of total amount it generated through its machines. While the data for each kiosk, its city, and monetary transactions was available, there was no mechanism in the kiosk to maintain a check on the choice of user language.

Table 10: Bill Payment via Kiosk Machine (2005 – 2007)

Year	2005	2006	2007
Number of Cities	3	26	31
Total Transactions	369	285,060	1,366,037
Amount Generated	Rs. 239,517	Rs. 236,935,964	Rs. 125,8064,224

Source: Own illustration (based on kiosk database)

Kiosk machines installed at NADRA's premises took a modest start in 2005 with only three cities (Islamabad, Karachi, Peshawar) where they were rolled out. The total transactions that were generated were also very few (less than 500). However, in 2006 and 2007 the total number of cities started to increase gradually and in 2006 the total transactions recorded in 26 different cities reached almost 0.3 million. These transactions exponentially increased in 2007 as the total transactions crossed 1.3 million whereas the amount generated via kiosk machines was almost PKR. 1.25 billion. The total transactions having crossed 1 million mark is an indication that the people have been looking for an alternative to the bill payment mechanisms and their reliance on the technology also indicates that new patterns of digital communication are slowly finding their acceptance among people. Similarly, the total transactions and amount generated via kiosk machines also implies that the kiosk oriented G2C solutions can assist in taking the burden off the banks, which had so far been the primary method of paying and accepting bills.

5.2.6.Kiosk for Utility Bill Payment

Kiosk machines became one of the first G2C service providers offered by NADRA which reduced the dependency on a computer, the internet or web portal and instead offered a mechanism through ATM-styled public machines. The kiosk project seemed a relevant and a decent alternative to web portal through which government could perform the task of information dissemination[610]. In 2007 when NADRA announced the franchise model of its kiosk machine project, a total of 33 machines went into operation. Initially the response seemed slow as only 36,821 transactions took place at these machines. However, the money generated via these machines amounted to PKR 1.8 million. In comparison to 2007, total franchise machines increased from 33 to 44 where more than one million transactions took place. This was a phenomenal uptake as the total transactions increased by 144%. These transactions also impacted the amount generated rising up to PKR 7.22 million. As these kiosks worked on franchise model, at the revenue share of PKR. 3 per transactions, the franchisees, thus, earned PKR 0.4 million from 144,479 transactions.

Figure 18: Utility bill payment via Kiosk franchise machines (2007-08)

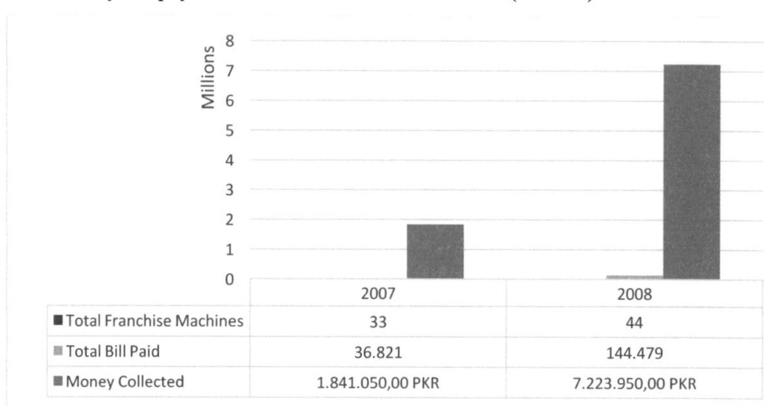

	2007	2008
■ Total Franchise Machines	33	44
■ Total Bill Paid	36.821	144.479
■ Money Collected	1.841.050,00 PKR	7.223.950,00 PKR

Source: Own illustration (based on kiosk database)

5.2.7. Geographical typology of Kiosk machines

In order to check the performance as per city, province, gender and transaction, the data does not automatically separate the city from their provinces. However, it is of keen interest to see what percentage or frequency of kiosk machine usage occurred in a particular province. This can help estimate the (a) kiosk's performance (b) the provincial citizen's attitude towards the kiosk machine usage.

In order to seek provincial results, each and every city was geographically coded into the map through an online geographical information system. The BatchGeo website, in this regard, provides a very useful solution in displaying spreadsheet

---

610 Prime Minister's Secretariat (2008). Report of the National Commission for Government Reform on Reforming the Government in Pakistan. Vol 1. Retrieved 04.05.2014 from https://www.pc.gov.pk/uploads/report/NCGR_Vol_I-1.pdf

results into graphical format with all the cities geocoded on the map[611]. For formal consent and email communication see Appendix IV.

In its first year of installation, starting with roll-out in three cities of Peshawar, Islamabad and Karachi, the transactions recorded were in their hundreds. This performance picked over the period of time from 2006 through 2009 when transactions went into thousands and then millions. This section provides chronology of kiosks' usage from 2005 through 2009 in the form of geographical maps.

As it can be seen from the figures in the following pages, most of the transactions that have taken place during 2005-2008 are in the Punjab province. The occurrence of kiosk usage in the Punjab province can also be interpreted that considering the most populous of all the five provinces there has been more demand for kiosk among the Punjab citizens.

Figure 19: Kiosk during 2005

Source: Author (through Nadra database / Batchgeo)

---

611   Batchgeo (2013). "About batchgeo". Retrieved 29.11.2013 from https://batchgeo.com/about/.

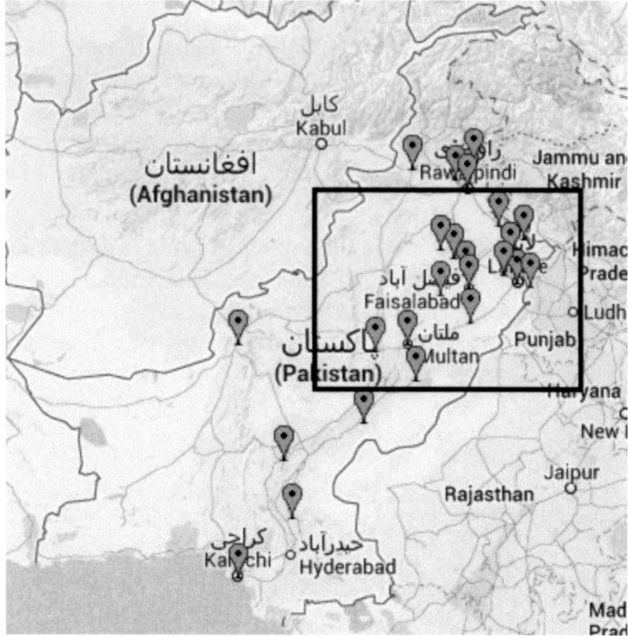

Source: Author (through Nadra database / Batchgeo)

Figure 21: Kiosk during 2007

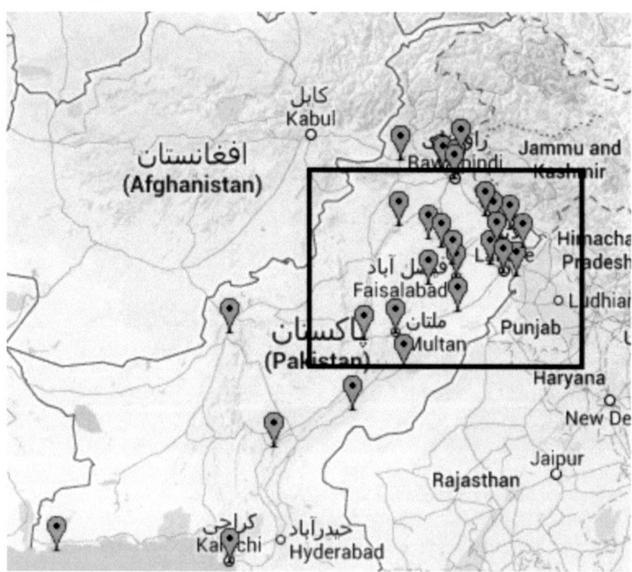

Source: Author (through NADRA database / Batchgeo)

156

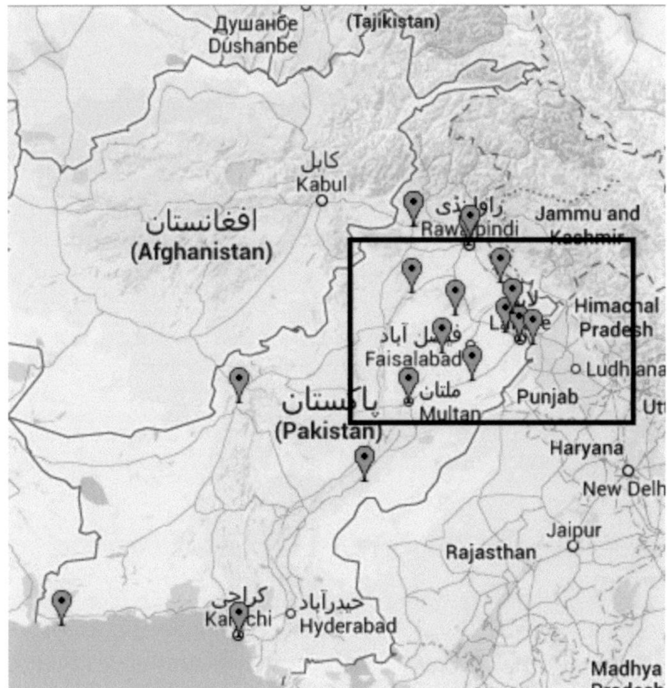

Source: Author (through NADRA database / Batchgeo)

5.2.8. City-wise performance of the kiosk machines

All the three variants of kiosk machines presented a unique utility and novelty for people. On one hand, it was helping reduce the work load from the banks, on the other hand it also introduced a new way of interactions with the government. People could pay utility bill either over the counter at the NADRA offices or via kiosk terminals or through franchised kiosk machines in the cities. The uptake and use of the kiosk, particularly through the kiosk machines (at NADRA) has been more prominent in the cities. While reviewing the Kiosk data received from NADRA, two cities were picked up from every year between 2005 and 2008 (Figure 23).

As it can be seen from Figure 23 maximum number of transactions are from the densely populated cities of Lahore, Multan, Karachi and Peshawar. In three out of four years through 2008, Karachi being a business hub, always made it as the top user of kiosk machine with transactions amounting to 259 (in 2005), 77,449 (in 2006), and 38,849 (in 2008).

Figure 23: City-wise performance of the kiosk machines from 2005-2008

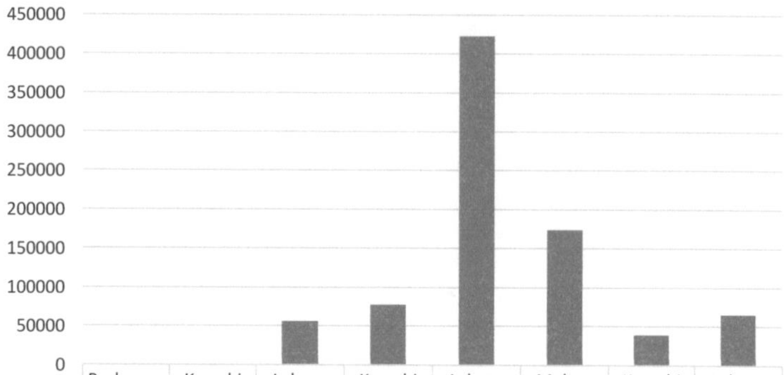

	Peshawar	Karachi	Lahore	Karachi	Lahore	Multan	Karachi	Lahore
	2005		2006		2007		2008	
■Transactions	40	259	56142	77449	422445	174146	38849	65592

Source: Author (through NADRA database)

5.2.9. Gender-wise Kiosk's usage for Utility Bill Payment

In order to reviewing and analysing the uptake of ICTs based on gender, a special request was made to NADRA if it is possible to parse the data on the gender variable. The parsing of data based on gender variable revealed interesting results. Between 2005 (launch of the kiosk machine) and 2007, the total number of transactions performed by male citizens started from meagre 337 transactions and by end of 2007 it reached up to 1.3 million (see Figure 24). These figures, however, in case of female citizens are extremely low. In 2005 only 32 female citizens used kiosk machines at the beginning whereas in 2006 there has been an increase of only 900 users. 2007 looks fairly decent in terms of numbers as the figure could walk past in thousands reaching up to 23929 transactions. Nevertheless, in comparison to male users of the kiosk machines installed at NADRA, the difference between kiosk uptake among the female citizens is by one million. These figures present a number of challenges for technology adoption among male and female citizens. The percentage difference between Pakistan's male and female population between 2000 to 2012 has remained 45 to 49% female and 51 to 55% male[612].

---

612   World Bank: 2013a

158

Figure 24: Kiosk adoption based on gender between 2005 and 2007

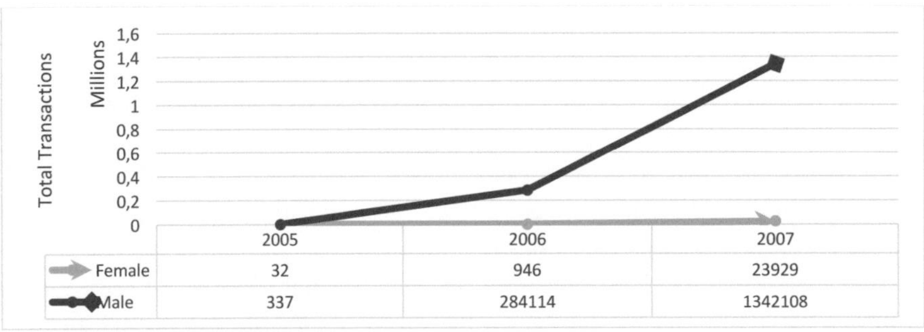

	2005	2006	2007
Female	32	946	23929
Male	337	284114	1342108

Source: Own illustration (based on kiosk database)

However, in case of kiosk's adoption the difference is not only big but also alarming. In terms of equal access between men and women is not only limited to information technology. In a patriarchal society of Pakistan, these differences can also be seen in terms of access to capital, freedom to work, or liberty of making life choices such as continuing education or be an entrepreneur[613].

Figure 25: City Wise Frequently Used Kiosks by female from 2005-2007

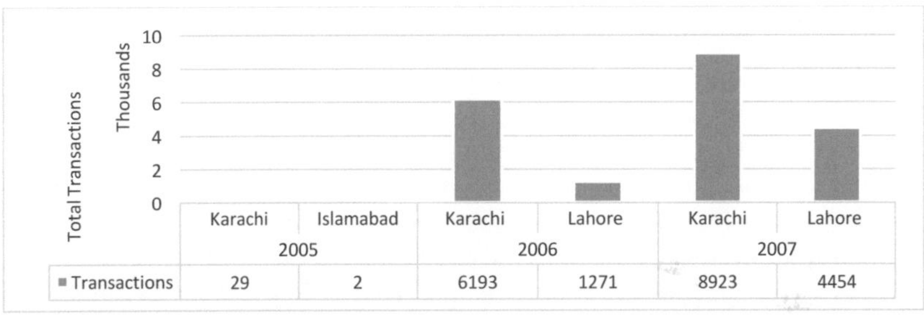

	Karachi	Islamabad	Karachi	Lahore	Karachi	Lahore
		2005		2006		2007
Transactions	29	2	6193	1271	8923	4454

Source: Own illustration (based on kiosk database)

Similarly, if top two cities are chosen based on the maximum number of transactions performed by female between 2005 and 2007, then the data reveal that most of the transactions are performed in metropolises. These cities include Karachi, Islamabad and Lahore where women more frequently use kiosk machines as compared to other. Apart from the issue of low technology adoption by females and the patriarchal system in Pakistan, these figures also reflect at the uneven developmental growth in rural and urban Pakistan as also discussed in Chapter 3.

---

613   Roomi, M. A. and Parrott, G. (2008). Barriers to development and progression of women entrepreneurs in Pakistan. The Journal of Entrepreneurship, 17(1), 59-72

5.3. NADRA's eSahulat platform

With its kiosk machine project, for the first time NADRA introduced an eCommerce platform in Pakistan. The distinguishing feature of the kiosk was the way it redefined the digital communication paradigm by relying on a technological artefact that defied the conventional G2C interaction with an eGovernment context. With kiosk machine, the citizens were able to interact with the government and formal finance sector with a machine in the local language without having to rely on web/Internet scenario[614].

NADRA not only revitalised the concept of traditional governance scenario for Pakistan but by developing database interactions and digital interfaces with other public-sector organisations, it also spruced up the concept of public service delivery. The economic opportunity for NADRA can of course not be ignored in this context. If NADRA was redefining the concept of electronic public service delivery in Pakistan, it was simultaneously also concerned about its commercial viability and organisational independence. NADRA's memorandum of understanding with the utility companies earned NADRA a profit of PKR 8/- per bill[615]. Out of these PKR 8 NADRA also paid a revenue share to its franchisees as well as the tax worth PKR 1.5 per bill. This means that NADRA's own profit in the end was PKR 1.5 per bill. However, the statistics in the previous section have shown how the transactions increased from thousands to millions over a span of two years. Therefore, if there are 1 million transactions per month then the revenue earned by NADRA is PKR 1.5 million per month.

This arithmetic dramatically changes when over-the-counter bill payment is also taken in consideration taking place at NADRA registration centres. As depicted in the previous section, with over-the-counter approach, NADRA created an alternative to its kiosk machine and offered citizens to pay bills via their NADRA's offices where its employees encash the bills using kiosk software installed on their computers. In this case, NADRA saved PKR 5 revenue share that it pays to its franchisees and resultantly only paid PKR 1.5 as tax per bill to the cooperating utility companies. Between 2005 through 2007, approximately 8 million bills were paid via kiosk's eCommerce platform. These 8 million bills generated PKR 5.7 billion taking NADRA's profit in millions of rupees[616]. Such business propositions opened doors to further entrepreneurial reckoning for NADRA to extend its kiosk's eCommerce platform based on market viability. Based on the profit it generated, kiosk as a franchise machine provided NADRA with a workable model. However, the cost of such machines, operations and maintenance involved for self-owned, dedicated machines as well as the technical

---

614   ICT Review / 2014 (2014e, December 16), Business Recorder. Retrieved 04.05.2015 from http://www.brecorder.com/pdf/ict2014.pdf

615   E-sahulat — a nightmare for Nadra franchisees (2011d, August 05), Daily Dawn. Retrieved 04.05.2014 from https://www.dawn.com/news/649717

616   NADRA ready to launch e-Sahulat (2008b, May 22), The Nation. Retrieved 04.05.2014 from https://nation.com.pk/22-May-2008/nadra-ready-to-launch-esahulat

infrastructure required for their upkeep and roll-out machines hindered NADRA's outreach and market share (for utility bills). In 2007, NADRA started to think about launching a fourth variation to existing kiosk machine which carries the essence of all three variants, that is, (i) an unassisted kiosk machine installed in NADRA premises, (ii) over-the-counter approach to pay bills at NADRA offices and (iii) a kiosk machine that can be franchised to small and medium business enterprises or persons. NADRA named its fourth variant as eSahulat.

5.3.1. eSahulat Application Context

With its fourth variant of kiosk's eCommerce platform, NADRA decided to franchise the software used in kiosks (which already existed for over-the-counter acceptance of utility bills at NADRA centres, second variant of kiosk (see figure 17)) instead of entire ATM-styled machines to small and medium business enterprises or persons. The basic idea behind this approach, apart from the business proposition, was also to counter the service-divide between the urban and the rural population that kiosks were probably not able to do or limited in their outreach[617]. The "e" in eSahulat stands for electronic whereas Sahulat is an Urdu language word (سُہُولت, pronunciation /sʊhu:lət/) which stands for facility[618]. A similar service was also initiated by the Indian State of Andhra Pradesh called as eSeva or electronic (public) service/help. eSeva was initiated in early 2000s as a one-stop source for providing government information and services via modern ICTs with citizen-centric G2C approach in mind[619]. NADRA formally launched eSahulat in February 2008 by inviting applications from interested business entities[620].

eSahulat's technical architecture is built upon the kiosk technology as NADRA has transformed the software used in kiosks into a computer software available in a Compact Disk and franchised the whole model for the general public. Any interested individual, small and medium enterprise has the possibility to apply for the software and use it for commercial purposes. Unlike franchise kiosk machines, for which NADRA's staff had to be responsible for its operations and physical location, eSahulat only came as a licensed CD for which the prospective business candidate needed to organise her/his own computer, internet connection and a printer. The computer specifications were kept to the minimum possible (in this

617   Nadra's e-Sahulat project woos huge response (2008a, March 14), Business Recorder. Retrieved 04.05.2014 from https://fp.brecorder.com/2008/03/20080314708668
618   Facility. [Def. 2]. Oxford Dictionary Online. In Oxford Dictionary. Retrieved from https://ur.oxforddictionaries.com/translate/urdu-english
619   India Filings: 2015
620   National Database Registration Authority (2008c). NADRA announces another public service(e-Sahulat). Retrieved 07.02.2009 from http://www.nadra.gov.pk/DesktopModules/top/topmore.aspx?tabID=0&ItemID=53&bID=0&Mid=3026

case Intel Pentium IV)[621]. eSahulat could be opened in any corner shop, shopping store or commercial/trade enterprise. According to this model anyone with a eSahulat license can advertise outside their business outlet and invite the general public to make their payment via NADRA's eSahulat software. People could just walk in these outlets running eSahulat software and pay their utility bills at their convenient time[622].

Figure 26: Development of NADRA's Kiosk machine (2005 through 2008)

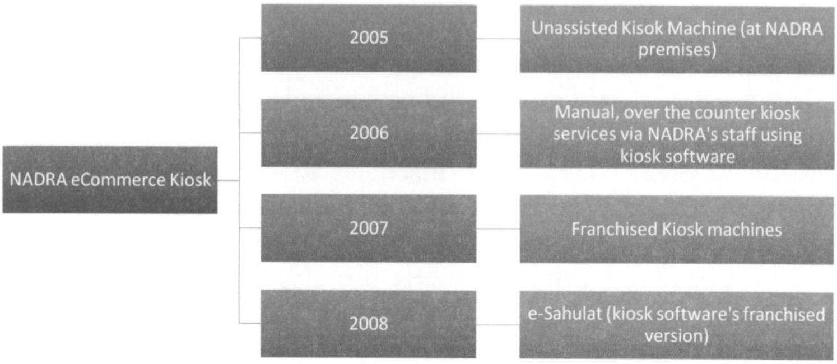

Source: Own depiction

eSahulat is based on a prepaid model. When it started in 2008 NADRA's kept the price of the eSahulat licensee at PKR 25,000. The price, however for the major/populous cities such as Lahore, Faisalabad, Gujranwala, Sialkot, Multan, Rawalpindi and Peshawar was PKR 100,00. On the basis of prepaid model a licensee had to first deposit cash in eSahulat's virtual wallet out of which payments would be paid whereas the walk-in customer would pay their utility bill price in cash to the eSahulat licensee. A prerequisite set by NADRA for a licensee's location was approx. 2500 to 3000 households in its vicinity. Following the first phase, NADRA shortlisted the applications, interviewed the prospective candidates and also surveyed the sites of eSahulat locations until April 2008 followed by award of licences, training and deployment of the software[623].

A major incentive for the owner of eSahulat software is the revenue share of PKR 5 per bill/transaction. eSahulat not only offered payment of utility bills but by 2008 it also included prepaid internet scratch cards, mobile phone top-up, calling

---

621 National Database Registration Authority (2012d). "e-Sahulat: Frequently Asked Questions". Retrieved 03.02.2012 from http://www.esahulat.com.pk/subpages/faqs.php

622 Over 100 more get Nadra's 'e-Sahulat' licence (2008, July 06), Daily Dawn. Retrieved 04.05.2014 from https://www.dawn.com/news/310440

623 National Database Registration Authority (2012a). "e-Sahulat: Individual Franchise Scheme". Retrieved 27.08.2012 from http://www.esahulat.com.pk/subpages/pop_up_individual_franchisee_scheme.php.

162

cards and a citizen verification system – regular feature in kiosk's platform (see Table 11 for exact revenue share).

Table 11: Licensees' revenue share in eSahulat payments

Nr	eSahulat Service	Licensees' Share
1.	Utility and Post-Paid Bills	Rs. 5 per bill
2.	Prepaid Internet Surfing Card	10% per card
3.	Prepaid Calling Cards	5% share per card
4.	Prepaid Mobile Calling Cards	10% per card
5.	Online Citizen Verification	Rs. 10 per verification

Source: Own illustration based on eSahulat Franchise Application Form[624]

5.3.2. eSahulat Results

Based on the data collected from NADRA, eSahulat seems to be attracting the public from its first year of inception in 2008. In comparison to kiosk machines' initial reception, a total of 1,620 people purchased eSahulat license all across Pakistan. Similarly, eSahulat programme also outdid the kiosk in the total number of transactions. At the end of 2008, over 7.6 million transactions took place in 1,620 different locations (no. of eSahulat licensees) in Pakistan whereas transactions via kiosks totalled 1 million in 2007, despite the kiosk technology being functional for two years.

5.3.3. Comparison of kiosk and eSahulat at respective launch years

When kiosk machine was launched in 2005, its purpose and functioning was not widely known. Similarly, the number of machines at the beginning, as noted in the previous section was limited to only few cities. Over the course of three years, kiosk machines and particularly their franchise variant received a good response from the general public. One of the primary reasons of eSahulat's uptake was that it banks on the public's awareness of the kiosk's eCommerce platform. This is the very reason that eSahulat's transactions in its first year were exponentially high as compared to the kiosk's transaction in its first year of launch. Total transactions recorded in 2005 at all the kiosk machines were merely 369 whereas the total transactions at eSahulat centres towered at over 7.6 million.

---

624   National Database Registration Authority (2013a). "e-Sahulat: Franchise Application Form". Retrieved 03.09.2013 from http://www.esahulat.com.pk/files_download/esahulat_application_form_urdu.pdf

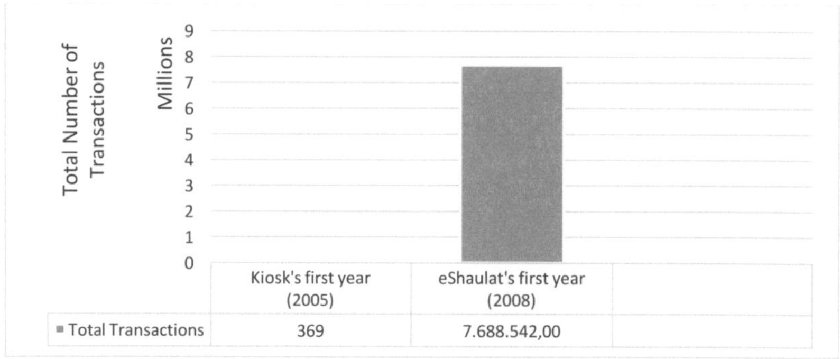

Source: Own illustration (based on kiosk database)

5.3.4. Comparison of Kiosk franchised machines vs eSahulat

In 2007 when NADRA franchised the kiosk machines to interested business entities, there was a considerable response by public and around 30 citizens showed interest in opting for the kiosk machines to be installed at their business premises. The transactions occurred at these machines totalled 0.7 million (755,906) with the payment of utility bills created a revenue for its franchisees PKR 3,779,530 (3.77 million). eSahulat in comparison with its franchised variant exhibited the scaling potential of its model. The transactions produced via eSahulat in comparison were 7.6 million, almost 10 times higher than the franchised kiosk variant. Considering that every licensee receives PKR 5 per bill as her/his commission, therefore, the revenue generated by eSahulat licensees in 2008 amounted to PKR. 38,442,710 (38.4 million).

Table 12: Transactions performed at eSahulat centres in 2008

Bill Payments via eSahulat	
Total Bills Paid	7.688.542
Money Collected via bill payment	Rs. 6.175.265.795,00
Revenue Generated for Licensees	Rs. 38.442.710,00
Citizen Verification via eSahulat	
Total Citizen Verifications	18.657
Money Collected via citizen verification	Rs. 932.850,00
Revenue Generated for Licensees	Rs. 186.570,00

Source: Own illustration based on kiosk's data

In this regard, eSahulat not only created a revenue for the businesses (see Table 12) who obtained the eSahulat license but it is also helping in sharing the load with the banks that had been sole authorities to accept utility bills in Pakistan, with the exception of the kiosks. In 2008, NADRA collected PKR 6,175,265,795 (just above 6 billion rupees), which is even greater than the cumulative amount of PKR 1,501,900,651 (a little over 1 billion rupees) from kiosk machines from 2005 through 2007. NADRA's own profit considering PKR 1.5 per bill as NADRA's commission amounted to PKR 11,532,813 (11.5 million rupees) only with the bill payments.

These figures do not include the commissions that licensees earned via citizen verification and sales of calling cards/top-up cards. In terms of business prospect, eSahulat is proving to be a favourable proposition for both the NADRA as well as for the license holders. It is interesting to note that a lot of people find it convenient to verify fellow citizens before getting into any business transaction. In this regard 18,657 took part in citizen verification whereas most of the citizen verification took place in the cities from Punjab province.

Figure 28: Adoption of franchised kiosk and eSahulat in their first years

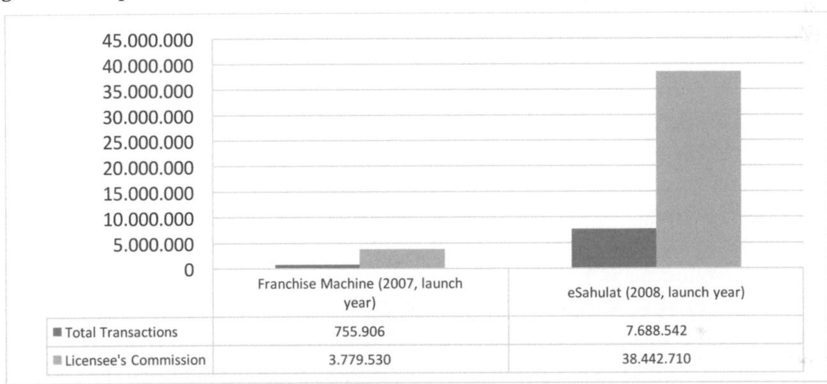

	Franchise Machine (2007, launch year)	eSahulat (2008, launch year)
■ Total Transactions	755.906	7.688.542
■ Licensee's Commission	3.779.530	38.442.710

Source: Own illustration (based on kiosk database)

5.3.5. Top Five Cities (most/least number of frequencies)

There is a stark similarity in the geographical trends of eSahulat and kiosk machines when it comes to the top cities where most and least frequent usage occurred. The top five regions where the maximum number of transactions took place in 2008 are Lahore, Faisalabad, Okara, Gujranwala and Multan. Similarly, the top five regions where minimum number of transactions took place are Thatta, Hangu, Chitral, Panjgur and D. I. Khan.

All the top five regions are in the province of Punjab which is not only the most populous province in Pakistan but it is also the political, agricultural, and industrial hub of Pakistan with the most population density. As pointed out in Chapter 4, Punjab contributes the most to Pakistan's agrarian, textile and cotton economy. It has over 110 million inhabitants which makes up for 55% of Pakistan's total population of 197 million. According to a study conducted by the Small and Medium Enterprises Development Authority (SMEDA) of Pakistan, Punjab hosts most of the start-ups and small and medium businesses in Pakistan[625].

Figure 29: Maximum and minimum transactions through eSahulat

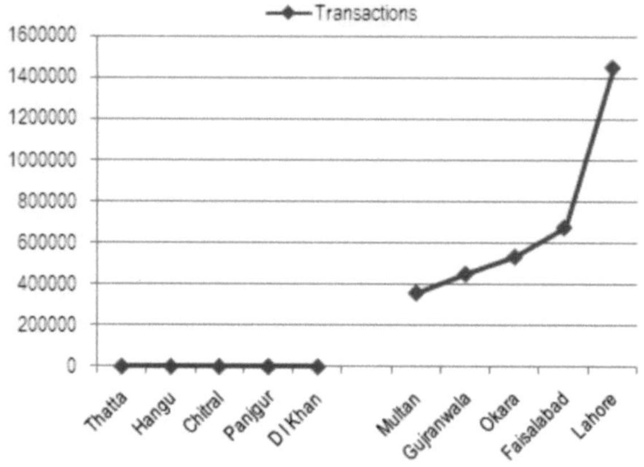

Source: Own illustration (based on kiosk database)

The tendency to take risks, launch new initiatives and business-oriented culture is comparatively stronger than rest of the provinces. This is one of the reasons that the other provinces such as KP, Sindh or Balochistan underperformed in the first year of eSahulat's launch. Karachi which is considered the most populous, industrial and port city in Pakistan ranked at number six with total transactions amounting to 295,692 (0.3 million) via eSahulat. It is noteworthy that eSahulat works entirely on voluntary basis. Some of the eSahulat licensees in Punjab hired daily wagers who advertised the eSahulat benefits within their vicinity. A network built on social capital resultantly let the people rely on more convenient ways of public service delivery.

Another interesting turn of events in 2008 was the number of transactions taking place via kiosk and eSahulat (see Figure 30). In 2007, when the transactions through kiosks started to increase from 285,149 in 2006 to 1,376,343, this number dramatically dropped down to 204,590 in 2008, which is even lesser than its performance in 2006. Even though kiosk machines were produced to provide an alternative to web/internet G2C model, however, eSahulat's model which is based

---

625 SMEDA. (2009). SME Baseline Survey. Retrieved 09.02.2011 from https://smeda.org/phocadownload/Publicatoins/SME%20Baseline%20Survey.pdf

166

on socio-technical capital and relies on human interface fared much better. When given an alternative in the form of eSahulat, the public apparently prefer the already existing oral communication pattern. The data indicate that kiosk might have been used only as an alternate to circumvent overloaded banks when no other option existed. eSahulat then took on the role of actively substituting kiosk as a more amenable, accessible and simpler way to access those same services which the kiosk had helped the banks to relieve of their burden.

Thus, NADRA not only created a G2C eService but also a new business model for the licensees of eSahulat based on the incentive involved, that is, higher the transactions higher the profit for the licensee. eSahulat seems different from other hybrid strategies in various ways. NADRA's eSahulat programme has tried to redefine the G2C scenario of eGovernment. The business model and an incentive for the licensee created via eSahulat is seldom witnessed when a government service such as utility bill payment and citizen verification mechanisms could be commercialised and scaled for the citizens.

Figure 30: Drop in kiosk machine's performance from 2005 through 2009

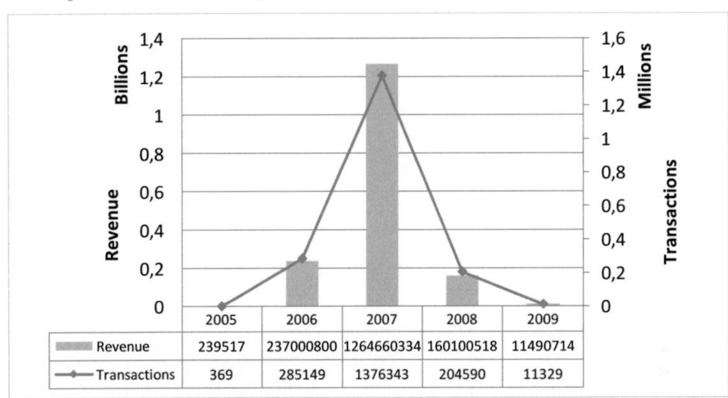

	2005	2006	2007	2008	2009
Revenue	239517	237000800	1264660334	160100518	11490714
Transactions	369	285149	1376343	204590	11329

Source: Author (through NADRA database)

Similarly, if we compare the generic eGovernment model discussed in Chapter 2, in any G2C eService, government interacts with its citizens via its official web portal. However, in eSahulat the government basically interacts with its citizen through a middle business entity, thereby enabling what may be Business to Citizen (B2C) service. eSahulat in this regard falls close to the concept of service centres where the computer operators act as middlemen between the general public while offering the G2C services. eSahulat removes the element of service centre and makes the business entities the middlemen, thereby bringing in private sector to bear the profit and efforts of the upkeep and expansion of its G2C. In comparison to Kiosk, eSahulat is different in the sense that the public does not have to interact with the kiosk, not be familiar with technology or to be personally present for the verification of the thumb impression.

5.4. Concluding Remarks

This chapter has focussed on the formation of NADRA and its role in the eGovernment application development in Pakistan. It has also highlighted the history of NADRA as to how it initially was developed as an NDO to help government with the census data of 1998. However, during Musharraf's military rule NDO's mandate was elevated and it was given a status of an attached department with the Ministry of Interior by promulgation of a presidential ordinance in 2000. This chapter has also focussed on the trajectory of eGovernment application development by NADRA explaining how initially its mandate to produce computerised national identity cards branched to develop several civil registry applications such as child registration certificates, family registration certificates, and machine-readable passports. All these products have now become mandatory and standard for Pakistani citizens.

One of the unique developments that took place in the field of eGovernment service delivery is the conception of kiosk machine and NADRA's eCommerce platform. Kiosk machine not only created a citizen-centric G2C platform but at the same time it also provided a reality check. Kiosk machine may not be the ultimate solution but it has at least triggered the realisation of persistent digital divide and how alternative channels of communication and devices are needed to be tapped in order to tackle eReadiness. In this regard, this chapter has identified kiosk and its other variants as one of the first G2C service in Pakistan that is developed by a public-sector organisation. One important aspect is also NADRA's growing influence in the field of G2C services, which has overshadowed the role and importance of another eGovernment initiative that is EGD (see chapter 4). In comparison to EGD, NADRA seems to have outperformed in terms of its G2C services and outreach. More detailed discussion on both the institutions can be found in chapter 7.

NADRA is the second institution along with the EGD that has been primarily established for eGovernment application development and G2C services. In this regard, NADRA has come under constant scrutiny by civil society and media with regard to integrity, privacy and the protection of citizen data. For instance, in 2007, NADRA's system detected an attempt being made by its own staff to produce fake identity cards[626]. Similarly, in 2010, NADRA again caught some of its own employees who were involved in making fake CNIC[627]. Though NADRA

626   Nadra employees caught in fake ID cards racket, November 17), Daily Dawn. Retrieved 04.05.2014 from https://www.dawn.com/news/276183

627   Corruption: Nadra staff caught making fake IDs (2010, December 10), The Express Tribune. Retrieved 04.05.2014 from https://tribune.com.pk/story/88181/corruption-nadra-staff-caught-making-fake-ids/

suspended 106 employees who were found involved in corruption and irregularities, it raises doubts and questions for society and citizens if they have been prepared for a revised social contract that their Government wants them to enter into; a contract where all data is maintained and mapped across government and private sector. Similarly, the question also arises as to how Pakistan would ensure the data privacy rights for its citizens and also the integrity of national identity card, which is now synonymous with citizenship in Pakistan. In 2009-10 the local media was abuzz with reports that the Ministry of Interior has secretly provided access NADRA's citizen database to American and British embassies[628]. NADRA's then deputy chairperson, Tariq Malik, maintained that there is no access provided to any agency[629]. Similarly, based on data protection and security, NADRA on another occasion also cancelled projects that involved collaboration with third/foreign companies[630]. There were also rumours of data theft and privacy that resurfaced when the founder of WikiLeaks, Julian Assange, interviewed former cricketer and politician Imran Khan who claimed that the Pakistani citizen database was compromised and data was stolen by foreign countries[631]. In 2012 questions also arose about the limits of data sharing even with other public-sector Pakistani organisations. For instance, upon the request of Federal Board of Revenue, NADRA allegedly helped prepare a list of 700,000 tax evaders based on the frequency of their passport renewal[632].

Such allegations have continuously brought NADRA's reputation under a penumbra particularly picked up by the media. NADRA's role in Pakistan is not only that of a repository of citizen database but equally as its digital gatekeeper. NADRA's data warehouse holds the most sensitive citizen data and the documents produced by NADRA (such as CNIC, SNIC, FRC, CRC etc.) are directly are linked to person's nationality, birth and family lineages. This data if mapped against persons' travel history, wealth and business dealings can raise serious questions about individual freedom and privacy. As NADRA has progressed over a number of years, it has also tried to pursue modern digital gatekeeping skills by employing technological tools or by following standardised set of rules to safeguard its database. For instance, as of 2012 NADRA is a CMMI

628 لٹی لٹی گل بونے اڑا اڑا گل رنگ (2010, [Faded stands the redness of rose; raped lies its fragnance!] March 18), Daily Nawaiwaqt. Retrieved 04.05.2014 from https://www.nawaiwaqt.com.pk/18-Mar-2010/106629

629 National Database Registration Authority (2009a). National Database is Fully Secured: Deputy Chairman NADRA. Retrieved 09.07.2009 from https://www.nadra.gov.pk/DesktopModules/top/topmore.aspx?tabID=0&ItemID=57&bID=0&Mid=3026

630 Nadra shelves project despite Malik's opposition (2013a, February 03), The Nation. Retrieved 04.05.2014 from https://nation.com.pk/03-Feb-2013/nadra-shelves-project-despite-maliks-opposition

631 Nadra conducts its own probe into 'UK visa scam' (2012d, July 24), Daily Dawn. Retrieved 04.05.2014 from https://www.dawn.com/news/736956

632 BBC News (2012). "Pakistan's experience with identity management". Retrieved 04.05.2014 from https://www.bbc.com/news/world-asia-18101385.

Maturity Level III organisation[633]. Similarly, it is also an ISO 27001 certified organisation[634].

On the other hand, over the years NADRA has also developed eServices such as kiosk and eSahulat that also indicates its entrepreneurial dispositions, which actually is the main area of exploration for the purpose of this research. The data collected as a part of empirical research from NADRA on kiosk and eSahulat platforms is neither publicly available nor does NADRA develops monthly or yearly reports about its business and operations. The data received from NADRA in this regard was only used for the purpose of exploring the digital trends by reviewing the frequency and patterns of usage between its two variants of G2C eService. No person data (such as names, age, address, CNIC number) was neither asked nor given by NADRA. The choice of language options chosen by the user at the start of kiosk machine operation could have been a helpful identifier in making projections about the use of language in technology adoption. However, the NADRA staff revealed that they do not store any information related to the choice of language. The kiosk data did have some erroneous entries which for the purpose of drawing statistical inferences was first cleaned up. How closely and minutely NADRA itself monitor and scrutinises the kiosk and eSahulat data itself remains beyond the scope of this study. However, within the context of this study, kiosk's data has helped understand adoption choices for digital trends. More importantly, the data has been helpful to realise that there is a need to tap into more alternative channels of electronic public service delivery. The channels that are more inclined to and/or culturally embedded with the existing patterns of communication tend to adopt faster. In this context NADRA has helped changed the G2C communication paradigm. With its ability to digitally access and link multiple citizen databases, standardize data collection and store it centrally with appropriate means to parse it meaningfully, it has opened the door to further innovation in digital service delivery. The next chapter will take this further to trace and analyse how NADRA's role is changing from a service provider to a service facilitator by engaging with other public and private sector organisations in Pakistan for the provision of eGovernment applications.

---

633   CMMI is short for Capability Maturity Model Integration and in past few years it has become an international standard in process improvement in the development of systems such as software and design engineering and programme management. Originally developed by at the Carnegie Mellon University, USA, there are five levels to CMMI namely, Initial, Managed, Defined, Quantitatively Managed and Optimising. NADRA in this case qualifies the Defined stage. For detailed overview on CMMI see: Chrissis, M. B., Konrad, M. and Shrum, S. (2011). CMMI for Development: Guidelines for Process Integration and Product Improvement. New Jersey: Addison-Wesley

634   NADRA information security system fully secured (2012a, September 10), Pakistan Today. Retrieved 04.05.2014 from https://www.pakistantoday.com.pk/2012/09/10/nadra-information-security-system-fully-secured/

Chapter 6

Interagency Collaborations: Election Commission of Pakistan, NADRA & Mobile Network Operators

The previous chapters have attempted to provide an overview of the provenance of eGovernment in Pakistan as well as its evolution in the first 12 years of the millennium. On the one hand, new institutions such as EGD, NADRA or ECAC have been established under different policies with different foundational vision while on the other hand we also saw in previous chapter how NADRA as a public-sector entity has taken the lead in developing eGovernment applications thereby consolidating its position in the field. Its eCommerce platform in the form of kiosk and eSahulat applications is not only the first ever public-sector financial platform, kiosk also happens to be one of the first G2C services. The most important feature of this platform is its realisation of digital divide and provision of alternative service delivery channel in the form of kiosk. Similarly, another important aspect discussed in the previous chapter is also NADRA's ability to enable G2C communication between the citizens and other public-sector organisations such as utility companies related to water, gas, electricity and telephone. NADRA has been able to develop its eCommerce platform by enabling the interaction of databases of different entities involved. In this regard, NADRA has been able to inspire other public and private entities to develop their own solutions.

This chapter particularly focuses on the aspect of interagency collaborations by reviewing the evolution of digital coordination at the level of G2B and G2G to provide G2C services. An important aspect that this chapter also focuses on is the changing patterns of service delivery channels. In this regard, this chapter will draw on two case studies (i) a private mobile network operator Mobilink's mobile commerce platform and (ii) an SMS application for voter verification developed by the Election Commission of Pakistan (ECP), a public-sector organisation. The key aspect of choosing these case studies lies in the fact that even though offered by a private and public entity respectively, both case studies use mobile phone as the front-end technology for G2C communication. Similarly, both Mobilink and ECP rely on NADRA as a digital authenticator, collaborator and partial-enabler of the service developed. Before getting into the details of Mobilink's mobile commerce platform, it is pertinent to briefly overview the history of mobile phone industry that led to it being an early player in electronic public service delivery and communication.

6.1. Overview of Mobile telecommunication in Pakistan

Among the major developments in the ICT sector that have taken place in the first decade of the millennium is rapid adoption and growth of mobile phones both in Pakistan and around the world[635]. In the beginning of 2000, mobile phones have mostly been associated with voice (calling) and SMS service. However, by the end of 2010, mobile network operators in Pakistan were able to offer their own indigenously developed electronic public services solutions[636]. In order to understand the journey of these ten years, the following section reviews mobile phone's role in triggering socio-cultural changes while also dominating socio-economic change(s) in Pakistan.

The first mobile network service licence in Pakistan was awarded to Paktel in 1989 which started its commercial services in 1990[637]. During this period two additional mobile network operators Instaphone and Mobilink also applied for licences and started their operations in Pakistan[638]. The higher costs of equipment (mobile phone) and telephone call rates kept mobile phones beyond the economic reach of the majority[639]. By end of 1996 there were only 63,038 combined mobile phone subscribers of Paktel, Instaphone and Mobilink[640]. These figures saw a very meagre increase on yearly basis and by end of 1999 a combined total of subscribers recorded by Pakistan Telecommunication Authority was 265,614[641].

Pakistan's telecom market saw a major change when PTA awarded three additional licenses to new mobile network operators at the beginning of the decade. These included Ufone in 2001, a subsidiary of Pakistan Telecommunication Limited (PTCL), UAE-based Warid Telecom and Norway-based Telenor both in 2004[642]. A presence of six mobile network operators changed the entire communication landscape in Pakistan and in 2005 mobile phone subscription rose to 33.9 million[643]. As the number of mobile network

---

635   Ansari, S. and Khan, A. (2009). Telecommunication trends in Pakistan. Market Forces, 4(4), 213-218

636   McCarty, M. Y. and Bjaerum, R. (2013). Easypaisa: Mobile Money Innovation in Pakistan. Retrieved 03.05.2014 from https://www.gsma.com/mobilefordevelopment/wp-content/uploads/2013/07/Telenor-Pakistan.pdf

637   World Bank. (1995). Telecommunications Regulation and Privatization Support Project. Staff Appraisal Report (Nr. 13298-PAK). Retrieved 04.05.2014 from http://documents.worldbank.org/curated/en/879571468774973666/pdf/multi0page.pdf

638   Looney, R. E.: 1998

639   Rafiq, A. and Gao, P. (2008). The transformation of mobile telecommunications industry in Pakistan. In The Pacific Asia Conference on Information Systems. (pp. 95). Retrieved 04.12.2010 from https://aisel.aisnet.org/cgi/viewcontent.cgi?article=1188&context=pacis2008

640   See Annex-9 in Pakistan Telecommunication Authority. (2005). Annual Report (2004-05). Retrieved 17.12.2010 from https://www.pta.gov.pk/annual-reports/ann-rep-05.pdf

641   ibid.

642   Mobilink continues to grow despite increasing competition (2005g, June 17), Business Recorder. Retrieved 03.04.2009 from https://fp.brecorder.com/2005/06/20050617282489/

643   Pakistan Telecommunication Authority. (2010a). Annual Report - 2009-10. Retrieved 03.05.2011 from https://www.pta.gov.pk/annual-reports/pta_ann_rep_2010.pdf

operators increased, so did the competition among these operators[644]. The increased competition among mobile network operators also opened new job opportunities for skilled workforce especially for young graduates[645]. All mobile network operators tried to offer lowest calling rates and inexpensive SMS packages[646]. These incentives were also supported by the global developments in mobile phone industry and from a luxury item, mobile phone started to become a necessity. The mobile network operators offered both prepaid and post-paid package plans for mobile users. Resultantly, Pakistani electronics market was flooded by a variety of mobile phones to suit budget, convenience and utility[647].

Throughout the first decade of 2000s, mobile phone shops started from bigger cities and equally spread in the rural areas of Pakistan. The mobile phone shops not only spread within the country but also within the cities as well[648]. Many bigger cities such as Lahore, Peshawar, Rawalpindi, and Karachi saw the development of shopping malls which only housed mobile phone shops, a very rare phenomenon. During the field visits to Pakistan in 2008 and 2009 this massive growth in mobile phone shops within shopping plazas could be witnessed for instance at Singapore Plaza (Rawalpindi), Bilour Plaza (Peshawar) or Hafeez Centre (Lahore). These shopping malls not only sell brand new mobile phones but also sell second-hand phones. The mobile phone prices vary from PKR. 800 (USD 8) up to PKR. 120,000 (USD 1,200) including feature to smart phones[649]. Apart from the sale of mobile phones, these shops and retailers offered additional services such as hardware repair, software update services, installation of caller tunes, selling mobile phone covers, and sale of prepaid calling cards.

Massive investment in mobile telecommunication sector, rapid roll-out of network infrastructure, coupled with a network of these small and big mobile shops and retailers helped establish a mobile phone culture in Pakistan[650].

644   Pakistan telecommunication Authority (PTA). (2007b, June 14). Pakistan Ahead of Regional Countries in Telecom Regulartory Environment. [Press release]. Retrieved 09.09.2009 from https://www.pta.gov.pk/en/media-center/single-media/pakistan-ahead-of-regional-countries-in-telecom-regulatory-environment

645   Zahra, K., Azim, P. and Mahmood, A. (2008). Telecommunication infrastructure development and economic growth: A panel data approach. The Pakistan Development Review, 47(4-II), 711-726

646   Pakistan Telecommunication Authority: 2007a

647   Malik, S., Chaudhry, I. S. and Abbas, Q. (2009). Socio-economic Impact of Cellular Phones Growth in Pakistan: An Empirical Analysis. Pakistan Journal of Social Sciences, 29(1), 23-37

648   End of an era: Pindi's Imperial Market is going cellular (2013e, July 21), The Express Tribune. Retrieved 04.05.2014 from https://tribune.com.pk/story/580119/end-of-an-era-pindis-imperial-market-is-going-cellular/

649   QMobile: Conquering the Pakistani market, one phone at a time (2013d, April 7), The Express Tribune. Retrieved 04.05.2014 from https://tribune.com.pk/story/532133/qmobile-conquering-the-pakistani-market-one-phone-at-a-time/

650   Delloitte's report for Telenor provides a good overview of investments in and economic impact of mobile communications. See Chapter 15 in Deloitte & Touche LLP (2008). Economic impact of mobile communications in Serbia, Ukraine, Malaysia, Thailand, Bangladesh, and Pakistan: A report prepared for Telenor ASA. Retrieved 05.05.2014 from

Therefore, deregulation in media and telecommunication policy during the decade from 2000 brought foreign and local investment in the telecom market making mobile phones affordable and thus bringing in even more mobile phone subscribers in Pakistan; thereby outperforming the total internet users[651].

According to the ITU data, from 2003 to 2013 there has been a growth of 9 internet users per 100 inhabitants in Pakistan[652]. Whereas in the case of mobile phone subscribers this figure has been phenomenal- mobile phone users increased from 1 per 100 in 2002 to almost 72 per 100 in 2013.

Figure 31: Mobile & telephone subscribers in Pakistan (per 100 inhabitants)

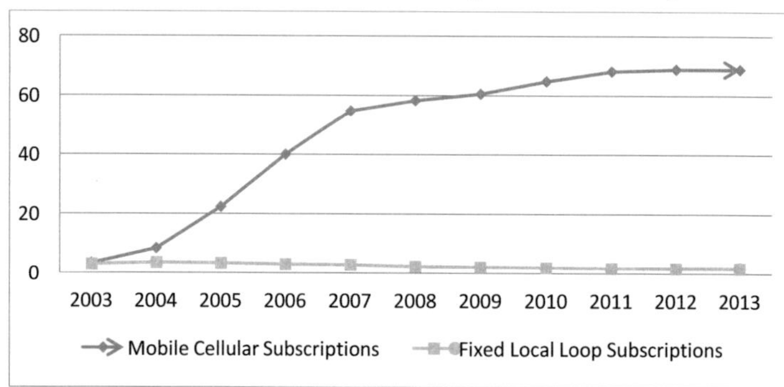

Source: World Development Indicators[653]

A number of factors help understand the wide gap between internet and mobile phones subscribers in Pakistan as depicted in Figure 31. To begin with, mobile phones' usage builds upon oral communication which people are acquainted with and vocal familiarity helps incentivize adoption[654]. Whereas the use of internet requires a completely different social and educational capital where a knowledge of web is foremost, followed by a possession or at least understanding of operational use of computers[655]. More importantly, there is a huge difference in prices of computers, mobile phones and internet rates. Internet required a considerable investment, first buying a desktop/laptop computer and, then a

https://www.telenor.rs/media/TelenorSrbija/fondacija/economic_impact_of_mobile_communications.pdf

651 Telecom sector attracts US$ 12bn investments in seven years (2014c, June 17), Business Recorder. Retrieved 30.09.2014 from https://www.brecorder.com/2014/06/17/179565/

652 International Telecommunication Union: 2013b

653 World Bank: 2013a

654 Raseroka, K. (2006). Access to information and knowledge. In Jørgensen (Ed.), Human rights in the global information society (pp. 91-105). Cambridge: The MIT Press

655 Koltay, T. (2011). The media and the literacies: Media literacy, information literacy, digital literacy. Media, Culture & Society, 33(2), 211-221; Bawden, D. (2008). Origins and Concepts of Digital Literacy. In Lankshear and Knobel (Eds.), Digital literacies: Concepts, policies and practices (pp. 17-32). New York: Peter Lang

connection and monthly subscription with an Internet Service Provider (ISP)[656]. The entire exercise also required a phone line connection. The economics of mobile phone on the other hand is not dependent upon so many variables, a simple mobile without a colour display would be charged in a few hours and can last sometimes up to 4 days depending upon the use. Particularly in the case of Pakistan, inexpensive mobile phones and a possibility to buy extra batteries (to overcome erratic electricity supply) on low cost has been one of the decisive factors in mobile phone uptake and spread[657].

Another factor for the wide gap as depicted in Figure 31 is also the lower call rates and package plans by the mobile network operators in Pakistan[658]. In order to tap into and exploit the mobile subscription market of Pakistan, all mobile network operators offer competitive call rates, sometimes as low as PKR. 1 per minute (USD 1 cent). Similarly, the mobile network operators also offer different call and SMS packages which users could subscribe to by purchasing pre- or post-paid calling cards. Mobile network operators seem to understand the social, cultural and economic situation of Pakistan as well as the purchasing power of a common person. Therefore, they introduced services that tried to focus on inclusion of majority population[659]. The mobile network operators made their prepaid calling cards available not only at their offices but at almost every retailer outlet, grocery stores, book shops, corner shops or gas stations throughout Pakistan. They have innovated services that are built around the economic and cultural nuances present in Pakistani society.

For instance, several mobile network operators introduced a service called as Balance Share[660]. Balance Share is a mobile phone service when the user's prepaid calling credit is completely exhausted or insufficient. In order to continue using mobile service, a new prepaid card would need to be purchased. The reasons for insufficient balance can be if the user cannot afford to buy another calling card due to financial constraints or if there are physical reasons (such as person's reduced mobility, unavailability of corner shop/grocery store). In such a situation,

---

656   Andonova, V. (2006). Mobile phones, the Internet and the institutional environment. Telecommunications Policy, 30(1), 29-45

657   PTCL losing customers to cheap cell phones (2002i, January 16), Daily Dawn. Retrieved 15.04.2009 from https://www.dawn.com/news/15176

658   See for example (particularly Traffic Trends, p. 24) in Pakistan Telecommunication Authority. (2009a). Telecom Quarterly Review. Retrieved 19.10.2010 from https://www.pta.gov.pk/media/tqr_dec_09_1.pdf

659   See for example Delloite's report prepared for GSMA on digital inclusion in Pakistan supported by telecommunication sector: Deloitte LLP (2015). Digital inclusion and mobile sector taxation in Pakistan. Retrieved 03.05.2015 from https://www.gsma.com/publicpolicy/wp-content/uploads/2015/02/GSMA2015_Report_DigitalInclusionAndMobileSectorTaxationInPakistan.pdf

660   Mobile network operators advertise credit sharing offers either through promotional SMS or via their websites. Similarly, they have also named them differently, either based on the name of their company or an entirely different name. For instance, Ufone calls their balance share mechanism as UShare, see: https://ufone.com/vas/utilities/Ushare/ and Telenor calls it Smart Share, see https://www.telenor.com.pk/value-added-services/smart-share/

mobile user could request her/his friend or family member to share a part of their balance on their pre- or post-paid subscription account. Resultantly another user transfers a certain amount of credit against a small fee thereby enabling another subscriber through resource sharing. It sounds like a financial/monetary transaction. In theory, the two users have only shared a part of money available in the calling card which can only be used for making calls or sending SMS.

Similarly, all mobile network operators have kept the rates of calling cards available in several denominations so as to reach out to citizens from all economic classes. For instance, the lowest denomination of the prepaid calling is available in the PKR. 100 (USD 1). Some of the prepaid call/text packages have been considered so inexpensive and readily available that the mobile network operators also received criticism for them[661]. In addition to the prepaid calling cards, network operators have also introduced a service namely 'Easy Load' for an economic class with even lower income level and that cannot afford the PKR. 100 top-up. With an Easy Load a mobile user can recharge an account with a value as low as PKR 10 (USD 10 cents). Different shopkeepers enter into a contract with mobile network operators to offer an Easy Load service and with every recharge, they earn a small percentage. With PKR. 10 balance in their account, several mobile phone users find various innovative ways to get in touch with their social network, for instance by giving others a missed call- an indication for the receiver to call back or another pre-negotiated message (behind a missed call)[662].

---

661 The omnipresent nature of mobile phone has ironically also received a criticism by the legislatures in Pakistan for the frequent use of mobile phone by youth. For instance, a member of Punjab's provincial assembly Mr. Sheikh Allauddin while addressing the assembly floor said, "Boys and girls talk the whole night on mobile phones and these packages are destroying the moral character of our youth". See for example: Jamil, F. (2013). "Pressured by lawmakers, Pakistan bans SMS and chat packages". In Aaj News. Retrieved 13.12.2013 from https://www.aaj.tv/english/technology/pressured-by-lawmakers-pakistan-bans-sms-and-chat-packages/; Ghosh, P. (2013). "Talk Dirty? Pakistan To Ban Youths From Obtaining Mobile Phone Package Deals To Prevent 'Immoral Behavior'". In International Business Times. Retrieved 14.12.2013 from https://www.ibtimes.com/talk-dirty-pakistan-ban-youths-obtaining-mobile-phone-package-deals-prevent-immoral-behavior-1401933; Pakistan ends cheap late-night mobile phone deals to protect young from 'vulgarity' (2012 The Telegraph. Retrieved https://www.telegraph.co.uk/news/worldnews/asia/pakistan/9692675/Pakistan-ends-cheap-late-night-mobile-phone-deals-to-protect-young-from-vulgarity.html

662 There is an increasing amount of research in innovative ways in mobile phone marketing, especially focused on the idea of missed call in developing countries. See for instance, Donner, J. (2007). The rules of beeping: exchanging messages via intentional "missed calls" on mobile phones. Journal of Computer-mediated communication, 13(1), 1-22; Spence, R. and Smith, M. L. (2010). ICT, development, and poverty reduction: Five emerging stories. Information Technologies & International Development, 6(SE), 11-17; James, J. and Versteeg, M. (2007). Mobile phones in Africa: how much do we really know? Social indicators research, 84(1), 117; Various countries have converted missed call feature into a profitable and functional business model. In India, a start-up named Zipdial calls their customers back who give them a missed call and in return they advertise different products, discounts and offers for their customers. See for instance (2013, August 17). Marketing a missed call. Retrieved from https://www.economist.com/schumpeter/2013/08/17/marketing-a-missed-call

With a number of mobile network operators and with their innovative ways of marketing and understanding of the market's economic needs and country's cultural communication, the mobile phone users have exponentially grown in Pakistan particularly in comparison with the internet users. In terms of economics, mobile phones are well within their reach- available in new and second-hand conditions and suit their geographic mobility. Similarly, for the prepaid calling cards, the calling rates have been kept extremely low by the operators to increase the subscriber base.

They rely on micro-entrepreneurs in communities, such as corner shops in villages or towns, and provide them a pre-negotiated percentage to sell their prepaid SIM/calling cards[663]. They use SMS alerts system to provide informational services such as latest news, stock reports, weather report, polyphonic ringtones and/or agriculture related news alerts for which smart phone is not even necessary. For instance, in 2011 alone a total of 237.58 billion SMS were generated in Pakistan[664]. In terms of telecommunication revenues, for the year 2012-13 they reached an all-time high of PKR 440.20 billion with a growth of 7.0% in a year, as reported by PTA[665]. Despite the fact that there were economic difficulties, power shortage, and constant threat of the terror attack on telecom infrastructure, such an increased number of revenue from the telecom sector indicates the strength and size of the market. With the arrival of 3G and 4G networks in Pakistan, mobile network operators may further explore new avenues and services, for instance, by reducing dependence on the voice communication channels[666].

With such a huge chunk of subscribers' share and innovation in the mobile phone services, a new platform for the provision of various electronic public services was set up in Pakistan[667]. Mobile network operators as such captured a huge market of Pakistani population to which they could offer variety of informational and financial services with regard to eGovernment[668]. This platform seems to be different from the one proposed by EGD and in its eGovernment policy for Pakistan (see Chapter 4). This platform would be readily available, does not

---

663 Long hours for little pay (2013d, June 02), Daily Dawn. Retrieved 04.05.2014 from https://www.dawn.com/news/1015552

664 Pakistan Telecommunication Authority (PTA). (2012). Proliferation of SMS & MMS in Pakistan with emphasis on Premium Rate SMS services. Retrieved 05.05.2014 from https://www.pta.gov.pk/media/sms_mms_report_130313_1.pdf

665 Pakistan Telecommunication Authority (PTA). (2014). Annual Report 2013-14. Retrieved 04.05.2015 from https://www.pta.gov.pk/annual-reports/ptaannrep2013-14.pdf

666 Telecom operators celebrate one million 3G, 4G users (2014d, November 05), 05.05.2015 from https://www.dawn.com/news/1142409

667 World Bank. (2013b). Leveraging Mobile Phones for Innovative Governance Solutions in Pakistan. Retrieved 04.05.2014 from https://www.worldbank.org/en/news/feature/2013/12/11/leveraging-mobile-phones-for-innovative-governance-solutions-in-Pakistan

668 Telenor Group. (2008, July 03). Telenor Pakistan connects the unconnected. [Press release]. Retrieved 08.09.2010 from https://www.telenor.com/media/press-release/telenor-pakistan-connects-the-unconnected

require dependency on the wired internet connection, detailed web/internet literacy and in comparison, with computer/internet costs, it is cheaper, and above all both voice and text-based. A similar trend can also be observed in other developing countries where mobile telephone subscribers have rapidly grown[669]. Most of the mobile phone subscribers in developing countries may not have a fixed line subscription but have more than one mobile phone in a household[670]. Thus, this leapfrogging to a cellular network is also providing a chance to bridge the gap between the marginalised and the privileged communities[671].

In several developing countries, mobile phones are now used for government to citizen services (G2C) as well[672]. Lower prices of telephone calls, unlimited SMS packages and cheap mobile phones are not only affordable for the large populations in developing countries but they also contribute in alleviating the digital divide[673]. From text communication to news subscriptions within the past decade, developing countries have gone a step further by establishing highly innovative mobile applications[674]. This has been accomplished by utilising the already existing human and technical infrastructures. Bangladesh, for example, has revolutionised its now famous microcredit scheme, which was made possible through Nobel Peace Prize winner Muhammad Younas's Grameen Bank and village phone programme[675]. One of the most innovative developments that took place in the field of ICTs and telecommunications has been in Kenya where its mobile network operator Safaricom entirely changed the financial services landscape with M-Pesa project[676].

M-Pesa created the opportunity for monetary transactions without the involvement of any bank in between. It eliminates the need for a bank and gives the subscriber an opportunity to conduct financial transactions through a mobile

---

669   A research study commissioned by UNICEF on 14 Case Study Countries provides a very refreshing read on (particularly Asian) countries pursuing innovative ways of digital and financial inclusion. See UNICEF. (2010). Mobiles for Development. Retrieved 23.03.2013 from https://www.unicef.org/cbsc/files/Mobiles4DeReport.pdf

670   The Economist (2009). "The power of mobile money". Retrieved 15.04.2011 from http://www.economist.com/node/14505519?story_id=14505519.

671   Castells, M., Fernandez-Ardevol, M., Qiu, J. L. and Sey, A. (2007). Mobile Communication and Society: A Global Perspective. Cambridge: Mit Press

672   Susanto, T. D. and Goodwin, R. (2011a). An SMS-Based e-Government Model: What Public Services can be Delivered through SMS? In Al Ajeeli and Al-Bastaki (Eds.), Handbook of Research on E-Services in the Public Sector: E-Government Strategies and Advancements (pp. 137-146). Hershey: IGI Global

673   James, J. (2013). Digital interactions in developing countries: an economic perspective. London: Routledge

674   Ming, A., Awan, O. and Somani, N. (2013). e-Governance in Small States. London: Commonwealth Secretariat. Retrieved 04.05.2014 from https://sscoe.thecommonwealth.org/wp-content/uploads/2018/11/eGovernanceinSmallStates.pdf

675   Rahman, A. (1999). Women and microcredit in rural Bangladesh: anthropological study of the rhetoric and realities of Grameen Bank lending. Boulder, CO: Westview Press, Inc.

676   Hughes, N. and Lonie, S. (2007). M-PESA: Mobile Money for the "Unbanked" Turning Cellphones into 24-Hour Tellers in Kenya. Innovations: Technology, Governance, Globalization, 2(1-2), 63-81

178

phone without a banking tier[677]. The case of Kenya vis-à-vis Africa could be witnessed by the fact that from the state of digitally unconnected in the 1990s, (quite similar to Pakistan) African mobile phone subscription tremendously grew in 2000s reaching up to 60 percent of African population with a mobile phone[678]. This level of subscription provided M-Pesa an opportunity to tap this potential. Within first year of M-Pesa's initiation in 2007, for about 1 million Kenyan population applied for the M-Pesa registration[679]. As a result, the amount that was transferred via mobile phones was estimated to be about USD 87 million using the M-Pesa system. Another initiative is from Brazil, where the national government remits social welfare payments through mobile platform referred to as Bolsa Familia. These examples have relied upon the indigenous communication patterns of oral communication as well as socio-economic factors to boost their subscriptions[680].

6.2. Mobile Banking in Pakistan

The wide operational capabilities of Keyna's "M-Pesa" initiative has also inspired the mobile network operators in Pakistan to take mobile services forward. Initially, Mobilink launched an application called Mobilink Genie in 2008[681]. The idea of Mobilink Genie was derived from the eCommerce mechanism with a motivation to introduce mobile commerce (mCommerce) platform in Pakistan in partnership with Inov8, CitiBank, KASB, Atlas Bank and Adamjee Insurance. With mCommerce, Mobilink wished to establish a payment platform using which people could pay their utility bills (such as electricity, gas or telephone) and buy pre- or post-paid mobile accounts[682]. For monetary transactions, Mobilink used its partner bank KASB bank's billing mechanism for the management and payment of utility bills. Although Mobilink's genie project was novel yet there were a few drawbacks to it. For instance, its software application was only able to run on mobile devices that supported internet connection (GPRS) and Java

677 Donovan, K. (2012). Mobile money for financial inclusion. Information and Communications for development, 61(1), 61-73

678 James, J. (2009). Leapfrogging in mobile telephony: A measure for comparing country performance. Technological Forecasting and Social Change, 76(7), 991-998

679 Mas, I. and Radcliffe, D. (2010). Mobile payments go viral: M-PESA in Kenya. In Chuhan-Pole and Angwafo (Eds.), Yes Africa Can: Success Stories from a Dynamic Continent (pp. 353-369). Retrieved from http://siteresources.worldbank.org/AFRICAEXT/Resources/258643-1271798012256/YAC_Consolidated_Web.pdf

680 United Nation Development Programme. (2012). Mobile Technologies and Empowerment: Enhancing human development through participation and innovation. Retrieved 04.05.2014 from https://www.undp.org/content/dam/undp/library/Democratic%20Governance/Access%20to%20Information%20and%20E-governance/Mobile%20Technologies%20and%20Empowerment_EN.pdf

681 Mobilink. July 08). Mobilink Genie pioneers Mobile Commerce in Pakistan. [Press release]. Retrieved from https://www.jazz.com.pk/media-center/press-releases/mobilink-genie-pioneers-mobile-commerce-in-pakistan/

682 Mobile market not saturated: Mobilink chief (2009, 16 November), Business Recorder. Retrieved 05.10.2011 from http://fp.brecorder.com/2009/11/20091116987256/

platform. This meant those mobile phone users who did not have a Java/Internet supported mobile phones could not use this service.

Then in 2009 Mobilink launched a pilot project called Mobile Money Order in collaboration with the Pakistan Post[683]. The service was only available for Mobilink customers, where they could have their virtual mobile wallets created for their mobile numbers, deposit money and send it to other Mobilink customers across Pakistan. Another mobile network operator, Telenor, has also introduced a broader spectrum of mobile services in partnership with Tameer Bank in 2009 under its brand name "EasyPaisa"[684]. This marked the first time in Pakistan that mobile services were offered to everyone in Pakistan, even if not a mobile phone user. Customers could walk into a Telenor franchise and pay bills for available utilities. Additionally, it became easier for customers to send remittances all across Pakistan. Mobile Wallets and international remittances were also added to the pool of services at a later stage. In late 2010, another mobile network operator Ufone established its first m-commerce product called "UPayments" which is only available to customers of Ufone and Habib Bank limited (HBL) [685]. Customers can pay their utility bills and buy airtime top-up using money available in their HBL account. In February 2011, Mobilink launched its first Over-The-Counter (OTC) mobile service, "Mobilink Utility Bills Payment" which is available for everyone and not only mobile users.

In all these mCommerce related applications offered by Mobilink, Telenor or Ufone, a common factor as we notice here is a partnership with a bank[686]. For instance, Mobilink partnered with KASB/Citi bank and Pakistan Post, Telenor partnered with Tameer Bank and Ufone cooperated with HBL. A number of factors are at play here. To begin with, is a partnership/cooperation with a bank necessary for every mobile network operator? Can the mobile network operators also conduct these mobile-based monetary transactions without formal banking channel partners? Similarly, in case of money transfer does a mobile subscriber not require a bank account? According to the regulations by the State Bank of Pakistan, any monetary transaction requires an involvement of a financial institution such as banks or postal services[687]. This regulatory clause is the very reason that the mobile network operators have partnered with banking institutions

---

683 Mobilink. (2008a, November 28). Mobilink and Pakistan Post Office to launch Mobile Money Order Service. [Press release]. Retrieved 28.10.2011 from https://www.jazz.com.pk/media-center/press-releases/mobilink-and-pakistan-post-officeto-launch-mobile-money-order-service/

684 Telenor Easypaisa. (2009a, October 15). Telenor Pakistan & Tameer Bank launch 'easypaisa'. [Press release]. Retrieved from https://www.easypaisa.com.pk/pressview/ep-launch.html

685 Ufone launches UPayments (2010, December 03), The Nation. Retrieved 03.04.2013 from https://nation.com.pk/03-Dec-2010/ufone-launches-upayments

686 State Bank of Pakistan (2012b). Branchless Banking Newsletter: Leveraging Technologies and Partnerships to Promote Financial Inclusion. Retrieved 04.05.2014 from http://www.sbp.org.pk/publications/acd/2012/BranchlessBanking-Jul-Sep-2012.pdf

687 State Bank of Pakistan (2007). Draft: Policy Paper on Regulatory Framework for Mobile Banking in Pakistan. Retrieved 09.03.2009 from http://www.sbp.org.pk/bprd/2007/Policy_Paper_RF_Mobile_Banking_07-Jun-07.pdf

that are responsible for monetary transactions although the user only sees results on their mobile phone and through their mobile phone operator. The banks only operate as an intermediary between the customer and a mobile network operator in case of money transfer. In case of utility bill payment, bank serves as an intermediary between mobile network operator and utility company (such as electricity or gas).

6.3. The era of Branchless Banking

However, the question remains if the mobile phone users need a bank account for any money transfer. There cannot be a definite no or a yes to this answer. Technically speaking, the mobile phone user in a typical setting does not walk into a bank and sign up for an account. The idea behind mobile phone banking is precisely based on catering to the unbanked population that cannot afford the bank accounts[688]. A citizen in a developing country is not always expected to have a bank account based on factors such as financial affordability and geographical access[689]. One of the prime factors behind the rapid popularity of M-Pesa is also based on purchasing power in maintaining the bank account within a developing country context. Similarly, it also needs to be understood that the mobile banking mechanisms currently in place in Pakistan are usable via simple feature phone without the need for a dedicated and fixed terminal (ATM) or a fixed and reliable internet access through a desktop/laptop. These online banking services, such as money transfer, payment of utility bills or other payments can be accessed via specialised free of charge mobile SIM menu provided by the respective mobile network operator[690]. The mobile phone users punch in the necessary information, perform monetary tasks and disconnects the call/session when done. Alternatively, mobile phone user can also walk into mobile network operator's authorised agent, such as its own franchises, corner shops or grocery stores (whosoever is a mobile network operator's authorised agent). Such a phenomenon has been referred to as branchless banking, where (a) the person does not need to own a bank account (b) the person and the bank are not directly involved for a financial transaction whereas the entire transaction takes place via mobile phone[691]. Mobilink's OTC approach, Telenor's Easypaisa and Ufone's Upayment

---

688   Branchless banking: Branching out (2014c, June 18), Daily Dawn. Retrieved 23.06.2014 from https://www.dawn.com/news/1113550

689   World Bank. (2008a). Finance For All? Policies and Pitfalls in Expanding Access. Washington DC: World Bank. Retrieved 09.12.2010 from https://siteresources.worldbank.org/INTFINFORALL/Resources/4099583-1194373512632/FFA_book.pdf

690   See for example Telenor Group's report on the uses of mobile phone (other than voice and text communication): Telenor. (2013). Reach. Retrieved 04.05.2014 from https://www.telenor.com/wp-content/uploads/2013/02/TG_Mag_2013-02-20_72dpi_spreads.pdf

691   (2011). The transformative role of Mobile Financial Services and the role of German Development Cooperation. Retrieved 09.12.2011 from http://www2.gtz.de/dokumente/bib-2011/giz2011-0068en-mobile-financial-services.pdf

mechanisms all follow and fall under the phenomenon of branchless banking for money transfers and utility bill payments.

In order to provide an enabling environment for the growth of mobile banking services in Pakistan, the two financial and telecommunication regulatory authorities- State Bank of Pakistan (SBP) and the Pakistan Telecommunication Authority (PTA)- work in close collaboration to develop a regulatory framework[692]. As of 2013, all the m-banking services in Pakistan are being governed by the SBP's Branchless Banking Regulations Act 2008, which was later amended in 2011[693]. Realising the market demand and potential, almost all the mobile companies in Pakistan are actively working on mobile banking initiatives in collaboration with different financial institutions[694].

Branchless Banking on one hand offers a good potential to the growth of electronic public services using ICTs whereas on the other hand it brings in the underbanked population from the informal to formal financial sector[695]. Branchless Banking can be particularly important for the population set is dependent on daily wages and cannot afford to open a bank account or later maintain it. A number of daily wagers travel big distances either within their city, district or country to find work and later send back money to their families. With branchless banking sort of mechanisms, they can skip the condition of sending and receiving money via banks or postal services and use their mobile phones as instantaneous tool for financial transactions[696]. However, there are few parameters as well as criteria that have to be adhered to while undertaking the branchless banking implementation. The following section discusses these criteria within the context of Pakistan.

---

692   Pakistan Telecommunication Authority (PTA). (2009b, November 19). PTA and SBP to formulate third party mobile banking regulations. [Press release]. Retrieved 28.08.2012 from https://www.pta.gov.pk/index.php/en/media-center/single-media/pta-and-sbp-to-formulate-third-party-mobile-banking-regulations

693   State Bank of Pakistan (2008a). Branchless Banking Regulations for Financial Institutions Desirous to undertake Branchless Banking. Retrieved 15.11.2011 from http://www.sbp.org.pk/bprd/2008/Annex_C2.pdf

694   Branchless banking – an effort to control cash, document economy (2013c, April 02), The Express Tribune. Retrieved 04.05.2014 from https://tribune.com.pk/story/529689/branchless-banking-an-effort-to-control-cash-document-economy/

695   Frankfurt School of Finance & Management's study prepared for the Asian Development Bank provides an insightful information in the form case studies covering Asian countries and critically explains how branchless banking can work as a successful instrument for reaching lower income groups, particularly those in the remote areas. See Asian Development Bank (ADB). (2014). Regional: Financial Sector Development in Central and West Asia: Making Mobile Financial Services Work for Central and West Asian Countries. Retrieved 04.06.2015 from https://www.adb.org/sites/default/files/project-document/173360/43359-012-tacr-07.pdf

696   Hussain, N. and Tahir, A. (2014). Financial Inclusion's Catalytic Role in the Urbanization of Pakistan's Rural Poor. In Kugelman (Ed.). Pakistan's Runaway Urbanization: What Can Be Done. (pp. 135-139). Retrieved 13.11.2014 from https://www.wilsoncenter.org/publication/pakistans-runaway-urbanization

6.3.1. Branchless Banking Regulations in Pakistan

According to the regulations defined by the State Bank of Pakistan (SBP) branchless banking services can only be provided by Financial Institutions (FIs) which cover Commercial banks, Islamic banks and Microfinance banks. Currently four models of branchless banking services are permissible[697].

- One-to-One: One bank partners up with one mobile network operator to offer branchless banking services.
- One-to-Many: One bank can network with multiple mobile network operators to offer branchless banking services.
- Many-to-Many: Multiple banks can join multiple mobile network operators to offer branchless banking services to nearly all bankable customers.
- Alternative Channels: Branchless banking services can also be provided using agents other than mobile network operators (e.g. post office, gas distribution stations) and using technologies not limited to mobile phone.

Based on these regulations, this means that any bank can offer branchless banking services or also collaborate with either one or more mobile network operators to utilise the mobile network operators' agent networks. Branchless banking tries to focus on population within informal economy by bringing them into a formalised economy, there is also an incentive involved for the participating mobile network operators or banks. Any mobile commerce application offered via a branchless banking platform by the mobile network operators in Pakistan offers a possibility of revenue with the sales of these services[698]. For instance, for the money transfer or for a utility bill payment, mobile network operators keep a certain margin from every transaction. While mobile network operator's share is not directly deducted for utility bill payments, for any money transfer, the mobile phone user is communicated an additional amount required to be paid on top of the money transaction[699].

Since many services are rapidly being provided via mobile network operators, it seems noteworthy to also briefly outline the permissible activities SBP has provided in its guidelines[700].

---

697   State Bank of Pakistan: 2008a; Hussain, N., et al.: 2014

698   Mas, I. and Kumar, K. (2008). Banking on Mobiles: Why, How, for Whom? Retrieved 10.12.2012 from https://www.cgap.org/sites/default/files/researches/documents/CGAP-Focus-Note-Banking-on-Mobiles-Why-How-for-Whom-Jul-2008.pdf

699   See for instance the section: Service Charges in Mobilink (2013). "Money Transfer". Retrieved 04.05.2014 from http://mobicash.com.pk/money-transfer.

700   State Bank of Pakistan: 2008a

- Opening and maintaining a Branchless Banking (BB) account.
- Account to account funds transfer: BB account holder can transfer funds to other registered bank accounts
- Person to person funds transfer: Funds can be transferred from one BB account to another BB account or bank account or to a mobile subscriber who is not a BB account holder. In this case the recipient will have to become a BB account holder before having access to the funds.
- Cash In and Cash Out: Customer may deposit or withdraw funds from their BB accounts
- Bill Payments: Bill payments can also be done under these regulations (e.g. Electricity, Gas, Water)
- Loan disbursement/repayment: BB accounts can be used by banks to disburse small loans while the same accounts can be used by customers for loan reimbursements.
- Merchant Payments: BB accounts can be used for making payments to buy goods/services.
- Remittances: BB accounts may be used to send/receive remittance subject to existing regulations.

From the list of permissible activities for Branchless Banking, we see that based on the SBP regulations, the entire system and functioning of financial sector seems to be completely revamped and revised in Pakistan[701]. Banks who have so far had an authority to perform all sorts of financial transactions have now the possibility to go through rapid transformations by employing the new ICTs[702]. The ubiquitous nature of ICTs has briskly influenced the financial sector in Pakistan. A second-hand phone for PKR 800 (US $ 8) or a brand new mobile phone PKR 1800 (US $ 18) can be served as a mobile bank account without having a bank account per se. The person's mobile number is also her/his bank account number[703]. A mobile phone in this case can be a recipient of payments, remittances and loans. At the same time the mobile phone can also be used for sending money, making payment of utility bills can also be used as an electronic payment mechanism for purchasing any goods[704].

701   Mustafa, R. (2015). Business model innovation: pervasiveness of mobile banking ecosystem and activity system–an illustrative case of Telenor Easypaisa. Journal of Strategy and Management, 8(4), 342-367

702   World Bank. (2009). Bringing Finance to Pakistan's Poor: A Study on Access to Finance for the Underserved and Small Enterprises. Retrieved 12.03.2011 from http://documents.worldbank.org/curated/en/358531468057878762/pdf/486720WP0Bring10Box 338916B01PUBLIC1.pdf

703   Analysis: Banking of the future (2015b, May 18), Daily Dawn. Retrieved 02.06.2015 from https://www.dawn.com/news/1182626

704   For an overview on mobile banking models in different countries, see International Telecommunication Union (2011a). Mobile Banking. Retrieved 12.05.2013 from https://www.itu.int/net/itunews/issues/2011/07/32.aspx

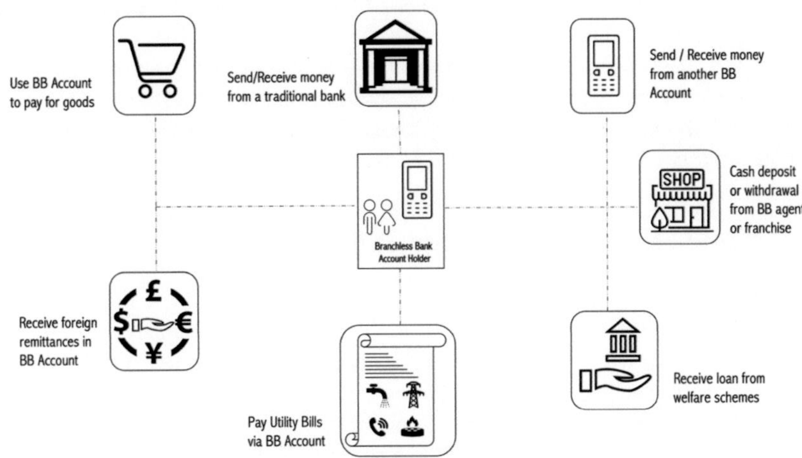

Source: Own illustration, derived from Branchless Banking regulations of State Bank of Pakistan[705]

Figure 32 depicts a branchless banking scenario in Pakistan. A mobile phone bank account holder can simultaneously send and receive money from other branchless bank accounts and pay utility bills. One of the important features of branchless banking particularly in Pakistan is its connectivity with the merchant accounts and with foreign reimbursements. This means that the mobile phone bank account holders can make payments for any goods purchased from various retailers, shopping malls, bookstores or grocery stores. Similarly, the mobile phone bank account holders can also receive foreign reimbursements directly in their phone accounts. Cash can be collected or deposited in the respective phone accounts via mobile network operator's authorised agent or franchise as depicted in Figure 32.

The financial inclusion of such a kind, as depicted in Figure 32, has introduced a digitisation of different form in Pakistan. For instance, there appears to be not one layer of digitisation only, rather several layers of digitisation with multiple checks and balances available. Any phone owner cannot have a SIM card without having a valid CNIC or foreign passport in Pakistan[706]. A SIM card owner needs to verify her/his identity at the mobile network operator's office by going through a CNIC biometric verification system developed by NADRA. If NADRA's computer approves of the thumb impression, only then a customer can purchase a SIM card[707]. Mobile network operators and NADRA systems are interlinked with each

---

705   State Bank of Pakistan: 2008a

706   Terrorists to 'lose big weapon' as Pakistan tightens mobile phone control (2015a, February 26), Daily Dawn. Retrieved 02.06.2015 from https://www.dawn.com/news/1166052

707   Pakistan Telecommunication Authority. (2014a, November 14). New sims are being issued through biometric verification system. [Press release]. Retrieved 03.05.2015 from

other for the verification of the person through CNIC. Similarly, utility bill companies such as electricity, gas, water or telephone have also digitised their data and its access is interlinked with banks and NADRA systems[708]. The following section takes a look at an example of branchless banking practiced by the mobile network operator Mobilink and the respective data sharing by different entities in Pakistan as a part of this example. The study of Mobilink's Branchless Banking mechanism was conducted as a part of this research.

6.4. Mobile service in use: Case of Mobilink

By 2011, Mobilink had reached 31 million subscribers making it a leading mobile network operator in Pakistan[709]. In Branchless Banking category, Mobilink did few experiments with Mobilink Genie and Mobile Money Order projects. However, both the projects had their own set of limitations which have been explained in earlier section. With an investment of over USD 3.3 billion and 8000 telecommunication towers all over Pakistan, Mobilink wanted to tap the potential of Branchless Banking built on inclusive model, that is, to include the segments of society that do not even possess mobile phone[710]. Mobilink wanted to develop a system of payment of utility bills that can be operated via mobile phone without involving internet banking mechanism. A service that can be operated through a SMS feature which is available in every mobile phone. It was with this motivation that Mobilink launched its service called "Mobilink Utility Bills Payment" in February 2011 available for every adult citizen in Pakistan through Mobilink's retailers' network. As a banking partner, Mobilink chose to collaborate with NADRA's eCommerce platform eSahulat[711]. NADRA's eSahulat platform itself carries a digital database of utility companies and it offers such a payment service through their own kiosk and eSahulat's retail network. However, NADRA transformed their eSahulat platform into a commercial entity to be available for banks and telecommunication operators (see Chapter 6). In this scenario Mobilink network acts as a client on behalf of their subscribers whereas NADRA's eSahulat platform performs as a server. That is, Mobilink's mobile platform and NADRA's eSahulat database are digitally connected to each other whereas this interconnectivity is irrelevant for the mobile phone user who receives the

https://www.pta.gov.pk/en/media-center/single-media/new-sims-are-being-issued-through-biometric-verification-system-

708   e-Sahulat Branchless Banking: Nadra facilitates financial institutions, telecom operators (2010a, December 23), Business Recorder. Retrieved 04.05.2014 from https://fp.brecorder.com/2010/12/201012231137422/

709   Mobilink. (2011b, February 22). Mobilink Launches Utility Bill Payment Solution. [Press release]. Retrieved from https://www.jazz.com.pk/media-center/press-releases/mobilink-launches-utility-bill-payment-solution/

710   Mobilink. (2011a, April 13). Mobilink & Ufone Pen Agreement for Tower Sharing. [Press release]. Retrieved 03.05.2014 from https://www.jazz.com.pk/media-center/press-releases/mobilink-ufone-pen-agreement-for-tower-sharing/

711   NADRA, Mobilink join hands for utility bills payment (2011b, January 13), The News. Retrieved 12.11.2013 from https://www.thenews.com.pk/archive/print/279693

information via SMS on her/his screen. Mobilink's utility bill service functions in the following manner:

- A person walks into Mobilink's retail shop with a utility bill(s).
- Mobilink retailer/franchise sends her/his consumer reference number (printed on the bill) along with utility company's name to Mobilink via SMS short code.
- SMS command reaches Mobilink mobile platform which forwards the query to NADRA eSahulat database.
- NADRA's server responds back with bill information which includes (i) due date, (ii) amount payable before due date and (iii) amount payable after due date.
- Retailer confirms the bill payment request via mobile phone; Mobilink's mobile commerce platform sends this confirmation to NADRA eSahulat.
- eSahulat responds back with a transaction ID which means confirmation of bill payment.

Mobilink's mCommerce platform deducts a specified amount from retailer's wallet and adds it into NADRA's wallet. Commissions and taxes are also credited to respective wallets in real time. It also sends out SMS notifications to the retailer and to the customer's mobile number (if provided) along with the NADRA transaction ID acting as a digital but physical receipt.

Mobilink's mCommerce platform is one among the innovative applications developed in Pakistan. Earlier there had been less or almost no digital interaction between/among the data banks of different agencies. Mobilink's example is one of the first few instances where three different entities are digitally interacting with each other in order to facilitate a citizen. This includes, a government agency (NADRA), a public utility company (government-owned electricity, water and gas companies) and a mobile network operator (Mobilink) collaborating to provide electronic services to the citizen via mobile phone. Mobilink's collaboration with NADRA's service can not only lessen the burden on a customer, but it is also able open a door of opportunity for potential small businesses which would like to act as a retailer/facilitator[712].

---

[712] The advantages for the retailers are manifold. As in the case of e-Sahulat, mobile network operators seem to be following the similar footsteps. For elaborate study on mobile phones influencing other sectors such as health, education and agriculture, see International telecommunication Union. (2015). m-Powering Development Initiative: A report by the m-Powering Development Initiative Advisory Board. Retrieved 04.05.2015 from https://www.itu.int/en/ITU-D/Initiatives/m-Powering/Documents/m-PoweringDevelopmentInitiative_Report2015.pdf

Figure 33 : Mobilink Service Mechanism

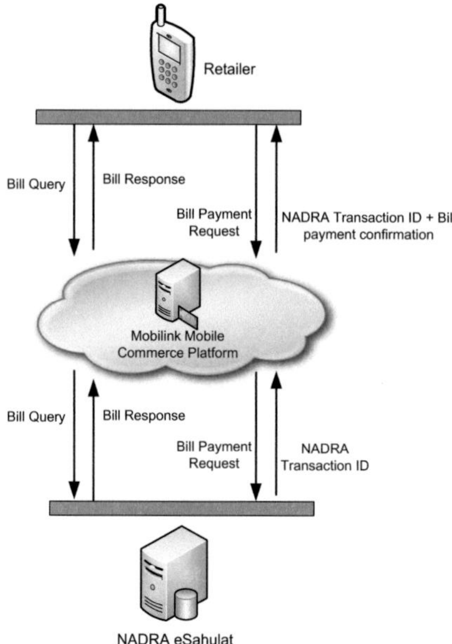

Source: Own illustration based on Mobilink data

Figure 34: Number of transactions in first 40 days of its launch

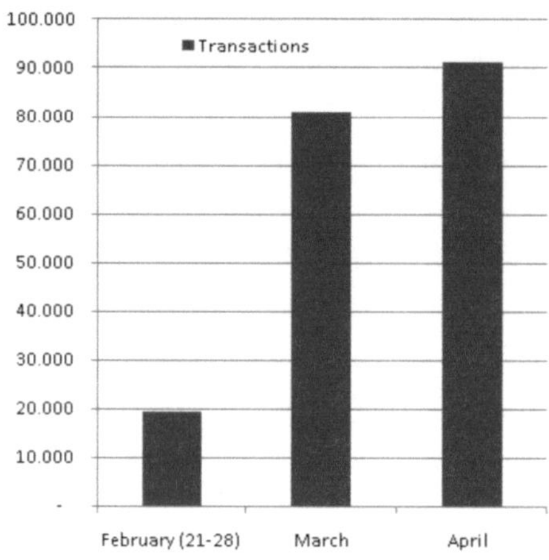

Source: Own illustration based on Mobilink data

188

Performance-wise, the service received a fair reception by attracting customers with more than 100,000 transactions in the initial 40 days of its launch[713]. Mobilink advertised its service through print and electronic media. Franchises, retailers and business centres providing the bill payment service indicated the procedures in Urdu and local languages to reach out to a wider audience. Brochures and fliers were also placed in these outlets to inform customers about how the service works and what they should know to avoid any fraudulent transactions. In a country like Pakistan, there is a huge reliance on person to person contact and social network that has its strong existence from community to community based on the language of the region[714]. This is further accentuated by the fact that majority of the population relies on these oral communications rather than written instructions.

6.4.1.Mobilink Retailer Network

Mobilink's utility bill payment mobile platform can also be seen as a business opportunity for its retailers. Mobilink has so far, a network of about 200,000 retailers and franchises, widely spread across the country, which sell airtime top-up. Utility Bills service was started with 10,000 retailers (out of these 200,000) which are evenly distributed in all regions. These retailers are selected on the basis of various factors such as their location or unavailability of banks in a particular area. Mobilink plans to provide all m-commerce services at all 200,000 retailers and franchises. Mobilink signs an agreement with a franchise, which allows them to carry out various business-related activities, such as subscribing new customers, SIM replacement, airtime top-up, post-paid bill payment and now utility bills payment as well. The franchise brings in a number of retailers as its POS (point of sale) agents, who only need a mobile phone to carry out transactions such as airtime top-up and utility bill payments. For utility bills payment, Mobilink has created wallets for all the retailers and franchises against their mobile number. A franchise can then buy credit in this wallet from Mobilink. After that, the franchise transfers some part of this credit to the targeted retailer of a transaction.

On every transaction, retailers and franchises receive some amount back in their wallets as part of their commission. For a single utility bills transaction, the retailer gets PKR 2.84, while a parent franchise receives PKR 0.7. Additionally, PKR 1 is received by Mobilink while PKR 3 is transferred to NADRA. Mobilink's mobile platform is rapidly attracting small and medium business enterprises that are welcome to assume the role of a retailer with very small investments. Additionally, transactions for domestic remittance, airtime top-up, post-paid bills,

---

713   This study was conducted in 2011 with the help of Mr. Ahsan Mansoor who has extensive experience of Pakistan's telecom market and worked for the Department for Value-Added-Services in Mobilink, Islamabad.

714   Kamran, S. (2010). Mobile phone: Calling and texting patterns of college students in Pakistan. International journal of business and management, 5(4), 26-36

cash in and cash out, is making m-commerce an attractive business opportunity for lower middle class as they use the liquidity afforded by the branchless banks in novel ways offering instalment options and loans into their physical networks.

Figure 35: Mobilink's Revenue share

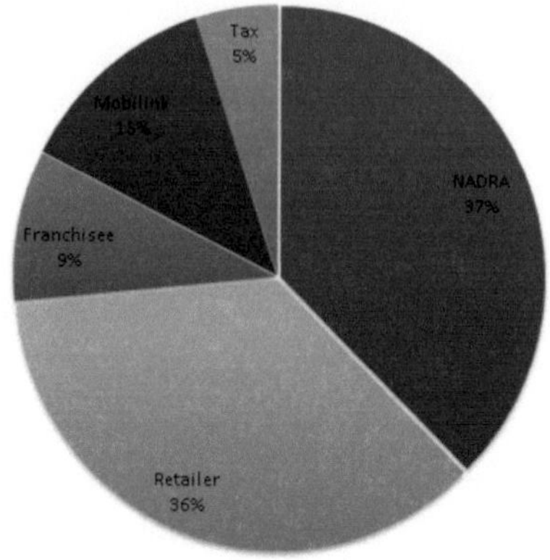

Source: Own illustration based on Mobilink data

6.4.2. Mobile applications as an alternate channel for G2C transactions

Plastic money has so far been failing to attract people in countries such as Pakistan due to factors such as lower income levels, undocumented economy, lack of trust and/or unavailability of cash machines[715]. A number of factors have contributed to this. Firstly, the majority of people in Pakistan do not have a bank account, which is a necessary prerequisite to own a debit or credit card. By 2015, there were a total of 41.7 million bank accounts in Pakistan while only 12% of adult population in Pakistan has an access to banking services in some form[716]. Secondly, the Point of Sale (POS) and Landline network does not have the penetration to a level where it could be adopted by the far-flung population in rural as well as in urban areas. The telecom sector's wireless network on the other hand is almost omnipresent and carries the potential to become the alternative medium for financial transactions. Currently there are more than 100 million mobile phone users in Pakistan that reflects the adoption level of this medium[717].

---

715 Offline transactions for online business (2015e, February 22), Daily Dawn. Retrieved 03.05.2015 from https://www.dawn.com/news/1164656/offline-transactions-for-online-business

716 Pakistan to have 100 million bank accounts by 2025 (2015, November 27), The Express Tribune. Retrieved 03.12.2015 from https://tribune.com.pk/story/999299/one-minute-bank-account-pakistan-to-have-100-million-bank-accounts-by-2025/

717 Telecom Indicators. Retrieved April 15, 2011 from http://www.pta.gov.pk/index.php?option=com_content&task=view&id=269&Itemid=658

190

Thus, mobile applications, in absence of other channels, carry the potential of becoming an alternative medium to carry out G2C transactions. The case of Mobilink serves as a workable example to understand how channels of public service delivery are undergoing a change. Similarly, NADRA's eCommerce platform serves as a gateway as well as collaborator in Mobilink's example. The process of such collaborations has already started with mobile network operators by partnering up with other businesses in order to offer a wide variety of G2C services via mobile phone.

6.5. eService for Election Commission of Pakistan

The previous section depicted the onset of changing G2C service delivery channels in Pakistan and the way they are rapidly changing and improving. The role of mobile phone is therefore no longer only a medium of voice communication or interaction via text messages. Its omnipresent utility has been tapped by the mobile network operators to bring innovation in G2C services in Pakistan where NADRA provided a platform of validation of the digital. This section further dwells on the interagency collaboration and analyses the role of mobile phone as a G2C medium. In this regard, this section studies the collaboration between the Election Commission of Pakistan (ECP) and NADRA which jointly developed an SMS application for voter verification for the elections held in 2013.

In Chapter 5 we looked at the origins of NADRA and its initiation and development from NDO in 1998 and eventually assuming a role of NADRA in 2000. In this regard, NADRA's role between 1998 through 2000 was to help the Government with digitising the 1998's population census data and digitally organising data for its assistance in socio-economic development indicators. NADRA later went on to develop computerised identity cards based on the census data as this study has pointed out in previous chapters. However, NADRA's core competence remained in handling millions of citizen data records and it is based on this expertise that ECP sought NADRA's assistance in digitising the electoral rolls[718]. The rationale behind digitising the electoral rolls lies in the problems of transparency in the ECP and also those encountered in the paper-based electoral rolls[719]. ECP relies on the same electoral rolls for the general and local elections. In a paper-based scenario it becomes difficult to easily verify any duplicates or substantiate the entry of new voters in the voter list. Historically, the voter list has remained an object of severe criticism in Pakistan and instances have emerged of

718   Need for electoral reforms in Pakistan (2010a, August 11), Daily Dawn. Retrieved 04.05.2014 from https://www.dawn.com/news/813373

719   Election Commission of Pakistan (2008). Delegation to Observe the Parliamentary Elections in the Islamic Republic of Pakistan (14 - 21 February 2008). Retrieved 04.05.2014 from http://www.epgencms.europarl.europa.eu/cmsdata/upload/b7862f31-0376-4959-a4b7-5b64e74a421c/Election_report_Pakistan_18_February_2008.pdf

electoral fraud and tempering of the voter list[720]. For instance, in 2007 ECP announced that for the then general elections the eligible voters were estimated to be 60 million, that is, 12 million less voters as compared to the election in 2002. Such errors in the voters list were either termed as technical inaccuracies whereas media and independent election observatories characterised it as a pre-poll rigging or election fraud that has been overlooked in the paper-based scenario[721].

The case of missing voters from the voters list in 2007 compiled by the ECP led the politician and two-time Prime Minister, Benazir Bhutto (1953-2007), to formally file a complaint in the Supreme Court of Pakistan (SCP)[722]. Upon the intervention of higher judiciary, ECP had to issue an additional/supplementary voters list with additional voters based on the election in 2002. ECP's explanation was based on the possession of old and new national identity cards. Even though the old ID cards were called obsolete in 2004, for 2007's voter list ECP had to rely on 2002's data as well. These discrepancies led to the idea of digitisation of voters list with a motivation to match it with (a) NADRA's citizen database and (b) to remove duplicate entries in the voters' list[723].

Election Commission of Pakistan (ECP), which operates as an autonomous government institution, has employed ICTs in various forms for facilitating the citizens during elections. The ECP first made use of ICTs to provide a G2C service in 2008. For the general elections in 2008, the ECP initiated a website through which citizens could verify their constituency as well as check the location of polling booths through ECP's digitised electoral rolls. The service was however criticised due to privacy concerns till the ECP called off their online database. However, post 2008, ECP decided to collaborate with NADRA in order to develop fresh electoral rolls. Considering the penetration of mobile phones in Pakistan, instead of relying on the web-based solution, ECP decided to make use of mobile phones through which voters could verify themselves with the help of simple SMS. Drawing on partnership with the NADRA, an SMS-based voter verification application was launched in March 2012 for over 85 million voters as

---

720   Kamran, T. (2002). Electoral Politics in Pakistan (1955-1969). Journal of Pakistan Vision, 10(1), 82-97

721   Free and Fair Elections Network. (2010). National Assembly Elections in Pakistan 1970-2008: A compendium of elections related facts and statistics. Retrieved 04.05.2014 from http://fafen.org/wp-content/uploads/2010/08/Compendium-National-Assembly-Elections-1970-2008-Pakistan.pdf; Also see European Union Election Observation Mission's detailed report on 2008's election: Election Observation and Democracy Support. (2008). Islamic Republic of Pakistan: Final Report - National and Provincial Assembly Elections. Retrieved 04.05.2014 from http://www.eods.eu/library/FR%20PAKISTAN%2016.04.2008_en.pdf

722   Haider, Z. (2007). Bhutto challenges Pakistani voter list in court. Retrieved 04.05.2014 from https://www.reuters.com/article/us-pakistan-bhutto-idUSISL18463420070626

723   Ansari, S., et al.: 2009

a part of the ECP's strategy to reinvent and redesign its functioning following a multi-stakeholder consultation process[724].

The entire process of organising electoral rolls shows how ECP as a public-sector organisation tried to undergo a major reform by heavily employing new technologies and revived its working by setting forth a multi-stakeholder agenda to form a five-year strategic plan[725]. The highlight in this context is a new voter verification service that was launched and its consequent effects on provisions for better validated, transparent and accountable elections. It is perhaps the first time that the ECP on behalf of the Government of Pakistan introduced an SMS service that enables citizens to verify and confirm themselves as registered voters for the general elections in 2013[726]. Alongside, the ECP also overhauled its strategic deliverables to foster digitisation of the electoral rolls, development of a more comprehensive website, translation of content in regional languages, a better Content Management System (CMS), adoption of Electronic Voting Machines (EVM), a mobile version of the website and online availability of all ECP information, results and notifications[727].

6.5.1. Interagency collaboration

Based on the discrepancies found in 2007's electoral rolls, the parliament passed a bill in 2011 to make it compulsory to possess a valid identity card for the voter registration as well as for casting vote[728]. This parliamentary bill paved way for ECP and NADRA to collaborate together to refine the electoral rolls[729]. In 2011, ECP and NADRA formally signed a contract for the revision of electoral rolls to update the Computerised Electoral Rolls System (CERS I) developed in 2006[730].

---

724 Election Commission of Pakistan (ECP). (2012, March 25). ECP inaugurates SMS facility tomorrow. [Press release]. Retrieved from http://www.ecp.gov.pk/ViewPressReleaseNotific.aspx?ID=1503

725 Demand to strengthen election commission (2010c, August 11), Daily Dawn. Retrieved 04.05.2014 from https://www.dawn.com/news/919073/demand-to-strengthen-election-commission

726 Election Commission of Pakistan. (2012b, February 28). ECP inaugurates SMS facility tomorrow [Press release]. Retrieved 04.05.2014 from https://www.ecp.gov.pk/PrintDocument.aspx?PressId=11053&type=Text

727 Election Commission of Pakistan (2012a). "Progress Report on Implementation of the Five Year Strategic Plan". Retrieved 24.03.2012 from http://www.ecp.gov.pk/Reports/SPStrategicPlanProgress.pdf.

728 Commonwealth Secretariat. (2013). Pakistan General Elections: Report of the Commonwealth Observer Mission. Retrieved 04.05.2014 from https://thecommonwealth.org/sites/default/files/project/documents/Pakistan%20General%20Elections%202013%20Commonwealth%20Observer%20Mission%20Report.pdf

729 ECP signs contract with NADRA for verification (2011, June 27), Aaj News. Retrieved 04.05.2014 from http://www.aaj.tv/english/national/ecp-signs-contract-with-nadra-for-verification/

730 ECP first attempted with the computerisation of electoral rolls in 2006 by involving a private software company KalSoft. See E-electoral rolls soon: ECP (2007a, January 28), Business Recorder. Retrieved 12.04.2014 from https://fp.brecorder.com/2007/01/20070128522704/; The

NADRA extended its services to the ECP by relying its citizen database. It appended and added supplementary information from their unique CNIC system for every voter[731]. The CNIC, as it has been discussed in an earlier section, carries a unique identification number with multiple biometric features, photograph and includes details of residence, contact information and family status. The idea of such an inter-sectoral and inter-agency collaboration was to map the citizen's database by ensuring unique identity within the system to prepare electoral rolls, which became the basis for CERS II.

The ECP established a project management unit comprised ECP and International Foundation for Electoral Systems (IFES) officials to steer and help monitor this task[732]. As a first step, ECP shared its existing electoral rolls data to NADRA for verification, for updating and augmenting the electoral rolls of 2006-07 on the basis of the CNIC database. Similarly, in April 2011, the Population Census Organization (PCO) which is responsible for conducting census in Pakistan was requested to collect the information of the head of family's data in a specially prepared form for preparing preliminary electoral rolls. This strategy assisted in revising the electoral areas on the basis of new census blocks as prescribed by the PCO[733].

NADRA used the data for assignment of new census block codes after verification and update it to existing electoral rolls with its CNIC data[734]. Resultantly, NADRA created a new database for the ECP by customising it along the lines of the newly defined electoral areas. The idea was to bring more clarity and precision of voter distribution. The ECP then conducted a country-wide door-to-door verification of these rolls from August through October 2011[735]. A pilot project was completed in 60 selected electoral areas in four districts that assisted them in identifying issues during a field exercise, which were corrected and accounted for while conducting the verification at the national level[736].

<hr>

results of the first attempt at the development of CERS, however, reportedly had serious discrepancies. See Electoral rolls are full of errors (2009, August 03), The News. Retrieved 04.05.2014 from https://www.thenews.com.pk/archive/print/665969-electoral-rolls-are-full-of-errors

731   Harnessing technology for electoral transparency (2011, October 17), The Nation. Retrieved 04.05.2014 from https://nation.com.pk/17-Oct-2011/harnessing-technology-for-electoral-transparency

732   International Foundation for Electoral Systems. (2013a). IFES Pakistan Fact Sheet: IFES in Pakistan. Retrieved 05.05.2014 from https://www.ifes.org/sites/default/files/ifes_in_pakistan.pdf

733   Nadra completes 80m draft electoral rolls (2011c, August 19), The News. Retrieved 04.05.2014 from https://www.thenews.com.pk/archive/print/616008-nadra-completes-80m-draft-electoral-rolls

734   ECP one step closer to 'one CNIC, one vote' goal (2011a, August 19), The News. Retrieved 04.05.2014 from https://tribune.com.pk/story/234741/ecp-one-step-closer-to-one-cnic-one-vote-goal/

735   2007 electoral rolls: 37.18 million unverified voters removed (2011, September 25), Business Recorder. Retrieved 04.05.2014 from https://fp.brecorder.com/2011/09/201109251234878/

736   Bokhari, H., et al.: 2012

Figure 36: Digitisation flowchart of electoral rolls by the ECP

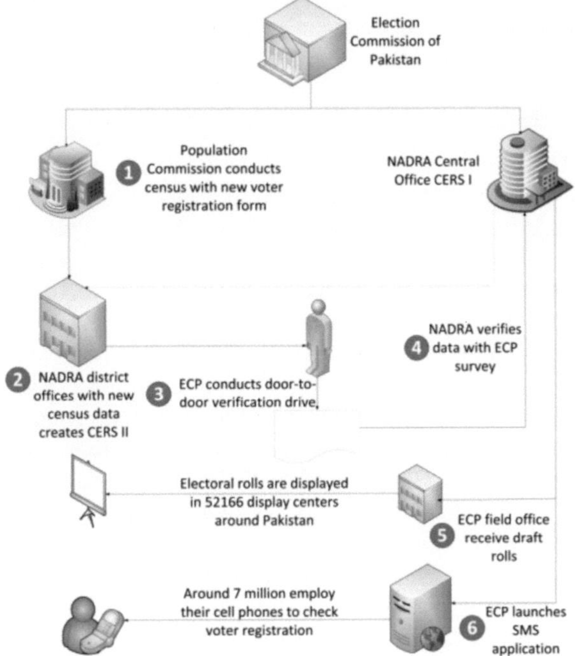

Source: Bokhari and Khan[737]

NADRA then entered the data changes on the basis of voters' data forms collected during door-to-door verification in light of guidelines, using double-blind data entry procedures[738]. Eventually, the preliminary electoral rolls were printed by NADRA and delivered to the ECP in 2012[739]. The ECP released the draft electoral rolls at 52,166 display centres across Pakistan for public display so that any corrections and objections could be rectified on the prescribed forms according to the legal framework[740]. In February 2012, ECP simultaneously launched both the paper-based electoral rolls and an SMS facility to verify citizens' voting details[741]. After making necessary revisions, the final electoral rolls were released on in July

---

737 Bokhari, H. and Khan, M. (2012). Digitisation of electoral rolls: analysis of a multi-agency e-government project in Pakistan. In Proceedings of the 6th International Conference on Theory and Practice of Electronic Governance. (pp. 158-165). Albany: ACM Press.

738 Gelb, A.: 2014

739 Preparing transparent and error-free electoral rolls (2012d, October 19), The Express Tribune. Retrieved 04.05.2014 from https://tribune.com.pk/story/453924/preparing-transparent-and-error-free-electoral-rolls/

740 International Foundation for Electoral Systems. (2013c). Electoral Rolls: Pakistan Factsheet. Retrieved 04.05.2014 from https://www.ifes.org/sites/default/files/the_electoral_rolls.pdf

741 Citizens use SMS to verify their votes (2012, February 29), Daily Dawn. Retrieved 04.05.2014 from https://www.dawn.com/news/699224

2012[742]. NADRA and ECP have planned to extend their cooperation by updating voter data simultaneously in the ECP database whenever voters would visit the NADRA offices for renewing or changing the personal details of their CNIC.

6.5.2.Improvements in electoral rolls from 2007 to 2013

According to CERS I, there were 81.21 million registered voters in Pakistan. ECP provided this data to NADRA for scanning and digitisation purposes. These brought to light some interesting discrepancies. The digital verification system at NADRA revealed that there were about 15 million entries which were identity-less, that is NADRA's citizen database did not have any record of such citizens. Similarly, there were about 9 million duplicate entries found along with the 13 million invalid voters (such as deceased) which were removed from the voter list[743]. Thus, due to NADRA's collaboration on this project, about 37 million of these were removed representing almost 45.5% of the total voters. These included multiple/duplicate entries and/or deceased voters. The door-to-door verification process assisted by the ECP to develop CERS II, increased transparency and district-wise voter distribution, equally assisted by data of family heads on 30.7 million forms during the Population Census 2011. The voter rolls were arranged according to details of city, region, streets and down to blocks to reduce time delays for the presiding election official during the voting process. For the first time, the Final Electoral Rolls (FER) 2013 included a picture of the voting citizen in order to ensure physical presence and verification while voting[744].

Around 11 million new voters were added as first-time voters. Similarly, the number of women voters amounted to 31 million[745]. NADRA made extensive use of its mobile registration vehicles to reach out to those geographically disconnected areas by relying on its local staff[746]. Similarly, to increase the female participation, apart from employing its mobile registration units, NADRA also established female only data procurement centres to encourage women to register for CNIC[747].

742   Daily Dawn (2012a). "ECP starts Rolls display, SMS service for voters verification. ". Retrieved 25.12.2012 from http://www.dawn.com/2012/02/29/ecp-starts-rolls-display-sms-service-for-voters-verification.html.

743   Nadra chief briefs EU team on steps to ensure fair polls (2013c, April 24), Daily Dawn. Retrieved 05.05.2014 from https://www.dawn.com/news/793880

744   Nadra, ECP working on electoral reforms (2013a, November 25), The News. Retrieved 04.05.2014 from https://www.thenews.com.pk/archive/amp/468792

745   Poll preparations: ECP sets up display centres for electoral rolls, February 20), The Express Tribune. Retrieved 04.05.2014 from https://tribune.com.pk/story/338951/poll-preparations-ecp-sets-up-display-centres-for-electoral-rolls/

746   Houreld, K. (2012). Pakistan ID cards remove ghost voters, target poor for aid. Retrieved 04.05.2014 from https://uk.reuters.com/article/uk-pakistan-identity/pakistan-id-cards-remove-ghost-voters-target-poor-for-aid-idUKBRE8AL0Y120121122

747   ECP directs NADRA to update FERs every month (2012c, September 04), Business Recorder. Retrieved 04.05.2014 from https://fp.brecorder.com/2012/09/201209041233785/

6.5.3.SMS Voter Verification Application

The ECP launched a widespread print and electronic media campaign to announce the electoral rolls. For geographically isolated and urban areas alike, the ECP produced voter awareness material and provided these to regional offices as well as to civil society and media organisations to help promote the electoral rolls and the SMS verification application[748]. The SMS application required a voter to text her/his CNIC number to a short code '8300'[749]. Irrespective of the Mobile Network Operator (MNO), the voter was charged, PKR 2 (USD 0.02) and a reply SMS in Urdu was received, showing the CNIC number, electoral area name, block code, tehsil/district and serial number[750]. In order to protect the privacy of the voter, the personal details such as name, parentage and home address were not included in the return SMS.

The block code in the return SMS is the assigned block or municipality number allotted after the census of 2011. The serial number replaces the old manual slip system that was previously issued to every voter with the serial number and name, at the time of registration. The idea of digitally replacing the paper slip system also intended to save cost and paper and to also reduce the chances of misplacing the slips[751]. With a digitisation of electoral rolls, voters only need to carry their CNIC and newly issued serial number to polling stations and cast their ballot. In order to ensure the reach of the voter rolls to all concerned, the physical display of electoral rolls continue side by side with the SMS service.

It would perhaps mark the first time in Pakistan that a number of governmental agencies namely, the ECP, NADRA and the Population Census Organisation have cooperated with one another over such a massive digitisation project for producing the electoral rolls[752]. The Pakistani voters' uptake on this service in both rural and urban areas appeared to be very receptive. Indicative of the demand shift in the new service delivery, voters did opt to use the SMS service. Within 17 days of the launch of the application, over 1.67 million people used this service

---

748   Election Commission of Pakistan (2015). ECP Newsletter: Issue I. Retrieved 04.05.2015 from https://www.ecp.gov.pk/Documents/Newsletter.pdf; Also see Election Commission of Pakistan. (2013a). Report on the General Elections - 2013. Retrieved 03.05.2014 from https://www.ecp.gov.pk/Documents/General%20Elections%202013%20report/Election%20Rep ort%202013%20Volume-I.pdf

749   Election Day: Expired CNIC holders would be allowed to vote (2013f, May 08), The Express Tribune. Retrieved 03.05.2014 from https://tribune.com.pk/story/546150/election-day-expired-cnic-holders-would-be-allowed-to-vote/

750   See Figure 37. Also see the section on Verification of voter in: 5-year roundup: By-polls scorecard puts PML-N ahead of PPP (2013g, March 22), The Express Tribune. Retrieved 04.05.2014 from https://tribune.com.pk/story/524655/5-year-roundup-by-polls-scorecard-puts-pml-n-ahead-of-ppp/

751   Khan, I. H. (2011). Electoral malpractices during the 2008 elections in Pakistan. Karachi: Oxford University Press

752   Nadra carries out data entry: ECP meeting reviews situation (2012a, February 10), Business Recorder. Retrieved 04.05.2014 from https://fp.brecorder.com/2012/02/201202101152327/

to verify their electoral details[753]. In this regard, most of the queries received were from the most populous province of Pakistan which amounted to 854,024 SMS. The table provides an overview of the initial response of the ECP's voter verification app.

Figure 37: Sample of return SMS with voter registration details

Source: Author

By end of the verification period in March 2012, the number of citizens who used this facility exceeded 7 million. These numbers rose to 16.3 million hits in April 2013[754]. In 2013, ECP announced that they received over 55 million voter verification requests[755]. The SMS service served as an auxiliary provision that enabled the registered voters in Pakistan within the verification period[756]. However, this level of collaboration can hint at the new areas of interagency collaborations, which actually could be considered as the third and fourth phase

---

753   Over 1.67m people used SMS service to verify voter list (2012b, March 02), Daily Dawn. Retrieved 04.05.2014 from https://www.dawn.com/news/699639
754   Kayani, D., & Rafi, F (2013). Pakistan Votes. Islamabad: Coffee Communications
755   55m text messages sent to ECP, companies make billions (2013b, May 15), The News. Retrieved 04.05.2014 from https://www.thenews.com.pk/archive/print/630483-55m-text-messages-sent-to-ecp,-companies-make-billions
756   SMS service was only one part of ECP and NADRA's collaboration on digitizing electoral rolls for citizen verification against the voters' list. In order to avoid fake voters, as has been the case in past, ECP/NADRA collaboration produced a voters list that also carried a photograph of every registered voter for the first time in Pakistan's electoral history. See section on General Elections 2013 (p. 12) in Election Commission of Pakistan (2013b). Second Five-Year Strategic Plan 2014-2018 (revised). Retrieved 04.05.2015 from https://www.ecp.gov.pk/Documents/sp%20revised/Strategic%20Plan%20Revised%20final.pdf

of horizontal and vertical integration, if it is compared with a eGovernment model (see chapter 2). Similarly, this SMS service also helped the ECP earn a name in the Guinness Book of World Records for having the world's largest voter verification SMS service for a voter density of 83.28 million[757]. The ECP used the same SMS service during the general elections on May 11, 2013 to announce and inform voters of their allocated polling stations[758].

Table 13: Initial Response to ECP-NADRA SMS App

Province	Total SMS received
Punjab	854,024
Sindh	421,247
Khyber Pakhtunkhwa	244,785
Baluchistan	95,735
FATA	18,945
Islamabad	44,799
Total SMS Received	*1,679,535*

Source: Own illustration based on Dawn[759]

The architecture behind the SMS application is a G2C informational exchange initiated by texting a short code. Similarly, on the front end, it uses text-based communication for any GSM cellular network in a local language and numbers. The middleware used to send SMS is a NADRA SMS Gateway. All the mobile network operators reserved the short code 8300 for the ECP[760]. Once a network operator receives an SMS, it passes the message to the NADRA SMS gateway. In return, if the message is appropriate, that is, if it contains 13 digits of CNIC, the voters' information is returned.

6.6. Concluding Remarks

In comparison to EGD's initiatives such as first official web portal and other automation projects (chapter 4) and NADRA's G2C projects such as kiosk and eSahulat (chapter 5), this chapter has tried to present the potential that various other ICTs can have in electronic service delivery. The use of mobile phone as a medium of G2C service provider is changing the eGovernment landscape in

---

757   ECP 'SMS voters' verification service may break World Record. (2012b, March 25), The Express Tribune. Retrieved https://tribune.com.pk/story/350452/ecp-sms-voters-verification-service-may-break-world-record/

758   International Foundation for Electoral Systems (2013b). Using a Gender Lens to Examine Pakistan's Historic Election. Retrieved 04.05.2014 from https://www.ifes.org/news/using-gender-lens-examine-pakistans-historic-election

759   ibid.

760   Election Commission of Pakistan: 2015

Pakistan. After NADRA's kiosk machine, we see that there has been a continuous search for and reliance on alternative service delivery channels. There appears to be a less focus on G2C services provided via computer/web mechanisms. The idea appears to focus more on the communication medium with which most public can be reached. In comparison to computer/web, mobile phones offer a better potential with wider outreach. Similarly, the choice of front-end technology that is, SMS is available as a basic function in every mobile phone which makes the G2C service accessible for every mobile phone user. In both Mobilink and ECP's SMS app, the mobile phone does not need to have any particular specification as a prerequisite. Neither do any of these two applications discussed in this study require the use of smartphone or internet.

Such level of innovation in G2C service scenario has also been possible due to the supporting regulatory frameworks. In this regard, State Bank of Pakistan seems to have played an active role in initiating a framework that is not only applicable and supportive of the banks but also to the mobile network operators who are regulated under by PTA. In the case of Mobilink, we saw how Pakistan is moving towards digital financing and experimenting with its own forms of mobile banking solutions. The case studies presented in this chapter have also tried to sum up the level of digitisation, intra and interagency collaboration to develop G2C services. Based on the case studies presented in this research, the question arises as to what kind of eGovernment model is emerging in Pakistan. How similar (or dissimilar) is it from the generic eGovernment models that have been perceived in other countries or the ones discussed in Chapter 2. Similarly, and more importantly, based on the case studies presented in this research, which organisation is mandated with eGovernment development in Pakistan? Is it EGD which was specially launched to advise on eGovernment adoption and development or is it NADRA which developed a number of G2C projects? The next chapter discusses the role of these institutions as well as emerging eGovernment approaches in Pakistan.

Chapter 7

Discussion

In this research, five different case studies have been presented to explain the role of ICTs in Pakistan and their use in the electronic public service delivery. These include the eGovernment initiatives by the Electronic Government Directorate (EGD), Ministry of Information Technology, electronic public services developed by the National Database Registration Authority (NADRA), Ministry of Interior, Election Commission of Pakistan (ECP)'s joint venture with NADRA, and private telecom operator Mobilink's Branchless Banking services. These case studies have chronologically guided this research to explore the foundations and growth of G2C eServices in Pakistan. They have also tried to explore the area of digital communication channels adopted by the eGovernment service providers. The present chapter discusses all the five case studies as a whole and tries to understand and present the model of eGovernment that has been evolved in Pakistan.

EGD was launched in 2002 as a premier institution by the Government of Pakistan to guide all the ministries and divisions at the federal and provincial level in the area of eGovernment. Its role was not only of a facilitator but also of an initiator to implement eGovernment projects at national and provincial level. This section reviews EGD's formation and performance over the years and tries to analyse and place the relevance of Pakistan's first eGovernment strategy prepared by the EGD. Similarly, this section also discusses Pakistan's first eGovernment strategy.

7.1. EGD's E-Government Strategy and 5-Year Plan

Since the beginning of EGD in 2002, EGD seems to have taken necessary steps that were mandated to it. If the government of Pakistan's notification of EGD's formation is reviewed, EGD was delegated five tasks. It was expected from the EGD (a) to implement eGovernment programme (b) provide technical advice and guidelines for implementation of eGovernment (c) plan and prepare eGovernment projects (d) provide standards for software and infrastructure in the field of eGovernment and (e) to undertake any other assignment that government may delegate the EGD[761]. The half a page of notification did not carry any other details, such as legal, financial or technical except that the EGD would function under the Ministry of IT. In October 2005 EGD conducted a briefing for focus persons of

---

761   Electronic Government Directorate (2002). Notification: No.3-16/2001/Coord. (Establishment of Electronic Government Directorate). Retrieved 10.10.2007 from http://pakistan.gov.pk/e-government-directorate/about/notification1.pdf

different federal ministries[762]. The briefing included presentation of EGD's mandate, its various projects, training programme and the eGovernment strategy and 5-year plan for the government of Pakistan. E-Government Strategy and 5-Year Plan for the Federal Government prepared by the EGD has so far been the singular strategy paper that has laid out an eGovernment plan for Pakistan. It is the sole official document that reflected on how EGD, and thus the Ministry of IT in Pakistan, defined eGovernment and how it wanted to pursue eGovernment implementation. It defined the goals and strategy as well as laid out the key success factors and essential measures that EGD perceived necessary for eGovernment implementation in Pakistan.

Table 14 depicts the salient features of Pakistan's eGovernment 5-year strategy. EGD defined the goals of eGovernment for Pakistan as to increase government's efficiency and effectiveness by employing the new ICTs to enhance the delivery of public services. EGD's eGovernment strategy was based on five key elements-providing basic IT infrastructure to all the ministries, establishment of common software applications, agency-specific eServices, formation of eGovernment standards and creating an enabling environment for all agencies involved who would then take ownership of their respective eGovernment programmes. EGD also constituted National eGovernment Council (NEGC) under the chairpersonship of the prime minister[763]. NEGC was expected to meet every quarter to review the progress on eGovernment implementation and to examine the legislations, rules and regulations.

As an essential measure for the implementation of eGovernment in Pakistan, EGD's strategy paper called for setting up a committee within EGD that would submit recommendation to the government of Pakistan twice a year. In addition, another essential measure was to ask every federal ministry to identify a focal person (a bureaucrat of joint secretary's level) who would represent her/his ministry while collaborating with EGD. The focal person would also work within her/his own ministry to identify three most important eServices for their ministries/departments in collaboration with EGD. One of the key aspects of interest in the eGovernment strategy paper is of EGD's autonomy. To carry out country's eGovernment agenda and establish strategic framework, EGD deemed its organisational and financial autonomy as an essential measure for the implementation of eGovernment in Pakistan.

In section 3.2 (of EGD's strategy paper), *Phases of E-Services to citizens*, EGD divides its eGovernment approach into four phases namely (a) informational (b) interactive (c) transactions and (d) collaborative[764]. These phases seem to have

---

762   Electronic Government Directorate (2005b). Briefing For Focal Persons About E-Government Strategy and 5-Year Plan. Retrieved 28.10.2009 from http://pakistan.gov.pk/e-government-directorate/presentations/briefing_for_focal%20Persons13102005Final.pdf

763   Kayani, M. B., Haq, M., Perwez, M. R. and Humayun, H. (2011). Analyzing barriers in e-government implementation in Pakistan. International Journal for Infonomics, 4(3), 494-500

764   Electronic Government Directorate: 2005a

been derived from existing phases of eGovernment that have been globally adopted for instance by international organisations such as UN, World Bank or World Economic Forum or as also mentioned in several academic studies. One of the most problematic aspects about these phases is EGD's apparently simplified view about them. The Informational phase was defined as below:

"Informational: This is the first phase and includes the provision of information alone. The quality, usability and currency of the content determine the value of this phase of e-government. This is the least complex of all the phases."[765]

Table 14: Salient features of Pakistan's 5 Year eGovernment strategy

Salient Features	Description
Goals	. Increase government efficiency and effectiveness . Increase transparency and accountability . Enhance delivery of public service
Key Success Factors	. Top-level sponsorship . Concerted effort of all the ministries, ownership and will to change . Capacity to deliver and absorb . Country wide coherent policy & strategic framework for eGovernment projects . Timely availability of funds . Strong relationships between public and private sector . Enabling amendments in laws, rules and procedures . Ability of citizens to access and use public service offered through E-Government.
Strategy	. Basic Infrastructure . Common Applications . Agency-Specific Applications & e-Services . Standards . Enabling Environment
Essential Measures	. Approval of eGovernment strategy by the federal cabinet . Set up committee to submit recommendations biannually . Ensure integration and interoperability by making compliance to EGD Framework . Compulsory IT Training federal government employees . Ministries' Divisions: Identify focal point at the level of Joint Secretary for eGovernment Identify 3 high impact G2C eServices in collaboration with the EGD Take ownership to accelerate the implementation within 24 months . Easy hiring and retention of contractual IT professionals . Organisational and financial autonomy for EGD

Source: E-Government Strategy and 5-Year Plan for the Federal Government[766]

---

765   ibid.
766   ibid.

The informational phase pertains to the availability of government's ministries information and regulation on official government website. This strategy paper released in May 2005 seems to have ignored the teledensity in Pakistan. Between 2000 and 2005, the total number of fixed telephone subscribers increased from 3 to 5 persons per 100 inhabitants. Similarly, in 2005 there were only 6 internet users per 100 inhabitants[767]. These figures vividly explain the teledensity ratio in Pakistan and the sluggish growth rate of internet users. EGD's eGovernment strategy seems to take these figures for granted and considers the informational phase as "the least complex of all the phases"[768]. It might have been the least complex phase for EGD to implement but the question that EGD had to answer was how it planned roll outs of phases of eGovernment services in a situation where less than 6% of the targeted Pakistani population had internet access.

Moving to the second phase of EGD's eGovernment strategy which is interactive.

"Interactive: In this phase, E-Government provides some degree of online interaction. For instance, citizens can enter complaints or job applications online. This phase does not include secure transactions such as financial or other transactions that require a high degree of authorization and audit."[769]

The problematic area observed here and through the EGD's eGovernment strategy is the vague use of terms like computer, internet and online. For instance, here EGD refers to "some degree of online interaction". However, it does not explain how exactly it means or defines the word online. With the simplistic use of the term 'online', it is assumed that the interaction is through a website, irrespective of the means and modes to be online with. Both of the first two phases show EGD's understanding or perhaps the misunderstanding of ground realities of Pakistan's public, its digital capacity, IT literacy, linguistic abilities and the capacity to understand content in a new or foreign context[770]. EGD's portal for the Government of Pakistan is entirely in English. The portal does not offer any facility for local content, making eGovernment for Pakistan as highly exclusive and centralised affair with no understanding for the majority of the adult public, which has little or no comprehension of English.

Key Success Factors, additional feature in EGD's strategy paper, is another aspect in eGovernment strategy of Pakistan that seems to point to EGD's erroneous judgement towards understanding eGovernment as a strategy. For instance, in order to implement successful eGovernment in Pakistan, EGD considered it

---

767   Pakistan Telecommunication Authority (2010b). Telecom Indicators. Retrieved 10.11.2010 from https://www.pta.gov.pk/en/telecom-indicators/2

768   Electronic Government Directorate: 2005a, p.8

769   ibid., p.8

770   Ameen, K. and Gorman, G. (2009). Information and digital literacy: a stumbling block to development? A Pakistan perspective. Library Management, 30(1/2), 99-112

'essential' to adopt the best practices from other countries. In this regard, EGD's paper takes a 'particular note' of few countries that have successfully implemented eGovernment programmes[771]. These include the USA, Singapore and South Korea. EGD's focus appears to be based on key success factors of the above-mentioned countries. Basically, EGD found these countries as a workable example to adopt eGovernment initiatives from these countries by adjusting and scaling them to Pakistan's eGovernment aspirations. From the choice of words and terminologies used in EGD's paper (for instance eGovernment has also been termed as eBusiness) and from the choice of countries (USA, Singapore, South Korea) to emulate, it appears as if EGD's understanding of eGovernment is limited to only bringing the government data on websites[772]. It seems to ignore other vital factors such as historical, cultural, political, linguistic, and of course technological aspects, that have a very significant role to play in the development and adoption of eGovernment projects[773].

Technologically speaking, Singapore, South Korea and the USA have not only been technologically well advanced but they are also the manufacturers and producers of communication technologies and devices[774]. If the digital divide among Pakistan, South Korea, Singapore and the USA is compared, as we see in the Figure 38 internet subscribers in Pakistan have not been able to cross 10 (users) per 100 inhabitants. In case of Singapore, USA and South Korea, their internet subscribers are well above 30 users per 100 inhabitants. From 2000 to 2005, the growth of internet users in all the three countries has rapidly increased. By 2005 the ratio in none of the countries was below 60 users per 100 inhabitants. In case of Pakistan this growth has only increased from 1 user to 6 users per 100 inhabitants[775]. The market for G2C services available via web/internet platform was ripe for these three countries. Web/internet as a medium of communication could serve as a primary, if not the only, mode of G2C communication[776]. EGD has also specified the list of challenges that these three countries faced in their respective eGovernment implementation. It is interesting to note that none of them mentioned digital outreach as a hurdle, which as the statistics indicate has been one of the important hurdles for Pakistan[777].

---

771   Electronic Government Directorate: 2005a, p.17

772   Alshawi, S. and Alalwany, H. (2009). E- government evaluation: Citizen's perspective in developing countries. Information technology for development, 15(3), 193-208

773   Aladwani, A. M. (2013). A contingency model of citizens' attitudes toward e–government use. Electronic Government, An International Journal, 10(1), 68-85; Stemberger, M. I. and Jaklic, J. (2007). Towards E-government by business process change—A methodology for public sector. International Journal of Information Management, 27(4), 221-232

774   Chung, Y.-I. (2007). South Korea in the fast lane: Economic development and capital formation. Oxford: Oxford University Press

775   International Telecommunication Union: 2010

776   Schware, R. and Deane, A. (2003). Deploying e-government programs: The strategic importance of "I" before "E". info, 5(4), 10-19; Gibbs, J., Kraemer, K. L. and Dedrick, J. (2003). Environment and policy factors shaping global e-commerce diffusion: A cross-country comparison. The information society, 19(1), 5-18

777   Electronic Government Directorate: 2005a, Appendix III

Similarly, EGD's eGovernment strategy paper also seemed to have overlooked the fact that basic G2C eServices are generally announced and outlined in any eGovernment implementation plan[778]. For instance, the strategy paper does not describe the basic G2C services that it wishes to develop/offer over the course of five years (considering that the strategy was meant to be a 5-year plan). What this strategy paper does offer is the identification of three high priority eServices that every ministry should identify in collaboration with the EGD (see Table 14, essential measures). In order to help identify basic eServices for different ministries, EGD hired the services of Singapore-based consultancy firm NCS in 2006 to conduct a survey for the 20 most important eServices for Pakistan[779]. The consultancy study was conducted in two different phases. The survey was a very detailed plan of action to gauge the government of Pakistan's readiness for eGovernment implementation.

Figure 38: Digital Divide among Pakistan, USA, Singapore & South Korea

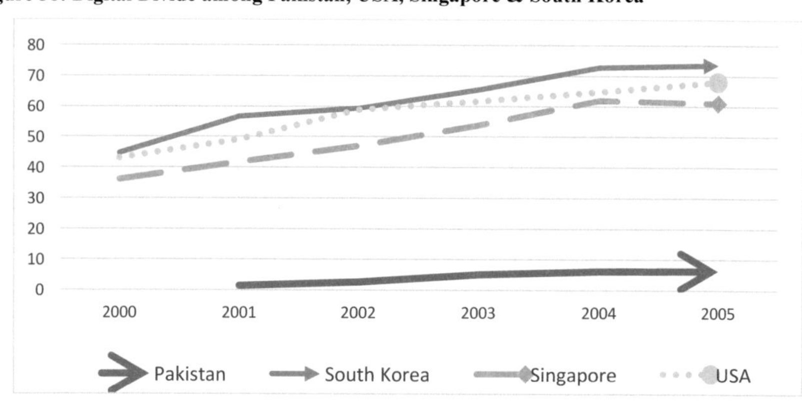

Source: Own illustration based on ITU[780]

The questionnaires developed by the consultancy firm expected each and every ministry and their attached departments to provide their detailed functioning[781]. It

---

778 infoDev's handbook released in 2002 in this regard provides a good overview for developing countries pursuing eGovernment implementation. See infoDev and Center for Democracy & Technology: 2002

779 NCS. (2009). e-Government Consulting – Shaping your e-Government Blueprint: Mapping the Future. Retrieved 09.07.2011 from https://www.ncs.com.sg/documents/20184/73726/eGov+Consulting+%28062009%29.pdf/941b f5b9-c760-4054-a6f2-995aae69ae0a

780 International Telecommunication Union. (2013a). ICT Statistics. Retrieved 20.05.2013 from http://www.itu.int/en/ITU-D/Statistics/Pages/stat/default.aspx

781 Electronic Government Directorate (2006a). Consultancy Study on 20 Most Important Services Assessment for Pakistan Government Services Inventory Questionnaire (Phase 1). Retrieved 09.08.2008 from http://pakistan.gov.pk/e-government-directorate/information/media/GSI_QUESTIONNAIRE_Phase_1_210306.pdf

asked the ministries to particularly identify the areas where the ministry and its attached department get in touch with the citizen and the kind of service they offer. In its second phase, the consultancy study also intended to ask details about ministries' IT budget, total staff as well as the number of days needed to attend to citizen services in a computer-less environment[782]. From questionnaires handed out to various ministries, it appears that EGD continuously tried to follow gradual and holistic approaches towards all ministries in order to document their working, outreach and service orientation just to hone the selection of eServices.

However, EGD's eGovernment strategy paper would have been better grounded if, for instance, the choice of countries to compare with for the critical success factors had been different. For a comparison, the USA, Singapore and South Korea would have worked well, had there also been an additional set of countries with ground realities similar to that of Pakistan. As elaborated in Chapter 2, eGovernment in developing countries has rapidly attained a special focus since 2000. A number of case studies, research articles, best practices, guidelines and toolkits have been regularly produced not just by the academia but also by the international organisations such as ITU, UN and World Bank[783]. Some of the important commentaries for developing countries include Heeks, Ndou, Basu, Bhatnagar, Wagner et al., De', Avgerou and Madaon or the eGovernment toolkits or guidelines developed by the UN or infoDev in 2001 and 2002 (see Chapter 2). From the review of EGD's Pakistan eGovernment strategy paper, the EGD doesn't seem to draw any inspiration from this useful and burgeoning eGovernment repository on developing countries. These studies and toolkits serve as a blueprint or a starting point to think about, plan, implement and practice eGovernment services[784]. Some of the core and common features that can be found in these studies include infrastructural development, digital divide, IT literacy, accessibility, eReadiness, benchmarking, local content and local context(s) and channels of communications- all which would have been pertinent to EGD's ambit and attempts at eServices.

Since 2001, United Nations' Division for Public Economics and Public Administration has been conducting international eGovernment surveys which do not just include the eGovernment rankings but also cover various issues, barriers

---

782   Electronic Government Directorate (2006b). Consultancy Study on 20 Most Important Services Assessment for Pakistan Government Services Inventory Questionnaire (Phase 2). Retrieved 09.08.2008 from http://pakistan.gov.pk/e-government-directorate/information/media/GSI_QUESTIONNAIRE_Phase_1_210306.pdf

783   In this context, a very useful handbook for the developing countries is prepared by Pacific Council On International Policy, see Pacific Council On International Policy. (2002). Roadmap for E-government in the Developing World: 10 Questions E-Government Leaders Should Ask Themselves (The Working Group on E-Government in the Developing World). Retrieved 04.05.2014 from http://unpan1.un.org/intradoc/groups/public/documents/caricad/unpan008526.pdf

784   See for example; Organisation for Economic Co-operation and Development: 2003

and the best of practices in eGovernment[785]. Since the first study, UN's report has developed a methodology of measuring a country's eGovernment performance based not only on aspirations and plans but also on its telecommunication indicators, web presence and human capital measures[786]. Over time these surveys have matured and become more inclusive of countries with lower levels of digital participation and infrastructures. Since 2008, UN surveys have acknowledged the problems of eGovernment services only accessible via web/internet and recognised the presence of and need for investment in alternative channels of electronic public service delivery[787]. If Pakistan's performance on UN's eGovernment eReadiness scale from 2001 to 2005 is reviewed, we see that the countries originally cited as benchmarks by EGD were on top of eReadiness ranking whereas Pakistan's position mostly fluctuated below 100.

Table 15: eReadiness: Pakistan, USA, Singapore, South Korea & Singapore

Countries	2001	2003	2004	2005
USA	1	1	1	1
Singapore	4	12	8	7
South Korea	14	13	5	5
Pakistan	94	137	122	136

Source: Own illustration based on UN Knowledgebase[788]

EGD's strategy paper does not seem to have taken these global surveys into consideration, nor has it highlighted the core areas of digital deficiencies that put Pakistan on 136th position in global ranking. In fact, eReadiness as a terminology has not even made it to the strategy paper along with benchmarking and local context. Local content is the only passing reference that finds way into the strategy paper. The first web portal and its subsequent versions developed for the government of Pakistan did not support local language script and were available only in English. With these figures (as in Table 15 and Figure 38) it should have perhaps been clear to EGD that its approach particularly with regard to the service

---

785   The survey is conducted biennially and helps assess the eGovernment development status of all 193 United Nations Member States, see United Nations. (2012a). Preparatory Process of the United Nations E-Government Survey. Retrieved 04.05.2014 from https://publicadministration.un.org/egovkb/en-us/About/-eGov-Surveys

786   United Nations. (2001). Benchmarking E-government: A Global Perspective (Assessing the Progress of the UN Member States). Retrieved 04.05.2014 from https://publicadministration.un.org/egovkb/Portals/egovkb/Documents/un/English.pdf

787   United Nations: 2008

788   United Nations. (2014). UN E-Government Knowledgebase. Retrieved 04.05.2014 from https://publicadministration.un.org/egovkb/en-us/About

delivery methods, may have to be diametrically different from the countries chosen as best practices references. However, the strategy paper keeps stressing upon online services via web/internet mechanism.

It is interesting to note that in March 2005 the Asian Development Bank Institute (ADBI) and the United Nations jointly organised a workshop on designing eGovernment in Tokyo with focus on developing countries[789]. The workshop invited the country representatives from Asian countries including Pakistan, which was represented by Joint Secretary Asad Sibtain from the Ministry of Information Technology (MoIT). Sibtain's proposal, made during the workshop, highlighted the fact of conducting focus group discussion that could help identify requirements for the implementation of eGovernment. Among the four questions he posed, the first two were related to the eServices, that is, what services the communities wish to see and what the delivery channels communities wish to use[790]. EGD, which has functioned under the ambit of the MoIT, did not seem to reflect such considerations about the service delivery channels in its paper, but senior officials at its parent institution (MoIT) were aware of the hurdles. From Mr. Sibtain's proposals that were made at the workshop, it appears that within the same ministry there already existed alternative views on the use of web/internet for the eGovernment service delivery.

7.2. Rise and Fall of Electronic Government Directorate (EGD)

Apart from drawbacks in the eGovernment strategy paper, there have also been issues that unfortunately reflected poorly on EGD's profile. The formation of EGD took place by abolishing the Information Technology Commission that was established by the Finance Division on August 9, 1997[791]. The EGD was not autonomous government entity rather the federal cabinet approved EGD as a special cell within the Ministry of Information Technology[792]. EGD's tasks were not only to develop eGovernment strategy for Pakistan but also to provide guidelines for ICT developments, installation of technical equipment at the ministries as well as to provide IT trainings to government employees. Majority of projects initiated by EGD were related to computerisation of ministries and government departments (see chapter 4).

It is interesting to note that since 1971 there already existed an institution called Pakistan Computer Bureau (PCB), whose primary responsibility was to advise the government and related public-sector organisations on computerisation[793]. PCB undertook projects that involved systems analysis, software development and

---

789   United Nations: 2005
790   ibid., p. 154
791   Electronic Government Directorate: 2002
792   Govt steps up efforts to introduce e-govt (2013c, September 02), The Nation. Retrieved 23.09.2013 from https://nation.com.pk/02-Sep-2013/govt-steps-up-efforts-to-introduce-e-govt
793   Pakistan Computer Bureau (2009). Functions of and Trainings provided by PCB Retrieved 24.10.2009 from http://www.pcb.gov.pk/Trainings.asp

implementation of computer based systems[794]. In addition to that PCB also provided IT trainings to officers and staff in public-sector organisations. PCB was initiated as an attached department of the Cabinet Division in 1971 but later it was transferred to Information Technology and Telecommunication Division, Ministry of Science and Technology in 1999[795]. After the establishment of the Ministry of Information Technology (MoIT) in 2000, PCB was shifted to operate under MoIT[796].

Keeping in view the roles and tasks of these institutions, ideally, all projects related to computerisation of ministries, training of the government staff and software related tasks were likely to be the priority of the PCB, whereas EGD could have assumed the role of policy making and eGovernment standardisation. However, there appears to be an overlapping of tasks and mandates in the case of EGD and PCB. Nevertheless, one of the important tasks that EGD was assigned to was the introduction and development of eGovernment services[797]. For that matter EGD initiated several automation projects with the foremost being the launch of Government of Pakistan's online portal. The portal only included basic information about the government ministries and there was no G2C service provided via the portal. While EGD continued to computerise ministries and departments, the fact remains that it could not develop any national level G2C service in the first eight years of its formation[798].

On the other hand, EGD's integrity as a fair and transparent institution also came under question[799]. In 2010, reports of data violations and theft in the EGD-managed projects started to emerge. EGD which took over digitisation of Prime Minister Secretariat and installation of Management Information System (MIS) thereof came under scrutiny when an EGD official was reportedly found copying sensitive government data on to his private laptop[800]. Even though the official was caught by the secretariat staff, it does highlight the fact how data breaches can happen on apparently secure computer systems. Over the period of time EGD continued to surround itself in further controversies and its integrity, output and performance were continuously probed. It started to become apparent that projects

794   Senate of Pakistan (2013). Report of The Senate Standing Committee on Cabinet Secretariat and Capital Administration & Development. Retrieved 04.05.2014 from http://www.senate.gov.pk/uploads/documents/1428589865_112.pdf
795   CCoR approves IT entities merger (2014, July 19), The Nation. Retrieved 04.05.2015 from https://nation.com.pk/19-Jul-2014/ccor-approves-it-entities-merger
796   Shafique, F. and Mahmood, K. (2008). Indicators of the emerging information society in Pakistan. Information Development, 24(1), 66-78
797   Ali, M. and Mujahid, N. (2015). Electronic Government Re-Inventing Governance: A Case Study of Pakistan. Public Policy and Administration Research, 5(2), 1-8
798   Iqbal, A. J. and Javed, H. (2013). Public Sector Project Management: Perception of Risks in IT. Journal of Strategy and Performance Management, 1(1), 24-32
799   Computerising land records: Contract awarded in violation of rules (2013b, October 06), The Express Tribune. Retrieved 04.05.2014 from https://tribune.com.pk/story/614128/computerising-land-records-contract-awarded-in-violation-of-rules/
800   Classified data stolen from PM`s secretariat (2010b, June 09), Daily Dawn. Retrieved 04.05.2014 from https://www.dawn.com/news/857783/classified-data-stolen-from-pm-s-secretariat

that had received millions in funding were not completed even after five year delays. Reports of corruption and embezzlement by the EGD were raised even in National Assembly[801]. In November 2012, National Assembly's standing committee on Information Technology took notice of embezzlement of PKR 1 billion that were paid by the EGD to its contractors[802]. Eventually the matter was also reported to the National Accountability Bureau (NAB) for the non-transparency cited in EGD projects[803]. Considering the administrative and financial irregularities, and failure of projects and the lack of support from the client ministries, it was decided that the EGD's mandate shall be siphoned off its present form and it shall be merged with the PCB[804].

The Ministry of Information Technology under which both the PCB and EGD had functioned since 1999 and 2002 respectively was prompted to restructure both the organisations. In July 2014, the government finally approved the merger of PCB and EGD by forming a new institution namely National Information Technology Board (NITB)[805]. The NITB's mandate primarily focusses on the provision of IT consultancy services to all government ministries, web development and hosting as well as advisory services[806]. NITB is also mandated to function under the Ministry of Information Technology and is led by an executive director[807]. EGD which was launched with high expectations in 2002 and which lauded and heralded for initiating Pakistan's first official eGovernment strategy was shut after a dilapidated run of 12 years. The website domain names which were registered by the EGD, that is, www.e-government.gov.pk have also been abandoned and never renewed by the Ministry of IT. EGD's projects and two eGovernment strategy papers that have been referred to in this study are also no longer available online[808].

801   EGD projects: National Assembly panel detects Rs 1 billion embezzlement (2012b, November 08), Business Recorder. Retrieved 04.05.2014 from https://fp.brecorder.com/2012/11/201211081256384

802   IT embezzlement: 13 E-services projects referred to FIA (2014e, March 21), The Express Tribune. Retrieved 04.05.2014 from https://tribune.com.pk/story/685362/it-embezzlement-13-e-services-projects-referred-to-fia/

803   Operational problems: Affiliate dept draining IT ministry, says Anusha (2014a, January 22), The Express Tribune. Retrieved 04.05.2014 from https://tribune.com.pk/story/661785/operational-problems-affiliate-dept-draining-it-ministry-says-anusha/

804   "National IT Board" being set up: PCB-EGD merger approved (2014a, July 19), Business Recorder. Retrieved https://fp.brecorder.com/2014/07/201407191204134

805   Technology merger: Govt merges bloated tech departments to form National IT Board (2014d, July 18), The Express Tribune. Retrieved 04.05.2015 from https://tribune.com.pk/story/737597/technology-merger-govt-merges-bloated-tech-departments-to-form-national-it-board/

806   National Information Technology Board (2014). Advisory Services. Retrieved 04.05.2014 from http://nitb.gov.pk/newnitb/index.php?section=functions&page=advisoryservice

807   Governance: E-office implementation reviewed (2014b, November 08), The Express Tribune. Retrieved 05.12.2014 from https://tribune.com.pk/story/787411/governance-e-office-implementation-reviewed/

808   Although the Ministry of Information Technology, Government of Pakistan has taken down the domain name www.e-government.gov.pk, however, the policy documents, EGD reports, strategy

Several factors seem to be at play in the rise and fall of EGD. Apart from the usual red tape and the administrative discrepancies in the EGD, there also appears to be two important factors that have resulted in the failure of EGD (projects). EGD was approved by the federal cabinet in 2002. This time period is of prime importance as there were other and similar steps also being taken for growth and development of eGovernment and G2C projects. Firstly, NADRA had already been launched in 2000 under the Ministry of Interior with a mandate to digitise the entire citizen database. NADRA did not stop its mandate with the computer identity cards instead it went on to take further steps within the field of eGovernment[809]. These steps include machine readable passports, drivers' licences and G2C financial solutions such as kiosk and eSahulat programmes. It is interesting to note that although EGD seemed to be the flagbearer of eGovernment standards, establishment of eGovernment strategy and a principal entity to guide and provide technical assistance to all ministries and divisions, however, there have hardly been any instances where EGD proposed any eServices for NADRA or consulted and conferred with them.

NADRA on its own part has continued to position itself as one of the pioneers in digital identity projects in Pakistan with its semi-autonomous structure[810]. In terms of digital capacity, NADRA relies on its own set of technical teams, equipment, and the ownership of its projects, products and hardware remains with NADRA[811]. Both the entities have not appeared to have cooperated or shown any interest towards any joint initiative. Ministries, divisions and other government departments mentioned from 2003 through 2010 on EGD's projects pages does not mention or highlight any cooperation or collaboration with NADRA either (see Chapter 4). NADRA, on the other hand, has collaborated, as shown in this study with other public and private entities in Pakistan such as Election Commission of Pakistan and mobile network operators[812]. Similarly, mobile telecom market has also seen an exponential growth and several telecom operators also initiated their own electronic public services. These parallel initiatives developed by digitally, politically and financially more astute entities (NADRA and telecom sector for instance), seemed to have made EGD's efforts redundant particularly in the field of G2C services.

Nevertheless, despite several deficiencies and drawbacks, E-Government Strategy and 5-Year Plan for the Federal Government will be considered as one of the first comprehensive strategy on eGovernment that was promulgated by Pakistan. After its merger with PCB and formation of the constituent NITB, the

papers, EGD's own presentation files that were available on EGD's website which were collected as a part of this research are available with the author.

809 Nadra set to introduce online verification system (2003h, August 11), Daily Dawn. Retrieved 04.05.2014 from https://www.dawn.com/news/134572

810 Federal Ombudsman Secretariat: 2015

811 Interview with Mr. Usman Mobin, February 2009.

812 Nadra ready to verify thumb impressions (2013c, May 16), The News. Retrieved 04.05.2014 from https://www.thenews.com.pk/archive/print/630507-nadra-ready-to-verify-thumb-impressions

eGovernment strategy paper along with web domains of the of EGD, have been taken offline. Although MoIT and NITB keep referring to the strategy document, neither MoIT nor NITB has kept a copy of this document on their websites[813].

7.3. NADRA's Kiosk project and its possible decline

Parallel to EGD, NADRA had also been developing its own set of eGovernment application, the foremost being the Computerised National Identity Card (CNIC). NADRA's kiosk project presented in this study may be regarded as one of the most innovative technological device and electronic public service delivery mechanisms developed in Pakistan[814]. Its navigation system is based on ATM-styled machine with touch screen and it came with biometric security features. It offers unassisted and round-the-clock support which makes it convenient for citizens to operate at a time of their convenience. Apart from ATM machines, it is perhaps the first time in Pakistan that an unassisted kiosk machine for electronic public service delivery has been launched[815].

Considering its unique features and a direct access to G2C services, the adoption of Kiosk machine could be expected to reach higher usage. The data received from NADRA also prove this fact. However, statistics also revealed that the number of transactions initially started to rise but later declined drastically (see chapter 5). As evident from numbers, the frequency of use of the kiosks started dropping soon after NADRA's launch of another bill payment mechanism called eSahulat. Although the kiosk's numbers started to increase from their previous position in 2005 to 2006, yet the drop of Kiosk usage is dramatic (see Figure 39). Given the ease that it offered in terms of avoiding long queues, quick and efficient payment, secure transfer and dispensing of funds and personalised service in terms of access to account details and past records- the Kiosk should have ideally seen a steady growth. This sort of a service that cut out middlemen, inconvenience and offered service through an alternate and faster channel, which initially seemed to fill a big gap in the market gradually started to underperform[816].

813   Interestingly United Nations' Public Administration Network (UNPAN) hosts a number of government documents and eGovernment studies from several countries. Pakistan's E-Government Strategy and 5-Year Plan for the Federal Government can still be found at http://unpan1.un.org/intradoc/groups/public/documents/apcity/unpan037732.pdf

814   [New facility to pay utility bills] سہولت نئی کی کرنے ادا بل یوٹلٹی (2006, October 12), BBC Urdu. Retrieved 04.05.2014 from https://www.bbc.com/urdu/pakistan/story/2006/10/061012_utility_bills_kiosks_uk.shtml

815   Urdu Point (2006). "نادرا کیاسک کی بڑھتی ہوئی مقبولیت، صارفین نے 10لاکھ یوٹیلیٹی بلوں کی ادائیگی کے لئے کروالی رجسٹریشن [Increasing popularity of NADRA kiosk: 1 million subscribers registered themselves for the payment of utility bills] ". Retrieved 04.05.2014 from https://www.urdupoint.com/weather/news-detail/live-news-17018.html.

816   Yasir, M. (2011). "Interview with Tariq Malik, Deputy Chairman, NADRA". Retrieved 04.05.2014 from https://www.codeweek.pk/2011/06/interview-with-deputy-chairman-nadra/.

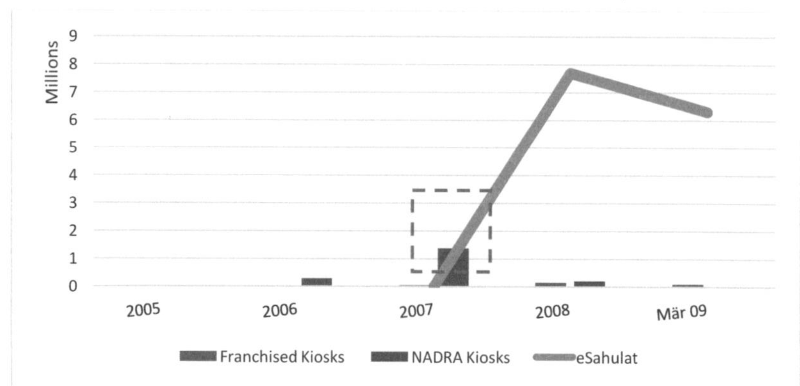

Source: Own illustration based on NADRA database

Apparently NADRA did not provide any public statement about this decline in usage as its emphasis and concentration shifted to franchising kiosk's software in its eSahulat project. However, during regular visits to NADRA during the research phase, the staff members revealed that the kiosk, despite its extraordinary value for money, also presented a number of challenges[817]. The market forces themselves decided against its future and favoured the use of eSahulat mechanism due to its licensing option. At a later phase, NADRA changed its focus from offering unassisted to assisted kiosks, but as the numbers of assisted machines also reveal, there has been a gradual drop when the eSahulat was launched. The software made for the kiosk, which is the core of eSahulat now dominates the market and more features are likely to be added to it[818].

At theoretical level, the kiosk's apparent underperformance leads to a number of observations. Can the theory of socially constructed technologies as discussed in Chapter 2 be applied to kiosk machines? Or could it be that digital literacy required for the use of kiosk was very limited? The previous chapters have tried to argue that Pakistan has experienced a massive digital leapfrogging from 2002 to 2008 particularly in case of mobile phone users which increased from hundreds of thousands to millions in a few years. However, in case of kiosk, there was no digital leapfrogging witnessed on the scale of mobile subscriptions. Kiosk appeared to have all the aspects that would have favoured its adoption. It is a publicly available machine, accessible to most, due to its bilingual offer of English and Urdu languages, no fees for usage and a coloured display screen. It is a good alternative to banks and above all it was the first ever eGovernment tool with

---

817   Interview with Ahmad Naseem, March 2009, Islamabad

818   NADRA launches Revolutionized Pension Disbursement Project (2013, April 13), Dunya News. Retrieved   04.05.2014   from   https://dunyanews.tv/en/Pakistan/168968-NADRA-launches-Revolutionized-Pension-Disbursement

214

which citizen can personally interact with the government, yet eSahulat based on the same architecture, continues to receive more preference[819].

The theory of social construction of technology (see Chapter 2) suggests that the technical/technological artefact has more tendency of adoption and acceptance if it is socially constructed. That is, if the artefact has the Social Construction of Technology (SCOT)'s elements of interpretative flexibility and closure. When the technology is iteratively constructed with constant and regular feedbacks between the users and producers then the technology produced tends to have higher maturity level and users are more adoptive towards technology and with that it can be said that the artefact eventually reaches the stage of closure[820]. In terms of interpretative flexibility, can we say that kiosk is a technological artefact that has different meanings and interpretations for various groups?

Different people hold different meaning and utility of a particular technology, in this case the kiosk. Based on the outcome of these groups' interaction with the kiosk- their apparent success or failure in utilising the kiosk for their intended purpose- will parse these groups according to their ability and intention[821]. Therefore, the feedback of these groups is necessary for a sustainable and inclusive technology. In case of kiosk, these groups can be divided in multiple categories. For instance, there can be groups who want to use kiosk and know how to use it but at the same time there can also be groups who want to use kiosk but don't know how to use it. Similarly, there can also be groups who are not even aware of it. However, in order to approach kiosk through SCOT's lens, the discussion focuses on those groups who know and who do not know the operational use of kiosk.

The group of people who could not use kiosk presented an important challenge to the debate on trust, Human-Machine Interaction and the technological experience[822]. Apart from ATM machines, it was perhaps the first time that a self-service public machine was introduced in Pakistan. The experience of ATM machines itself in Pakistan has been limited and known to result in errors[823]. There is a huge majority of people who own ATM/debit cards but face difficulties in using them. Many a time, the ATM machines are out of order or people/users might approach the staff inside the bank in order to seek their assistance in

---

819   Business Recorder: 2008a, March 14

820   Bijker, W. E. (2009). How is technology made?—That is the question! Cambridge journal of economics, 34(1), 63-76

821   Pinch, T. (2009). The social construction of technology (SCOT): The old, the new, and the nonhuman. In Vannini (Ed.), Material culture and technology in everyday life: ethnographic approaches. (pp. 45-58). New York: Peter Lang

822   Baecker, R. M. and Grudin, J., (Eds.) (2014). Readings in Human-Computer Interaction: toward the year 2000. San Francisco: Morgan Kaufmann Publishers

823   Out-of-order ATMs creating problems for customers (2012b, April 6), Pakistan Today. Retrieved 04.05.2014 from https://www.pakistantoday.com.pk/2012/04/06/out-of-order-atms-creating-problems-for-customers/

withdrawing cash or pensions[824]. A number of banks in Pakistan regularly report customer complaints related to debit cards being 'swallowed' by ATM machines, either due to technical errors or because of electricity failure. Yet ATM machines are generally known to function on the principle of inserting an ATM card in machine and drawing cash. However, in case of kiosk, it was not the case.

First, Kiosk did not expect the person to insert the ATM card or draw money from it. In fact, to use kiosk one has to deposit money. For the first-time use, a person enters her/his CNIC in the kiosk and chooses a language (that is, Urdu or English) and completes the registration by agreeing to terms and conditions of the kiosk. In the meantime, the person also goes through the navigation system for either bill payment, citizen verification or purchase of mobile top-up cards. To begin with, the initial registration process with the machine also reflects an ongoing debate on digital natives and digital migrants[825]. With the arrival of internet, the concept of online registration may have become a self-understandable and easy process. A number of web/internet users sign up for different email accounts and social media websites. Finding a link on a webpage such as 'Sign Up' or 'Register' can be an *obvious* method for digital natives. The idea of sign up itself follows a usual pattern that web users have over the years become acquainted to[826]. Choosing a user name and password, entering person information, street address and postcode are some common signup procedures. However, for a group of people who had never been a web/internet user and never had to go through online registration processes, NADRA's kiosk machine may have appeared to be a completely alien mechanism. Therefore, kiosk's first step of initial registration (even if it is an understandable requirement as first-time user) may have detrimentally effected its adoption, as experienced in other parts of the world in terms of ICT adoption[827].

Similarly, since kiosk functions on the concept of depositing money directly into the machine, the idea of handing over one's cash to the machine was also very alien. Even though many ATM machines have started accepting cash deposits since 2010, the history of ATM machines in Pakistan is very young. It was only after 1999 that State Bank of Pakistan approved electronic cash transfer and established the digital national switches necessary for ATM machines[828]. However, ATM machines formally started to arrive after 2002. Also, the ATM

---

824   Several ATM machines go out of order (2009, November 28), Business Recorder. Retrieved 04.05.2014 from https://fp.brecorder.com/2009/11/20091128991571

825   Jones, C. and Czerniewicz, L. (2010). Describing or debunking? The net generation and digital natives. Journal of Computer Assisted Learning, 26(5), 317-320

826   Thomas, M., (Ed.) (2011). Deconstructing Digital Natives: Young People, Technology, and the New Literacies. New York: Routledge

827   Dyson, L. E. (2004). Cultural issues in the adoption of information and communication technologies by Indigenous Australians. In Proceedings cultural attitudes towards communication and technology.   (pp. 58-71). Murdoch University Perth. Retrieved 04.05.2014 from http://citeseerx.ist.psu.edu/viewdoc/download?doi=10.1.1.198.7661&rep=rep1&type=pdf

828   Pakistan's e-banking – past, present and future (2013c, April 23), Business Recorder. Retrieved 04.05.2014 from https://fp.brecorder.com/2013/04/201304251178401/

machines network also slowly spread across Pakistan and they were commercially available after 2004. Therefore, between 2004 and 2005 (kiosk's launch) there is a very small amount of time during which the common public may have heard, learnt or used the ATM machine. Their reference point here and the lived experience with the technology adoption is very small[829]. For instance, in case of mobile phone, the landline telephones provided a good reference point and landline phone have existed for decades prior to the arrival of mobile phones. Therefore, depositing cash in a machine with which people cannot historically or experientially associate with, casts doubts and hinders trust in a machine that is publicly available without gate-keeping and human chaperoning[830].

The kiosk's gradual underperformance also matches with the experiments made with different ICTs in other developing countries. Heeks for instance has noted that many eGovernment projects fail when the design-reality gaps are too big[831]. While referring to computer kiosks in rural India, Heeks argues that while supporting the ICT projects, investments in hardware and software alone are not sufficient[832]. There are other factors such as people, politics, emotions and/or culture that are continuously at play while developing ICTs[833]. The theory of design-reality gaps also points to the inconsistencies that exists between the technological artefact and the given worldview that can cause failure(s) either in partial or total[834]. A superficial assumption or expectation based on the ground reality can trigger a gap between the producer (of technology) and the user. The chances of a failure of such an artefact are higher if a gap is bigger. Could we say that the entire design of kiosk had a big design-reality gap? The idea of ATM-styled machine, when ATM themselves were not very common, the mechanism of cash deposit directly into the machine (when the only known mechanism of cash deposit in Pakistan has been through a bank directly to the cashier against receipt), the process of first time registration with the machine (when the web/internet users were extremely low) and above all the entire human machine interaction itself makes the case of design-reality for kiosk very challenging to its sustainable adoption.

This also leads to the fact that kiosk was perhaps not a socially constructed digital communication device. Its purpose, its use and its outreach though had a number of advantages but it does not seem to entirely qualify the theory of SCOT. The

829   Curran, J. M. and Meuter, M. L. (2005). Self-service technology adoption: comparing three technologies. Journal of services marketing, 19(2), 103-113

830   Vance, A., Elie-Dit-Cosaque, C. and Straub, D. W. (2008). Examining trust in information technology artifacts: the effects of system quality and culture. Journal of management information systems, 24(4), 73-100

831   Heeks, R.: 2002a

832   Heeks, R. (2005b). ICTs and the MDGs: On the Wrong Track? Retrieved 30.11.2010 from http://hummedia.manchester.ac.uk/institutes/gdi/publications/workingpapers/di/di_sp07.pdf

833   Hornung, H. and Baranauskas, M. C. C. (2007). Interaction Design in eGov systems: challenges for a developing country. In Proceedings of the 34th Seminário Integrado de Software e Hardware, Porto Alegre, Brazil. (pp. 2217-2231).

834   Heeks, R.: 2006a

relevant social groups in this case are citizens of Pakistan whose lived experience, digital capacity, her/his approach of payment of the public utility consumption were supposed to be iteratively carried over in the continuous development of kiosk. From a stand-alone and unassisted device, kiosk started to become an assisted device. NADRA also realised that such kind of machine may require human support. Therefore, NADRA also announced in a public message for the citizens to register themselves with the kiosks that were installed on NADRA's premises where personnel would walk every kiosk user through the entire registration/navigation process[835].

There are other set of factors as well that also limited the employment of further kiosk machines. Installation of a kiosk machine required a digital infrastructure setup, purchase and acquisition of necessary equipment (touch-screen systems and machine with money recognition, depositing features), constant maintenance of the hardware, marketing and dispatching the project software and subsequent upgrades was a sizeable investment. Wherever unassisted kiosks were deployed to common public places, power supply and security measurements also needed to be considered, particularly taking Pakistan's acute energy crisis in consideration[836].

In order to promote and increase the use of kiosk, NADRA transformed its service centres by transforming kiosk's software on normal desktop-computers (for kiosk's variants see Chapter 5). This proved to be a way to significantly reduce the cost of the project and this opened a new stream for delivery of this service. NADRA's customer service centres then usually contained a kiosk machine and counters where people could pay their bills as well. This experience proved that the initial rationale of fulfilling a market vacuum for a G2C service, the utility bill payments, which had saturated the lines in public banks that struggled to service the burgeoning mass of the unbanked population. However, this growth also showed the potential of scaling this project further as the NADRA centres had their primary service of registering citizens and facilitating the identity services rather than collecting utility bills. Scaling of kiosk machine to far-flung areas also required infrastructural investment and sustainability of a sophisticated and expensive machinery is also cumbersome in the long run. Therefore, NADRA came up with its third variant of kiosk by franchising the kiosk machines to business entities which contributed to both an income option for NADRA franchisees and security for the kiosk machine[837].

Launched in 2007, eSahulat was the fourth variant of kiosk machine that was made operational in 2008. It utilised the same software of the kiosk but could be

---

835   Business Recorder: 2005d, August 30
836   Vaid, M. and Tourangbam, M. (2014). Pakistan's Energy Crisis. Retrieved 30.08.2014 from
         https://foreignpolicy.com/2014/08/13/pakistans-energy-crisis/
837   Nadra plans to franchise kiosks (2007, March 10), The News. Retrieved 04.05.2014 from
         https://www.thenews.com.pk/archive/print/46012-nadra-plans-to-franchise-kiosks

franchised to small shops and businesses rather than just the NADRA centres[838]. Designing a financial system to supplement the cash inflows to turn it into a successful business model for small shop-owners was engineered. A training system for utilising this eSahulat programme was also designed and licensees of the service had to undergo this mandatory training[839]. The important aspect that eSahulat addressed however, was the (re)introduction of a human agency in the utilisation of this software and service. The saturation point that the kiosk machines experienced was primarily because of the limited ability among the common public that was not used to such employment of technology. Kiosk required a sophisticated understanding of the transaction that ranged from identification mechanism, registration process, machine reading of the bill, input of correct payment amount and finally deposition of cash with the kiosk[840].

This in itself was a leap forward for a lot of individuals who were used to standing in queues, handing over the receipt of the bill and being told to pay a specific amount which they handed over and received a stamped customer copy in return[841]. The kiosk machine needed much more self-reliance and experience with technology. Although it had provisions for Urdu, scaling this beyond cities would also be a problem as it was not the regional and main language in the rural places. Literacy is also a major challenge in Pakistan and a lot of people who could not read were automatically excluded from using the kiosks singularly[842]. Issue of trust is also the reason for first-time users as the idea of putting money into a machine that completed most of its work on a screen and only returned a printed copy of the transaction, which they could not decipher, was difficult to imagine and adapt to. Suspicion of technology and wariness to change are often cited as critical factors that shape trust in any eGovernment initiative, especially so if it involves money[843].

The eSahulat fulfilled this *deficiency* by re-introducing the human agency and catering to the trust factor[844]. What it most ardently provided was the factor of an oral communication, which did not need further registering with machines, instead it involved face to face communication with the retailer (eSahulat

---

838   The Nation: 2008b, May 22

839   National Database Registration Authority: 2012d

840   As Bakardjieva reminds us, human agency sits at the center of technological development and in case of kiosk, people's choice was for more prevailing communication patter, that is oral, based on human/social network. See Bakardjieva, M.: 2005

841   As argued in previous chapters, the lack of technology adoption by Pakistan's banking industry along with the unbanked population has severely affected, among others, the utility bill payment process. It was only in 2008 that Pakistani banks decided to form a consortium named 1Link that initially brought 29 member banks to support the interbank funds transfer. See 1LINK building the e-Banking infrastructure in Pakistan (2008c, April 23), Business Recorder. Retrieved 04.05.2014 from https://fp.brecorder.com/2008/04/20080423726993

842   Ameen, K., et al.: 2009

843   Mazhar, F., Jam, F. A. and Anwar, F. (2012). Consumer trust in e-commerce: A study of consumer perceptions in Pakistan. African Journal of Business Management, 6(7), 2516-2528

844   Yu, F.-L. T. (2008). Uncertainty, human agency and e-government. Transforming Government: People, Process and Policy, 2(4), 283-296

licensee) who also provide bill's receipt, required amount of money and a stamped return of the customer's bill[845]. This was similar to the transactions committed at the banks and did not require a drastic change from the pattern of exchanges that were existent. Convenience was multiplied as people could turn up to small businesses like corner shops that had purchased the licence and pay their utility bills within minutes. Such sort of partial reliance on technology and partially on social network enabled the trust and acceptance of the eSahulat[846]. Banking with a physical human agency, people felt comfortable handing their money to the local/chosen shopkeeper that they had interacted with over the years or was physically present. The external identity, presence and reputation of the local business aided the trust factor of the eSahulat program.

Training and equipment that was provided to the franchisees also played its due part. For a small fee, licence owners learnt to operate the software in order to maintain a secure identity on their system (licensee account), enter information of the transaction, verify a successful payment and even troubleshoot the system in case of errors. In several areas where eSahulat service is available, the licensees have publicised its services on their own. However, what is more interesting in this scenario that the licensees have even engaged their co-workers to spread a word about the availability of eSahulat services in their vicinity. Therefore, a new form of engagement is emerging in Pakistan where citizen services such as utility bill payment, citizen verification and/or money transfer are being offered at the door step. eSahulat in this regard is providing new meaning to the social capital theory which is primarily composed of elements such as relationships, trust, and norms[847]. The interpersonal ties that people have developed over the period of time in terms of relationships are foundations of social networks. The participatory approach by citizens leads to the fact that eSahulat has exploited the very social capital in regrouping the people in their vicinities thereby establishing a sociotechnical capital[848].

The transition from kiosk machine to eSahulat showed the evolution of a G2C service that proved its utility by adapting to the common communication pattern of the people. Understanding the limitation that technology places on

---

845　The use of oral communication in order to support the eServices is even affective in several developed countries such as New Zealand where the government uses such assisted patterns and employ multichannel services for new immigrations, rural sector and their indigenous population. See for example Cullen, R. (2005). E-Government, A Citizens' Perspective. Journal of E-Government, 1(3), 5-28; Cullen, R. and Hernon, P. (2004). Wired for well-being: Citizens' response to e-government. Retrieved 28.10.2010 from http://unpan1.un.org/intradoc/groups/public/documents/APCITY/UNPAN017975.pdf

846　Shipps, B. (2013). Social networks, interactivity and satisfaction: Assessing socio-technical behavioral factors as an extension to technology acceptance. Journal of theoretical and applied electronic commerce research, 8(1), 35-52

847　Resnick, P. (2005). Impersonal sociotechnical capital, ICTs, and collective action among strangers. In Dutton, Kahin and O'Callaghan (Eds.), Transforming enterprise: the economic and social implications of information technology (pp. 399-412). Cambridge: The MIT Press

848　Mandarano, L., Meenar, M. and Steins, C. (2010). Building social capital in the digital age of civic engagement. Journal of Planning Literature, 25(2), 123-135

inclusiveness of the technically challenged populace, important gaps in the market can be fulfilled by such eGovernment services. Developing countries can potentially take up new technology but only if the consideration of trust, human agency, technical ability and sustainability of the technology model employed is ascertained. The case of eSahulat illustrates that identifying the factors that can hinder any technical initiative are most fundamentally those of technical acceptance and human awareness of technology. NADRA's own staff believed that in comparison to eSahulat, kiosk's ATM-styled physical structure, sophistication of use, continuous maintenance and hardware costs did not seem economically feasible[849]. Kiosk as a technological artefact also highlights the fact that even though it supported the provision of local language, nevertheless, it still carries the essence of technological determinism as also discussed in Chapter 2. For kiosk to be more socially inclusive technology, perhaps different hardware variations could have been tried, which are more compact and economically feasible. NADRA's kiosk as an alternative mode of G2C communication can well serve as a good alternative to web portal. It could very well serve as a point of information to retrieve government information and could be used to print various documents such as government forms "online".

7.4. NADRA's consolidating position as an eService producer/provider

Since its inception in 2000, NADRA has been continuously consolidating its position with its CNIC project as a major player in ICT/eGovernment category in Pakistan. Its role in Pakistan's current digital communication scenario works as a public organisation which plays a dual role of specialisation in digital identity management and digital gatekeeping[850]. Any financial and legal activity has to go through NADRA's digital approval/verification mechanisms[851]. NADRA's Verisys software acts as a digital gate keeper and without its approval any financial and legal activity cannot take place[852]. The Verisys software is a mandatory piece of digital approval software at every bank and mobile network operator's outlets. Any person who wishes to open a bank in Pakistan or buy a SIM card must present her/his CNIC and verify her/himself through a fingerprint

---

849   Interview with Ahmad Naseem, March 2009, Islamabad

850   Nadra to participate in card technology conference in Egypt from May 6 (2007d, May 01), Business Recorder. Retrieved 04.05.2014 from https://fp.brecorder.com/2007/05/20070501558069

851   SBP stresses photo identity for bank account (2005f, August 31), Business Recorder. Retrieved 04.05.2014 from https://fp.brecorder.com/2005/08/20050831319362/

852   Since 2013, police departments have also started using NADRA's verisys system in order to facilitate the filing of citizen complaints. See for example: Online FIR registration procedure announced (2013b, July 06), Daily Dawn. Retrieved 04.05.2014 from https://www.dawn.com/news/1023147

reading machine available at every bank, mobile telecom office and their franchisees.

To a large extent NADRA's role in digitising public service has been multifaceted. While its initial mandate included database creation and management, it has not been limited to only development and issuance of ID cards. Since the entire database for Pakistani citizen is already available in digital form, it became possible for NADRA to build other identity related eServices around it, such as related civil registration systems (see Chapter 5)[853]. In this regard, NADRA developed another/additional system for local/provincial governments for the issuance of birth, marriage, divorce and death certificates. This civil registry system was only manually possible prior to NADRA and every city municipality and the lower courts was responsible for issuing such certificates. Instead we see that because of the entire wave of digitisation, the public-sector organisations such as municipalities at all levels in Pakistan have also gone through a reform. The older and manual systems and procedures that have previously caused hindrances and delays are now more ICT-based and conveniently available for the common public[854].

NADRA further consolidated its position in the field of eGovernment by initiating kiosk and eSahulat projects. With kiosk machine project NADRA tried to exhibit its mandate, outreach and ambitions for further developing G2C services. NADRA tapped that empty vacuum of eServices that was so far not filled by any other public-sector or private organisation[855]. It tried to swipe at people's problems in the financial sector such as money transfer and bill payment and built kiosk and eSahulat around it. Similarly, it also tried to address the gaps in Pakistan's digital communication landscape by initially not limiting itself to web-based services. Kiosk's immediate rise, its underperformance in the wake of its more decentralised use in eSahulat is an attestation to NADRA's employing eServices based on alternative channels (in addition to and other than web-based services).

Since bill payments and money transfer incur transaction costs, NADRA found a feasible business model that could also generate money for it. NADRA has been continuously strengthening its own financial autonomy by being a more self-sufficient public-sector organisation, not greatly reliant on the national exchequer. For instance, any individual opening a bank account in Pakistan needs to be digitally authenticated by banks making use of NADRA's Verisys software for which NADRA charges PKR 35 (USD 0.35) per authentication[856]. Over the number of years, NADRA's financial position has been further consolidated after

853   National Database Registration Authority (2014a). Civil Registration Management System. Retrieved 04.05.2015 from https://www.pc.gov.pk/uploads/crvs/CRVS.pdf

854   Nadra to register births, marriages (2007f, August 13), Daily Dawn. Retrieved 04.05.2014 from https://www.dawn.com/news/260982

855   Gallup Pakistan (2014). Pakistan ICT Indicators Survey - 2014. Retrieved 04.05.2014 from http://gallup.com.pk/wp-content/uploads/2017/01/Pakistan-ICT-Indicators-Survey-2014-1.pdf

856   Gelb, A.: 2014

it registered a public limited company under the name of 'NADRA Technologies Limited'. The company is registered with the Security and Exchange Commission of Pakistan (SECP) since 2004[857]. As a limited company NADRA has been bidding for international projects and for that matter it received contracts from various governments[858]. These include, for instance, Bangladesh (digital driver's licence system), Kenya (passport control system and biometric passports), Sudan (Civil Registration System) and Nigeria (identity management system). The technology solutions that NADRA had specialised in are now also being used in several countries bringing in a huge revenue to NADRA. By 2013 NADRA's revenues totalled to $ 17.6 million (USD) which made it possible to pay for than 18,000 of its workers, technological hardware and equipment services[859].

Table 16: List of NADRA's international eGovernment projects

	Country	eGovernment Service
1	Bangladesh	National Driver's License System
2	Kenya	Passport Issuance & Control System
3	Kenya	Electronic Passport System
4	Sudan	Civil Registration System
5	Nigeria	National Identity Management System

Source: Own illustration based on NADRA's official website (www.nadra.gov.pk)

On the home front, NADRA continues to partner and collaborate with both private and public-sector organisations such as banks, telecommunication companies, universities and ministries[860]. In comparison to EGD, which was considered as a premier eGovernment institution in Pakistan, NADRA's trajectory in Pakistan's eGovernment scenario has been entirely different from EGD. Even though both EGD and NADRA started their functions as an attached department of Ministry of IT and Ministry of Interior, respectively. However, over the years EGD's role continued to diminish and its services were extended to a few ministries based in

---

857 Security and Exchange Commission of Pakistan (2014). Company Name Search. Retrieved 19.10.2014 from https://eservices.secp.gov.pk/eServices/NameSearch.jsp

858 National Database Registration Authority (2014c). "NADRA's International Clients". Retrieved 04.05.2014 from http://nadra.gov.pk/index.php/clients-a-partners/international-clients.

859 'The Nadra story' (2014d, November 12), Business Recorder. Retrieved 04.05.2015 from https://epaper.brecorder.com/2014/11/12/7-page/466533-news.html

860 HBL partners with Nadra for branchless banking (2013b, November 01), The Nation. Retrieved 04.05.2014 from https://nation.com.pk/01-Nov-2013/hbl-partners-with-nadra-for-branchless-banking; Virtual University of Pakistan. (2015, August 05). Virtual University Signed MoU With NADRA Technologies. [Press release]. Retrieved 30.08.2015 from http://vu.edu.pk/NewsDetails.aspx?type=&NewsId=2383

Islamabad, whereas NADRA continued to grow in authority, national digital outreach and in electronic public service delivery.

NADRA's products such as CNIC, civilian registration management system, biometric passports or kiosk/eSahulat programmes have rapidly changed not just the way public sector organisation functioned in Pakistan but they have also tried introduce culturally and commercially viable electronic public services[861].

7.5. Mobile phones as an emerging eGovernment service provider

Along with the NADRA and EGD's experiments with eGovernment implementation in Pakistan, the mobile phone services also grew in parallel. At the beginning of 2000, when mobile phone market was going through several experiments with bulkier and slimmer phones, it was difficult to assume if mobile phones could also serve as a medium of electronic public service delivery in developing countries[862]. With world-wide deregulation in telecommunication sector, mobile penetration rapidly increased and superseded the internet in terms of subscribers. Kenya's M-pesa initiative contributed to using mobile phone not just as a device for voice communication. It helped raised SMS standard as a carrier of eGovernment services[863]. Mobilink's example in this study has showed the potential and utility of simple mobile phone into a portable electronic public service device. One of the profound and groundbreaking policy initiatives in this regard that took place in Pakistan is the initiation of Branchless Banking (BB) Regulation by the State Bank of Pakistan in March 2008[864]. This initiative opened the door to financial institutions including commercial and microfinance banks to increase their outreach using ICTs[865].

Mobile telecom sector proved to be a perfect playground for banking the unbanked population of Pakistan. By 2011 there were 109 million mobile

---

861 In its annual report, UNICEF's Pakistan chapter highlights NADRA's contribution towards birth registration of households particularly in landlocked Pakistan where government's outreach has been difficult. UNICEF in collaboration with NADRA cooperate local government in subsidising the cost of birth certificates. See UNICEF. (2014). Annual Report 2013 - Pakistan. Retrieved 04.05.2015 from https://www.unicef.org/about/annualreport/files/Pakistan_COAR_2013.pdf

862 Barwise, P. (2001). TV, PC, or Mobile? Future Media for Consumer e-Commerce. Business Strategy Review, 12(1), 35-42

863 Barrett, M., Davidson, E., Prabhu, J. and Vargo, S. L. (2015). Service innovation in the digital age: key contributions and future directions. MIS quarterly, 39(1), 135-154

864 A.Shaikh, A., Glavee-Geo, R. and Karjaluoto, H. (2017). Exploring the nexus between financial sector reforms and the emergence of digital banking culture – Evidences from a developing country. Research in International Business and Finance, 42(2017), 1030-1039

865 See for instance Pakistan's premier development institution's review on branchless banking and growth of microfinancing in Pakistan: Pakistan Poverty Alleviation Fund. (2012). Branchless Banking and Savings. Retrieved 04.05.2014 from http://www.ppaf.org.pk/doc/microfinance/27-Session%20on%20Branchless%20Banking%20and%20Savings.pdf

subscribers in Pakistan as compared to 25 million banking accounts[866]. Telecommunication sector tried to harness this regulation and initiated multiple projects with utility bill payments being on top. Since mobile telecommunication companies are not a financial institution, therefore they established partnerships with different banks in order to initiate their mobile banking solutions. Pakistan Telecommunication Authority (PTA) and State Bank of Pakistan (SBP) in this regard collaborated with each other to draft a regulatory framework to facilitate mobile banking in Pakistan[867]. With necessary policy in place, mobile operators realised an opportunity into Pakistan's predominantly cash-based economy[868]. They realised the needs and financial limitations of a small and medium enterprises as well as the monetary concerns of a common person, which includes a combination of daily wagers as well as low salaried people who can neither afford the prerequisites nor the maintenance of traditional bank account[869]. Mobile phone operators initially introduced the electronic bill payment mechanisms and gradually moved on to offer money transfer mechanisms[870]. For telecom operators, this is an opportunity to earn money through transaction charges whereas for partnered banks they provide necessary electronic payment gate along with the service charge per transaction.

Initially, the mobile network operators collaborated with multiple banks to initiate bill payment and money transfer mechanism. However, realising the lucrative revenue stream on per transaction basis, many banks are also initiating newer services themselves with telecom operators. The market seems so attractive to mobile phone operators that they even started buying banks in order to have full control of the nature of financial services and the revenue stream that they generate. For instance, the Norwegian operator Telenor which initially collaborated with Tameer Bank for its Easypaisa project has now completely purchased the Tameer bank and renamed it as Telenor Microfinance Bank[871]. Any citizen can open a bank account with Telenor Bank with an amount as low as PKR 1 (USD 1 cent). Similarly, Mobilink after partnering with Waseela Bank and National Bank of Pakistan bought the entire Waseela Bank and now operates it

---

866   State Bank of Pakistan (2011). Branchless Banking Newsletter (Issue 1). Retrieved 04.05.2014 from http://www.sbp.org.pk/publications/acd/2011/BranchlessBanking-Jul-Sep-2011.pdf

867   Pakistan Telecommunication Authority (PTA): 2009b, November 19

868   Consistant policies can tap Pakistan's economic potential (2013d, March 16), Business Recorder. Retrieved 04.05.2014 from https://www.brecorder.com/2013/03/16/110972/consistant-policies-can-tap-pakistans-economic-potential

869   Mobile banking with 1.4mn accounts showing remarkable surge (2013a, April 14), Business Recorder. Retrieved 04.05.2014 from https://www.brecorder.com/2013/04/14/115161; The outreach of mobile banking (2009a, November 02), Daily Dawn. Retrieved 04.05.2014 from https://www.dawn.com/news/838971/the-outreach-of-mobile-banking

870   Easypaisa launches funds transfer facility to and from banks (2014a, June 13), Daily Dawn. Retrieved 04.05.2014 from https://www.dawn.com/news/1112521

871   Telenor buys Tameer Bank (2016, March 17), The Express Tribune. Retrieved https://tribune.com.pk/story/1067643/telenor-buys-tameer-bank

under the name of Mobilink Microfinance Bank[872]. Pakistan's national telecom provider Ufone followed the same pursuit and purchased Rozgar Microfinance Bank and currently functions under the brand name of U Bank[873]. Chinese mobile operator Zong has so far not initiated its own bank but in partnership with other banks it also offers similar services as those of other operators[874].

With BB regulations, mobile phone operators are at the same time also owners of Pakistan based banks regulated by the SBP. This also means that Mobilink, Ufone and Telenor now also offer all other services that a traditional bank does, that is, issuance of ATM card, international money transfers, current and saving accounts or loans. At the same time due to BB option, mobile phone operators' vast network of retailers spread across Pakistan serve as their agents through which people can deposit and withdraw money or pay utility bills[875]. A retailer is any shop who has the license to operate on behalf of the mobile phone operator for BB services. People who do not even own a mobile phone can perform any banking activity (bill payment, money transfer, deposit or withdrawal) through these agents over the counter by presenting their CNIC (see Table 17).

Table 17: Banking services by the mobile network operators in Pakistan

	Bank Ownership		Bill Payment		Money Transfer		
	Current Name	Previous Name	Over the counter (OTC)	via Mobile Phone	Mobile to Mobile	Mobile to CNIC	CNIC to CNIC
Mobilink Jazz	Mobilink Microfinance Bank	Waseela Microfinance Bank	✓	✓	✓	✓	✓
Telenor	Telenor Microfinance Bank	Tameer Bank	✓	✓	✓	✓	✓
Ufone	U Microfinance Bank Ltd	Rozgar Microfinance Bank Ltd	✓	✓	✓	✓	✓
Zong	-	-	✓	✓	✓	✓	✓

Source: Own Illustration

872   State Bank of Pakistan (2017). Quarterly Compendium: Statistics of the Banking System. http://www.sbp.org.pk/ecodata/fsi/qc/2017/Dec.pdf

873   U Bank is a wholly owned subsidiary of Pakistan Telecommunication Company Limited (PTCL), See: http://ubank.com.pk/about-us/

874   As mobile banking grows, Zong, Askari Bank join the race (2012c, November 20), The Express Tribune. Retrieved 04.05.2014 from https://tribune.com.pk/story/468247/as-mobile-banking-grows-zong-askari-bank-join-the-race/?amp=1

875   Haider, H. (2014, March 10). Branchless Banking Section: Banking Review 2013. Business Recorder. Retrieved 04.05.2014 from https://www.brecorder.com/pdf/banking-review-2013.pdf

According to the State Bank of Pakistan data, the number of BB accounts, BB agents, cash deposits between 2011 and 2015 have exponentially grown. At the end of 2011 total BB accounts in Pakistan were 500,453 (see Table 18). These figures reached up to 15 million at the end of 2015. Similarly, the amount deposited through the BB accounts by end of 2015 amounted to PKR 8,827 million. These numbers particularly in the case of BB accounts and cash deposits also include people who were prior to BB regulations unbanked. Due to mobile banking mechanism in place they are able to either use their mobile phones or use OTC approach via retailer.

The role of mobile phone based eServices is changing the dynamics of service delivery in Pakistan, as the examples, initiatives and figures stated above indicate. Pakistan seems to follow the similar pattern as also adopted by other developing countries with lower rates of internet adoption in comparison to mobile phone[876]. Susanto and Goodwin argue that the user of ICTs in developing countries hugely depend upon the familiarity with the communication channels[877]. If the technological channel and device is low cost and also supported by a native language then it makes it convenient both for the eService provider and consumer. In case of Pakistan, network of retailers spread all across Pakistan attempts to bring the eServices to areas for people who may not even own a mobile device[878].

Table 18: Branchless Banking (accounts and transactions) from 2011-2015

	2011	2012	2013	2014	2015
Total BB Accounts	500,453	2,112,052	3,475,458	5,414,655	15,322,171
Total BB cash Deposits (PKR. In millions)	Rs. 503	Rs. 1,055	Rs. 2,639	Rs. 6,668	Rs. 8,827
Value of Transactions (PKR. In millions)	Rs. 138,120	Rs. 492,282	Rs. 802,496	Rs. 1,352,517	Rs. 1,872,451

Source: Own Illustration based on State Bank of Pakistan[879]

The mobile-based services, as noted in this study mostly fall under financial services category. Even though they enable people to pay utility bills and transfer

---

876   Susanto, T. and Goodwin, R. (2013). User acceptance of SMS-based e-government services: Differences between adopters and non-adopters. Government Information Quarterly, 30(4), 486-497

877   Susanto, T. D. and Goodwin, R. (2010). Factors Influencing Citizen Adoption of SMS-Based e-Government Services. The Electronic Journal of e-Government, 8(1), 55-71

878   Microsave Consulting (2015). "Over the Counter (OTC) in Pakistan: Why It Works". Retrieved 30.06.2015 from https://www.microsave.net/2015/06/10/over-the-counter-otc-in-pakistan-why-it-works/.

879   State Bank of Pakistan Branchless Banking Newsletters. Retrieved 31.12.2015 from http://www.sbp.org.pk/publications/acd/branchless.htm

money using their mobile phones, there is, however, an important observation that these services do not include informational services. For instance, if the current scenario of mobile-based eServices in Pakistan are compared to the models of eGovernment (see Chapter 2), these services do not include provision of information about government procedures or laws. The emphasis appears to be focused mostly on earning revenues through payment mechanisms. Mobile network operators in Pakistan present a unique opportunity for G2C services which can be readily used for informational purposes as well but perhaps they might need to be commercially incentivised. Here it might still be the prerogative of the government to release, maintain and catalogue their services and processes for the benefit of citizenry.

There are several cases in other countries such as in Indonesia where health ministry uses SMS based service to announce disease outbreaks or health care updates[880]. In Istanbul, for example, the municipality of Sisli has taken the SMS-based services from informational level to a transactional level, that is, SMS cannot only be used as a broadcast but citizens can also engage in two-way communication. The SMS-based eService in Sisli enables the citizens to pay their taxes using their Tax ID[881]. Since SMS-based services do not require any internet and can also work in areas with limited coverage/signals, therefore the outreach of SMS is far higher and cost-effective than a web/internet apparatus such as a computer or NADRA's kiosk machine. In Philippines for instance, government uses SMS-based service through which citizens can communicate with their garbage collection services in order to report any additional garbage collection[882].

7.6. Conceptualising eGovernment model for Pakistan

As the case studies explained in this research have highlighted, the first twelve years of the new millennium have seen an array of technological initiatives in Pakistan in terms of new IT policy, eGovernment strategy by EGD, NADRA's digitisation initiatives, G2C services by mobile operators as well as other collaborative eServices such as those of the ECP and NADRA. The eService elaborated in this research as well as the more receptive ones in Pakistan are mostly contributing to the financial sector. If these case studies are reviewed in comparative perspective, particularly in the light of Layne & Lee eGovernment model (see Chapter 2), then EGD's web portal does not even properly qualify the

880   Al-Hujran, O. (2012). Toward the utilization of m-Government services in developing countries: a qualitative investigation. International Journal of Business and Social Science, 3(5), 155-160

881   Kervenoael, R. d. and Kocoglu, I. (2012). E-Government Strategy in Turkey: A Case for m-Government? In Bwalya and Zulu (Eds.), Handbook of Research on E-Government in Emerging Economies: Adoption, E-Participation, and Legal Frameworks. Vol.1 (pp. 351-373). Hershey (PA): Information Science Reference

882   Susanto, T. D. and Goodwin, R. (2006). Opportunity and overview of SMS-based e-government in developing countries. In Morgan, Brebbia and Spector (Eds.), Advances in Education, Commerce and Governance second International Conference on the Internet Society Advances in Education, Commerce and Governance (pp. 255-264). South Hampton: WIT Press

catalogue stage of generic eGovernment models or frameworks available. The interaction level for the citizens in the government's web portal as the analysis in Chapter 4 has shown, remains limited. On the other hand, if NADRA or mobile applications are being reviewed, they seem to have achieved the vertical and horizontal levels of integration in order to offer financial services. However, neither NADRA nor mobile operators currently offer information on government services that could be categorised for a catalogue stage. Thus, it is difficult to place present eGovernment scenario in Pakistan in one particular category. Market forces coupled with financial incentives drive the eGovernment services currently in practice in Pakistan[883].

Access to government information is still a major issue and web portals at the federal or provincial levels have put up information in English. The present scenario gives rise to a new category of information elites[884]. This in return creates a new divide within a national eGovernment discourse in Pakistan. This form of digital othering could prolong if the status-quo prevails. The Pakistan Telecommunication Authority (PTA) is perhaps among the few organisations that actually offer up-to-date information in Urdu as well as English[885]. Similarly, prior to the general election in 2013, the Election Commission of Pakistan created bilingual information brochures to explain the voting procedure. This information was also duly provided on the website. However, these are individual initiatives largely dependent on decisions made by bureaucracies. Had the EGD's strategic framework been stronger, there could have been a homogenous approach in order to develop the government website. If Pakistan's digitalisation approach with reference to web portal is compared with a digitally established country such as Germany, it provides a good example of how web portals need to be homogenised at regional, national and international level. For instance, Germany's web portal for Foreign Ministry's function and strategy is homogenous towards web content. The same web portal functions in the same way for all its related institutions such as Germany's embassies and missions abroad. For every German embassy abroad, the domain name www.diplo.de works as a parent domain. If the website of any embassy needs to be visited, the domain name can be preceded by the country's name. For example, Germany's embassy in Pakistan would have a domain name www.pakistan.diplo.de and same holds true for other country level domains of www.diplo.de. Similarly, all the embassies follow a similar design for the web portal and house the same information containers, albeit different data.

883   Gallup Pakistan (2013). Use of Mobile Money in Pakistan: Findings from FITS Study. Retrieved 04.05.2014 from http://gallup.com.pk/wp-content/uploads/2016/02/Aug-03-20131.pdf

884   Suoranta, J. (2003). The world divided in two: Digital divide information and communication technologies, and the'youth question'. Journal for Critical Education Policy Studies, 1(2), 1-31

885   Pakistan Telecommunication Authority (2014b). "   [Urdu Website of Pakistan Telecommunication Authority]پاکستان ٹیلی کمیونیکیشن اتھارٹی اردو ویب سائٹ : پی-ٹی-اے.پاکستان حکومت". Retrieved 04.05.2014 from https://www.pta.gov.pk/ur.

Former PTA employee, Naveed ul Haq, thinks that there is a lack of political will in government quarters towards eGovernment[886]. The clerks, officers and bureaucrats are at times incapable of handling modern ICTs. Expecting them to make ICTs usable for the public good may be a far cry. He shares similar opinion as NADRA's technology specialist Mr. Usman Javaid who thinks that Pakistan's eGovernment scenario also needs to look beyond financial services[887]. Both point to the same conclusion; that the most penetrated technological tool, which in Pakistan's case is the mobile phone, needs to be contextualised for better eGovernment service delivery. Information dissemination is among the first priority as is noted in the definitions of eGovernment[888]. As Bhatnagar and Basu have repeatedly emphasised, the developing countries pursuing eGovernment may need to employ both manual as well as partial electronic services to begin with[889]. The government, instead of cataloguing its information at the website level, shall ensure information dissemination in a paper-based environment as well. The eGovernment's motive is not just to put all information online; its objective is to also streamline government's administrative processes[890]. As the infoDev's guidelines for developing countries suggest, the level of automation must not be strictly followed[891]. In this context, Pakistan's eGovernment model seems to fall in a somewhat hybrid state, which constitutes few highly, sophisticated eServices (that of NADRA or ECP) or the less sophisticated ones (such as one by the EGD).

7.7. Towards hybrid model of eGovernment?

The G2C eGovernment services presented in this study have highlighted a number of key points and concerns for the provision of electronic public service delivery in Pakistan. Table 19 presents a tabular categorisation of the eServices discussed in this study based on their nature, front end technology, communication channel, language, initiator and the citizen-technology interaction. The question, based on the eServices presented in this study and drawing upon their tabular representation below, can Pakistan's eGovernment model or strategy be categorised? Can Pakistan's eGovernment model be compared or evaluated against the models presented in theoretical section in Chapter 2?

A number of corollaries can be drawn from the Table 19. Out of five eServices reviewed in this study, there are only two of informational nature, whereas the

---

886   Interview with Naveed ul Haq, November 2009, December 2010

887   Interview with Usman Javaid, February 2009

888   Schwester, R. (2009). Examining the barriers to e-government adoption. The Electronic Journal of e-Government, 7(1), 113 - 122

889   Bhatnagar, S. C.: 2002; Bhatnagar, S.: 2009; Bhatnagar, S.: 2004; Basu, S.: 2004

890   Nygren, K. G.: 2009

891   World Economic Forum. (2001). The Global Information Technology Report (2001–2002): Readiness for the Networked World. Retrieved 04.05.2014 from http://unpan1.un.org/intradoc/groups/public/documents/un/report.pdf

rest three eServices are of financial nature. Here informational service refers to a G2C eService through which either a citizen can attain/verify information about him/her from the government databases or a citizen could access information such as a departmental information, details about ministries, download government policies or laws, relevant forms, or contact information. Rest of the three services are financial in nature through which either a citizen can pay utility bills or transfer money. One of the important aspects in Table 19 is also the front end, that is, the technical medium through which a citizen can perform/access the G2C service. In the front-end category, only one eService, i.e. Government Portal is offered via web-based service either through mobile or personal computer whereas rest of the services are either accessible through mobile phone or via kiosk.

Table 19: Nature, Front End, Language, Channel & Initiator of eServices

eService	Nature	Front End	Language	Communication Channel	Initiator	Citizen-Technology Interaction
Government Portal	Informational	Computer Mobile	English	Internet	EGD	Yes
Utility Bill Payment	Financial	Mobile	English Urdu	Mobile Network	Mobilink	Yes
Voter Verification	Informational	Mobile	Urdu	Mobile Network	ECP NADRA	Yes
Kiosk	Informational Financial	ATM-styled machine	English Urdu	Internet	NADRA	Yes/No
e-Sahulat	Informational Financial	Computer software	English Urdu	Internet	NADRA	No

Source: Own Illustration

The categories of Nature and Front End in Table 19 are quite revealing in several ways and they also hint at the model or pattern of eGovernment emerging in Pakistan. While on one hand the Nature of eServices provides information about the government's priority towards eGovernment strategy, on the other hand it also hints at the citizen demand that these eServices are trying to cater to. Can it be deduced that there are more financial services options because they are more in demand than informational services? Or is the monetary aspect associated with financial transactions that service providers have been attracted by? NADRA's kiosk and eSahulat whereas Mobilink's eService reviewed in this study are all associated with the digitisation of utility bill payments. Similarly, it is not only NADRA or Mobilink that offers these financial eServices. There are other mobile network operators in Pakistan that have also initiated their own utility bill payment mechanisms[892]. However, with these financial services, it appears that there is less focus on informational services which happens to be the first criterion for any eGovernment model.

Another important factor in Table 19 is that of a front-end, that is, how the service provider wants citizens to access its eService or in other words through which communication technology the citizen will access the eService. Will it be web-based accessible via computer or mobile? Will it be mobile only or will there be an entirely different mechanism other than computer or mobile phone? We see that there is only one eService, government of Pakistan's web portal accessible via web and computer or mobile whereas rest of the eServices are either offered through mobile phones or via kiosk machine or the eSahulat software available at different retail stores or NADRA offices. This particular trend takes us back to the comparison of mobile and internet subscribers in Pakistan (see Chapter 3 and 4). While we see an exponential rise in mobile phone subscribers, the internet users between 2003 through 2012 have not been able to cross 10 users per 100 inhabitants in Pakistan. The eServices by NADRA, ECP and Mobilink seems to tap the very potential of mobile phone penetration in Pakistan whereas the government on the other hand supported such initiatives through revised regulations[893].

Figure 40: Pakistan's eGovernment scenario based on this study

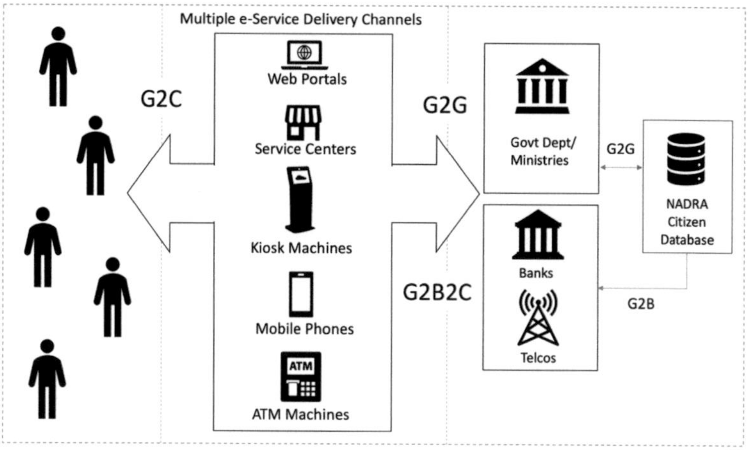

Source: Own Illustration

---

892 Mobile Banking is a growth story in Pakistan (2013b, March 14), Business Recorder. Retrieved 04.05.2014 from https://fp.brecorder.com/2013/03/201303141163171/

893 Pakistan Telecommunication Authority. (2009b, December 16). All Banks to be Connected with Mobile Operators for Mobile Banking Soon. [Press release]. Retrieved 04.05.2014 from https://www.pta.gov.pk/en/media-center/single-media/all-banks-to-be-connected-with-mobile-operators-for-mobile-banking-soon

Based on the corollaries mentioned above, what kind of eGovernment does Pakistan eventually follow? If all the eServices mentioned in Table 19 are compared with the eGovernment model presented in Chapter 2, such as the Layne & Lee model, a number of differences can be pointed out. For instance, in the light of Layne & Lee model, Pakistan's official web portal developed by the EGD (see chapter 4), has so far not been able to provide fully fledged eServices. Fully fledged here refers to the fact that (majority of) public services are available through web portal and all the government departments are digitally linked to provide not only departmental eServices but also cross-departmental services as well. In case of Pakistan's official web portal, most of the information and services provided in that web portal only qualify up to the catalogue and/or transaction stage, that is, the first two stages in Layne and Lee model. However, this qualification (to catalogue and transaction stage) cannot be considered as complete, instead it seems a partial qualification.

For a number of reasons, it amounts to EGD portal's partial qualification to Layne & Lee's model. Firstly, in multilingual countries particularly Pakistan here, Urdu is the national language, but most webpages on EGD's portal carry information in the English language. This aspect particularly limits the access of those citizens who are not conversant with English[894]. Similarly, apart from downloading government forms or contact details of the government officials, there are no further eServices provided on government web portal (see for example Chapter 4). For instance, the services offered by kiosk machine, eSahulat or by mobile companies are not available on government's official web portal (at least up to 2012). While the mobile and internet subscriber figures do explain this lack of interest in further developments in G2C services for web/internet users, but it also highlights the fact that there seemed to be a silent admission by the EGD during 2002 through 2012 to the rise of other eService competitors and perhaps also a lack of commitment or resources to pursue similar projects for web/internet users.

NADRA and mobile network operators have initiated their own G2C eServices. Their eServices are of a financial nature and they do not provide any government related information as is the case with web portals. This brings us to fourth aspect in Table 19, that is of Communication Channel. The service delivery channels in the case of ECP, Mobilink, Kiosk and eSahulat are entirely different from the web portal. For instance, in the case of election commission's voter information mobile application (see Chapter 6), service delivery mechanism is a simple mobile phone. Even a smart phone is necessary in that case. Similarly, in the case of NADRA's kiosk machine, the service delivery mechanism is a standalone ATM-style touch screen. NADRA's eSahulat programme does not even require a use of any electronic gadget on citizen's part and entire request processing is handled by a person who owns the eSahulat franchise. Service delivery channel in case of

---

894   Permal maintains that digital market can heavily benefit and further flourish shall the frequent use of native language be adopted in Pakistan. See Permal, A. J. (2014). Going native. Retrieved 04.05.2015 from https://aurora.dawn.com/news/1140683/going-native

eSahulat is NADRA's computer software whereas the communication medium is internet but that keeps the citizen out of personal interaction with technology[895].

If ECP, NADRA and Mobilink's eServices are evaluated against the Layne & Lee model, we find an interesting trend in Pakistan's eGovernment scenario. Layne & Lee's eGovernment model's stage four and five are referred to as horizontal and vertical integration[896]. These stages imply the electronic public service delivery's attainment of a maturity level. The public services in the vertical stage are completely digitised and accessible to citizens via digital communication medium such as web/internet. Whereas at the horizontal stage different eServices are also able to interact with each other to provide an accumulated reply to the citizen. Although EGD's web portal only partially qualifies to Layne & Lee's first two stages of catalogue and transaction, the eServices initiated by the ECP, NADRA and Mobilink qualify the fourth and fifth stage of vertical and horizontal integration stage. All the ECP, NADRA and Mobilink's eServices are not only completely digitised services (vertical integration) but they also interact with each other (horizontal integration). For instance, in case of ECP, NADRA's citizen identity (CNIC) database helps in citizen verification whereas ECP's database provides the voter registration information to the citizen. Similarly, Mobilink's utility bill service uses NADRA's eSahulat platform for billing information and money transaction. Both the ECP's service and Mobilink's service rely on the databanks of NADRA which corresponds to the horizontal integration stage in the Layne & Lee model.

There are however three important considerations while claiming ECP, NADRA and Mobilink services to be of horizontal integration stage. Firstly, Pakistan's eGovernment scenario as a whole cannot be entirely considered qualifying the horizontal integration stage. The horizontal integration stage as in Layne & Lee's model refers to the digitisation of all (or majority of) public services as well as their interaction with each other. In case of Pakistan's eGovernment scenario, there are only a few services that so far interact with one another. Similarly, as of Layne & Lee's model the services are accessible via web portal whereas in case of Pakistan the services are available on a wide variety of platforms which includes mobile phones, kiosk machines and some are over the counter, that is, in case of eSahulat where the citizen uses a franchisee's shop to perform the particular task (such as payment of utility of bill) on her/his behalf.

The case of Kiosk and particularly eSahulat brings us to the fifth aspect in Table 19, that is, Citizen-Technology Interaction. Based on the definition of eGovernment as a concept and the use of ICTs public services are supposed to be available via digital communication media. Similarly, and more importantly the ICTs were supposed to make access to public services more efficient and effective thereby giving citizens a faster and better access to public services as compared

---

895   National Database Registration Authority (2009c). "About e-Sahulat". Retrieved 14.09.2009 from http://www.esahulat.com.pk/subpages/about_esahulat.php.

896   Layne, K., et al.: 2001

to the manual scenarios. The eServices reviewed in this study as also listed in Table 19 present a slightly different scenario when Citizen-Technology Interaction is being analysed. While in the case of ECP, Mobilink and EGD's web portal, the citizen has complete and independent access to the service, communication channel and the front-end. Of course, the citizen could, for example seek assistance in the use of and access to public service. However, s/he is in control of all the three aspects, that is, the front-end, communication channel and the service itself. In case of Kiosk, the citizen can either independently access the kiosk machines or there is assistance available. However, eSahulat completely keeps the citizen out from her/his interaction with technology. The front-end and communication channel is completely in control of the eSahulat's franchise owner.

With the above-mentioned developments in public service delivery channels such as mobile phones, kiosk machines and eSahulat, it appears that a new pattern of eGovernment is emerging in Pakistan. This pattern seems to have realised that for the provision of eServices, web portals may not be among the most effective options for electronic public service delivery for Pakistan (at least at this stage). Considering the small number of web/internet subscribers in Pakistan, the ECP, Mobilink and NADRA eServices have exploited and engaged with multiple channels of electronic public services delivery that (for now) avoid the use of web portals and personal computers and offer eServices via alterative communication media such as mobile phones, kiosks and eSahulat service centres. In this context, it appears that there is a variety of mixed and/or hybrid strategies of manual and automated systems are employed to provide electronic public services.

The use of hybrid strategies in eGovernment systems is a known practice especially in several under-developed countries where technological capabilities for government and citizens are limited in nature[897]. These capabilities are not limited to only government and citizens but these also include a number of other factors such as internet access, digital divide, language barriers, e-literacy or affordability of gadgets that contribute opting for hybrid strategies[898]. With the eServices presented in this study, Pakistan appears to be following the same hybrid strategy of eGovernment where it uses a mixed approach of offering manual and digitised public services.

Figure 41 attempts to review the hybrid model of eGovernment in Pakistan in the light of Layne & Lee model (as depicted in Chapter 2). The difference in Pakistan's hybrid eGovernment model and the Layne & Lee model is at catalogue and transaction stages of eGovernment and vertical and horizontal integration stages seem to be working independently. Ideally and as per Layne and Lee model these stages build upon each other where front-end is a web portal whereas the communication channel is the internet. In Pakistan's case, instead of having a web

---

897 Bhatnagar, S.: 2009

898 Ramli, R. M. (2012). Hybrid approach of e-government on Malaysian e-government experience. International Journal of Social Science and Humanity, 2(5), 366-370

browser/personal computer to access the eServices via website, eServices are provided through multiple channels that may include web portals, service centres, kiosk machines, mobile phones, landline telephones and/or banks.

Figure 41: Layne & Lee model's comparison with eServices of this study

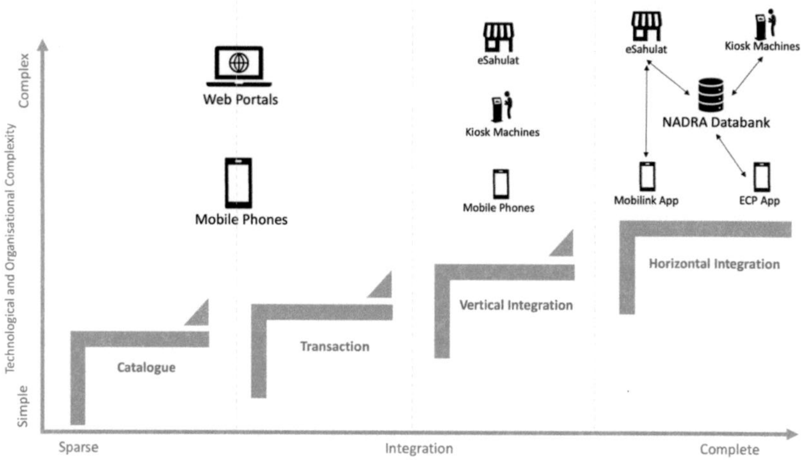

Source: Own illustration

A number of countries, such as Brazil, Malaysia, China or India are also taking initiatives that involve a plethora of hybrid eGovernment initiatives that works as an alternative eGovernment model[899]. In hindsight, hybrid strategy of eGovernment application changes G2C scenario in eGovernment debate and opens up new or revised ways of electronic communication between government and citizens. The prevalence of the digital divide and the emergence of hybrid strategies, particularly in technologically less developed countries, is providing a new opportunity of government to citizen interaction that can and will result in not only electronic public service delivery, but also in creating new jobs and more decentralised approach to the service delivery mechanism[900]. However, if these hybrid strategies wish to reach out to majority subscribers then recognition of communication patterns and information channels is highly important to access majority public[901]. Similarly, the language of these channels will also play a decisive role, otherwise it is very likely that the fate of these hybrid strategies and communication channels meets the similar reception as is the case of web portal.

---

899 ibid.

900 Heeks, R.: 2002a; Lau, T., Aboulhoson, M., Lin, C. and Atkin, D. J. (2008). Adoption of e-government in three Latin American countries: Argentina, Brazil and Mexico. Telecommunications Policy, 32(2), 88-100; Rocheleau, B.: 2007

901 Reddick, C. G., et al.: 2012

236

The following section takes a look at some of those communication patterns that can be employed for the eGovernment applications in Pakistan.

7.8. Government Information via existing communication channels

There is a growing consensus among technology consultants and experts in Pakistan that the emphasis on information access and dissemination shall be given a similar importance as is currently given to financial applications[902]. In this regard, it needs to be seen that in order to put hybrid strategies into practice, which communication channels are being employed. The case of the web portal makes it clear that in the presence of new communication devices such as mobile phones, web browser and internet as sole communication medium have a limited reach in numbers. In the case of mobile and internet users in comparison, it seems that people's preference towards the choice of communication medium is based on traditional and existing communication pattern, which in this case is oral communication. Therefore, it will be important to investigate and adopt those existing preferences of communication in Pakistan to have a wider public outreach. These communication channels are supposed to be more prevalent and omnipresent than the Internet. These channels share wider appeal and people's trust in these channels may be more enhanced than in new technologies. This section takes a look at few existing communication channels in Pakistan that people have used for decades. Such channels may complement the growth of G2C services and help improve the efficiency and effectiveness of public services in Pakistan.

7.8.1. Post Offices as a potential government service delivery

Post offices are considered as the most indigenous communication channel that has the potential to work as a catalyst for eGovernment applications[903]. In South Africa, for example, throughout the country, the government has installed public information terminals in about 800 post offices[904]. These post offices together with other multipurpose community centres aim to increase the adoption and use of ICTs in order to provide inclusivity in building an information society. Historically, the role of postman in Pakistan is not limited to the delivery of post. The postman in his personal capacity and out of his sheer respect in community performs more roles than he is required to or paid for. Due to lower literacy standards, the postman in many rural settings in Pakistan is often asked by the

---

902   Interview with Deirdre Williams, November, 2009

903   Rose, R. (2005). A global diffusion model of e-governance. Journal of Public Policy, 25(1), 5-27

904   Mutula, S. M. and Mostert, J. (2010). Challenges and opportunities of e-government in South Africa. Electronic Library, The, 28(1), 38-53

citizens to read out their letters to them[905]. It is also common that the postman writes letters on people's behalf as they dictate their content to him. Similarly, the post office's multiple roles - in providing money investment facilities and money transfer schemes - make its case as an indigenous communication channel very plausible, its capacity needs to be utilised for the dissemination of information at a wider level[906]. The network of post offices in Pakistan is among the largest after mobile communication. The post offices even operate in areas where at times mobile signals do not reach. If the post offices are equipped with software made on the pattern of eSahulat, then a huge rural populace of Pakistan can be served with information for which they may not have to travel distances and spend money on information retrieval.

Figure 42: Depiction of SIM menu of Mobilink's on-the-air eServices

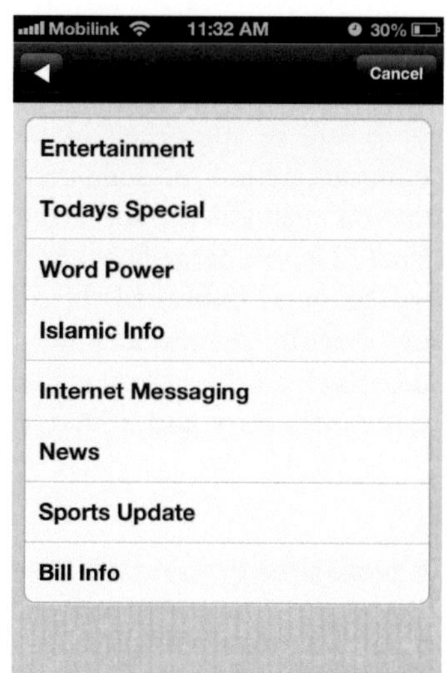

Source: Author

7.8.2. Mobile Phone SIM menu

Similarly, mobile phone's operational capabilities are also considered as a potential communication channel for not just financial applications but also for government information delivery[907]. Therefore, mobile phone's SIM menu could

905   VOA Urdu (2011). ڈاکیا : ماضی کا مضبوط معاشرتی کردار زوال پذیر ہے [Postman: The strong social character of the past is on the decline]. Retrieved 04.05.2014 from https://www.urduvoa.com/a/pakistan-postal-service-21apr11-120382089/1132690.html

906   Beynon-Davies, P. (2007). Models for e-government. Transforming Government: People, Process and Policy, 1(1), 7-28

907   Trimi, S. and Sheng, H. (2008). Emerging trends in M-government. Communications of the ACM, 51(5), 53-58

be channelised for this purpose. Currently every mobile phone in the world, regardless of being smartphone or not, has the ability to open SIM menu[908]. Mobile operators according to their own vision specially design the SIM menus. This information is sent on air where the SIM operates as a middleware between the user and the SIM menu that mobile operators use. Generally, the SIM menu contains information such as stock reports, news, sports, weather and/or bill payment. This is the mechanism through which Mobilink's Bill Payment functions or Telenor's Easypaisa initiative[909]. If the government realises or takes interest in using this medium as a tool to disseminate government information, then this could function as a widely reachable information service. The example of such a SIM menu may include information that a citizen could seek about procedure to open a bank account, information about a job sector, birth certificates, agricultural-related announcements, commodity prices or any other government to citizen information that are otherwise a part of a web portal[910].

7.8.3. Speech to Text Solutions

Utilisation of local and social capital towards developing citizen centric eServices based on existing patterns of communication can also help produce innovative eServices. The idea of speech to text solutions is also based on oral communication models. A number of call centres already use speech to text approaches in many parts of the world[911]. An aspiring customer or citizen calls up the call centre in order to seek information. In normal call duration s/he goes through a number of voice menus that either ask in return to press a certain button or to speak. Bajwa hints at the mobile subscriber in Pakistan who sometimes is not even able to operate the mobile but is aware that if they press a green button twice, this would eventually link them with the last call made[912]. He points out a more inclusive approach towards government to citizen services via speech to text or speech-to-speech mechanism that could read out the information to the user[913]. The information received through this procedure would eventually include those

---

908   Susanto, T. D., et al.: 2010

909   Telenor (2014). "About Easypaisa". Retrieved 04.05.2014 from http://easypaisa.com.pk/index.php/en/about/about-easypaisa.

910   Singh, A. K. and Sahu, R. (2008). Integrating Internet, telephones, and call centers for delivering better quality e-governance to all citizens. Government Information Quarterly, 25(3), 477-490

911   Power, D. and Power, M. R. (2009). Communication and culture: signing deaf people online in Europe. Technology and Disability, 21(4), 127-134; Fuglerud, K. S. (2009). Universal Design in ICT Services. In Vavik (Ed.), Inclusive Buildings, Products & Services: Challenges in Universal Design (pp. 244-267). Trondheim: Tapir Academic Press; Venkatraman, S. and Hughes, S. (2009). The development of an information systems strategic plan: an e-government perspective. International Journal of Business Excellence, 2(1), 50-64

912   Interview with Fouad Bajwa, Sharm el Sheikh, November, 2009 and Lahore, 2010

913   Watling, S. (2011). Digital exclusion: coming out from behind closed doors. Disability & Society, 26(4), 491-495

illiterate citizens who cannot read or write but regularly use mobile phone on a daily basis[914].

7.8.4. Collaboration, intermediaries and Public Private Partnerships

Most of the interview and conversation partners for this study have emphasised the fact that the government cannot be left solely responsible in creating capacity, providing digital literacy, building IT infrastructure and/or spreading awareness for the provision of electronic public services. The private sector in this regard needs to be engaged to closely work in association with public sector. The role of Universal Service Fund (USF) has repeatedly been highlighted for the inclusion of telecommunications in the rural sector[915]. The USF, as also discussed in this study involves all the telecom operators for the growth of telecommunication in Pakistan. The USF which is built on the Public Private Partnership model involves no funding from the government and runs on the basis of 1.5% revenue contributions of the telecom operators[916]. In this regard, the role of the USF is to use the revenue collected from the operators' contributions and to further invest in the telecom services for the un-served or under-served areas in Pakistan. Such a public private partnership overseen by the Government of Pakistan has seen the spread of telecommunication to over 1600 districts in Pakistan's rural sector.

The role of public private partnerships has been considered a valuable contribution to the ICT and eGovernment projects[917]. While working at the Information and Communication Technologies for Development section in 2005 at the GIZ headquarters in Eschborn, Germany, the author witnessed the interest in public private partnerships that sees to dominate Germany's development agenda as well. GIZ in collaboration with the Federal Ministry of Economic Cooperation and Development has cooperated in many projects in developing countries that have brought together local partners together with the national governments. Similarly, as part of a working group during GIZ organised conference on eGovernment for development, the author observed strong reciprocating interests from officials of African and Asian countries who wanted

---

914   William McIver, J. (2003). A Community Informatics for the Information Society. In Girard and Siochrú (Eds.), Communicating in the Information Society (pp. 33-64). Geneva: United Nations Research Institute for Social Development (UNRISD)

915   Siegmann, K. A. (2010). The Gender Digital Divide in Rural Pakistan-To Measure and to Bridge it. Retrieved 17.08.2012 from https://repub.eur.nl/pub/22393/SiegmannGenderDigitalDivide.pdf

916   International Telecommunication Union (2013d). Universal Service Fund and Digital Inclusion For All Study. Retrieved 04.05.2014 from https://www.itu.int/en/ITU-D/Conferences/GSR/Documents/ITU%20USF%20Final%20Report.pdf

917   Norris, D. F. (2008). E-government research : policy and management. Hershey: IGI Pub; Keogh, D. and Wood, T. (2005). Village Phone Replication Manual: Creating sustainable access to affordable telecommunications for the rural poor. New York: United Nations Information and Communication Technologies Task Force. http://www.infodev.org/infodev-files/resource/InfodevDocuments_14.pdf; Nair, K. and Prasad, P. (2002). Development through information technology in developing countries: experiences from an Indian state. The Electronic Journal of Information Systems in Developing Countries, 8(1), 1-13

to move beyond the red tape and invest more in such partnerships[918]. The conference concluded that the donors needed to develop an understanding about the private sector products, and they must be aware of the costs, trade-offs in order to help their clients in making the best choices for their public-sector management needs. In order to create digital capacities, as most of the conversation and interview partners of this study argued, the intermediaries' roles should be further streamlined. In this regard, internet cafes and public call offices are crucial and are currently being used by the public which requires internet use on day to day basis but based on their limited knowledge about the computer and the internet, they rely on the owner or the assistant of these café/offices in order to fulfil their needs[919].

7.8.5.Role of Mass Media in Government Information dissemination

One of the most important factors in Pakistan's eGovernment scenario is that the provision of electronic services such as identity card, electronic passport, utility bill payments, birth, marriage, and death certificate issuance mechanisms have been digitised. There are multiple procedures and multichannel services for citizen verification (such as through NADRA kiosk and eSahulat platform), mobile SIM verification (SMS-based service) or voter verification.

However, for the provision of government information data there are so far very limited options available such as the government website. The government procedures, rules and information about ministries or departments is only available via Pakistan's official government portal which is so far only available in English. From the eServices that are presented in this study, more focussed on financial services, that is, of a transactional nature. This also indicates that on one hand the service providers, that is Government of Pakistan as well as mobile telecom companies want to exploit the financial benefit by offering services such as bill payment, tax payment, money transfer etc. However, they do not seem concerned about the fact that information dissemination is as crucial to eGovernment debate and in fact a prerequisite for further adoption and better service delivery. From the website analysis through the Layne & Lee method (see chapter 4), Pakistan's official portal does not even qualify for the first stage (catalogue stage). The government information about the laws, regulations, and procedures is either incomplete or redirects to the invalid/dead webpages. There are no multichannel services for the catalogue stage neither do they seem to be a priority as seen in the status quo.

---

918 GIZ. (2008). eGovernment for development: The Promise and the Practice (A Conference Documentation). Retrieved 04.10.2010 from http://www2.gtz.de/dokumente/bib-2009/gtz2009-0504en-e-government-conference.pdf
919 Interview with Mr. Ahmad Naseem, March 2009; Interview with Shaheryar Idress, December 2010.

Pakistan's mass media, which has expanded at an exponential pace, in the past ten years has so far not been employed to do this task. With Pakistan's over 80 television channels, cable televisions and radio services reaching every remote corner of Pakistan offer a unique opportunity[920]. The mass media provides a potential to be employed as one of the eGovernment communication channel which can be used as a formal method to guide the citizens about government procedural information as an alternative way to compliment Pakistan's official web portal.

7.8.6. Traditional institutions and multipurpose community telecentres

Among the rural communities and countries with limited technology diffusion and teledensity, village kiosks and telecentres are also considered as a major player in bridging the digital divide[921]. Such telecentres are being used not only for the dissemination of knowledge but the citizens can also have access to eServices such as livestock rates, weather reports, crops and land records[922]. In Pakistan where the majority of the population lives in rural areas, many among them are living below the poverty line and cannot afford to have a telephone or computer. In such circumstances, community telecentres have the ability to help the rural population in gaining access to ICTs. In his study Mahmood has referred to three initial multipurpose community telecentres that started in Gwadar (south-west Balochistan), Mithi (south-east Sindh), Usterzai Payan (Khyber Pakhtunkhwa) from 2001 onwards[923]. These community centres were initiated by the Sustainable Development Networking Program (SNDP) in collaboration with the local NGO in order to provide assisted access to the internet, training courses and telephone call facility.

In a country such as Pakistan, the telecentres may well be a useful solution for initial eServices however, traditional institutions such as mosques in Pakistan play a more important role when it comes to information dissemination among close-knit communities in the rural areas of Pakistan[924]. The communication pattern that exists through such traditional institutions can also complement the catalogue stage of eGovernment for the information broadcast. In Iran and Nigeria, such traditional channels have already been employed for the use of government

---

920   Pakistan's next media revolution (2013h, February 14), The Express Tribune. Retrieved 04.05.2014 from https://tribune.com.pk/story/506846/pakistans-next-media-revolution/

921   Madon, S. (2005). Governance lessons from the experience of telecentres in Kerala. European Journal of Information Systems, 14(4), 401-416

922   Kumar, R. and Best, M. L. (2007). Social impact and diffusion of telecenter use: A study from the sustainable access in rural India project. The Journal of Community Informatics, 2(3), Special Issue: Telecentres

923   Mahmood, K. (2005). Multipurpose community telecenters for rural development in Pakistan. The Electronic Library, 23(2), 204-220

924   Rawan, S. M. (2002). Modern mass media and traditional communication in Afghanistan. Political communication, 19(2), 155-170

information. Unegbu and Haliso in their study have highlighted traditional rulers, mosques and churches in Nigeria which play effective channel in local government financial information communication[925]. In Iran's East Livan village of Golestan province example, one such initiative of involving community telecentre together with the mosque, rural clinic, schools, local community and government office has been experimented[926]. Built with the cooperation of UNESCO and World Bank, these telecentres aim to provide education for all ages and for all people, cultural activities, e-learning modules and general information centres for health and agriculture information. In Malaysia, a telecentre has been placed in the mosque compound in order to use the mosque as a potential vehicle of digital information dissemination[927]. In Pakistan, a traditional institution such as mosque can also help in becoming an alternative communication channel through which G2C interactions can be supported either in the form of kiosk or community centre.

925   Unegbu, V. E. and Haliso, Y. (2012). Local Government Financial Information Communication, Citizenry Awareness and Empowerment in Nigeria. Journal of Economics and Sustainable Development, 3(12), 116-128

926   Jalali, A. (2006). Socio Economic Impacts of Rural Telecenters in Iran. In World Bank Seminar on Women's Economic Empowerment and the Role of ICT [presentation slides].Retrieved 24.10.2012 from http://siteresources.worldbank.org/INTGENDER/Resources/AliJalali.pdf

927   Razak, N. A. and Malek, J. A. Education for All: Meeting Millennium Development Goals via the Telecenters. In Proceedings of the 5th WSEAS/IASME International Conference on Educational Technologies (EDUTE' 09). (pp. 218-220). Retrieved 04.05.2014 from http://www.wseas.us/e-library/conferences/2009/lalaguna/EDUTE/EDUTE-33.pdf

Chapter 8

Conclusion

In this study, an attempt has been made to understand the role of modern ICTs, their significance, applicability along with challenges in electronic public service delivery in Pakistan. In this context, this study focussed on a time period of 2000 – 2012. This time period is, historically as well as politically, important due to substantial and decisive developments that took place in the field of eGovernment in Pakistan. While approaching the central research question, how far modern ICTs affect the public service delivery mechanism in Pakistan, this study chronologically followed the institutional and legal advancements in the country as well as attempts in the realm of eGovernment. The academic explorations pursued in this study reveal how modern ICTs have led to the formation of new public institutions (such as NADRA and EGD) and also their subsequent successes and failures (at least in the case of EGD). In pursuance of a second query, understanding patterns of digital communication to channelise and navigate electronic public services, this study explains and analyses the official eGovernment strategy adopted by the Government of Pakistan. While exploring and chronicling the evolution of ICT-based electronic services, this study came across a number of eGovernment projects initiated by public sector organisations such as EGD (eGovernment portal), NADRA (kiosk machine and eSahulat programme), joint public-sector digital collaborations (between NADRA and ECP), and some initiatives by the private sector institutions such as Mobilink.

There are a number of key findings that have come across as part of this research in understanding the eGovernment scenario of Pakistan, and these include:

♦ Internet is an important tool and backbone of most modern communication. However, this study found that there are limited number of internet users in Pakistan where mobile phone users have exponentially increased. EGD's government web portal in this regard has a limited audience and there are also other discrepancies on government's part, for instance, the portal does not offer the content in local language(s).

♦ The official eGovernment strategy for Pakistan does not underscore the digital divide and eReadiness factors neither does it highlight establishment of any digital literacy projects in Pakistan. Most of the EGD projects, instead catered to the automation of manual processes.

◆ NADRA's kiosk and eSahulat initiative seems to have revamped the entire eGovernment landscape in Pakistan and the focus from information-based services has shifted to financial services.

◆ The mobile network operators in Pakistan are following the same pursuit as NADRA and while exploiting opportunities for increasing their customer base, they have initiated electronic public services through their digital and retailer networks.

◆ Kiosk machine project and eServices via mobile phones reveal that there is a growth of new (electronic) service delivery mechanisms in Pakistan that are socially, traditionally and culturally more viable.

◆ The new regulations such as Electronic Transaction Ordinance (ETO) and Branchless Banking are making it possible for public- and private-sector organizations (NADRA, ECP, banks, mobile network operators) to digitally collaborate with one another to produce new eServices.

This study has looked in detail at the formation, functioning and output of institutions that successive governments of Pakistan specifically launched to further the eGovernment implementation. For that matter, Pakistan constituted Electronic Government Directorate (EGD) in 2002 as an attached department under Ministry of Information Technology to develop eGovernment standards and Pakistan's official eGovernment strategy. Between 2002 and 2012, EGD undertook several tasks which included computerisation of ministries, identification of basic eServices and development of Pakistan's first eGovernment strategy which was launched in 2005. The problem with most of the projects that EGD undertook was that they lacked the understanding and identification of public services and their delivery, as pointed out in Chapter 4 and 7 in this study. Similarly, the eGovernment strategy developed by EGD is so far, the only officially documented strategy. The document among other aspects, advocated the development of common software applications for ministries in order to introduce standardised paperless office procedures. However, the fact remains that during the span of 10 years, EGD could not develop any national level G2C service and instead it continued to surround itself with a number of controversies including corruption and embezzlement of money (see Chapter 7). These controversies badly affected EGD's reputation and led to its merger with a sister organisation, Pakistan Computer Bureau (PCB), to constitute National Information Technology Board (NITB). EGD which was initiated with a lot of expectations and was supposed to carry Pakistan through eGovernment implementation process, could only survive for 10 years before getting stripped off its mandate.

Another aspect that also (indirectly) contributed to EGD's dissolution was the Government of Pakistan's launching of different institutions with similar mandates. For instance, two years before EGD's formation, Government initiated National Database Registration Authority (NADRA) as an attached department to the Ministry of Interior in 2000. This agency which was tasked to document, digitise and create a citizen database preceded EGD's output by five years and was well on its way to digitally mapping citizen identities, civil registry system, biometric passports and vehicle registration mechanism. The family-linkage system, which NADRA improved later on, also brought forth multiple mapping options for trading, property management and legal issues (see Chapter 5 and Chapter 7). This meant that by the time of implementation of the EGD's eGovernment strategy, NADRA was well ahead of EGD and most of its applications qualified the transaction stage (compared with eGovernment frameworks).

eGovernment and most of its propagation models as discussed in Chapter 2 elude to a stage- by-stage dissemination of the service which builds on logical succession to widespread use of eGovernment. The very first stage of any eGovernment model, that is, catalogue or informational phase advocates that any country pursuing an eGovernment implementation shall, in first place, work towards public dissemination (or cataloguing) of governmental procedures using modern ICTs in order to facilitate citizens. This stage is followed by transaction stage which enables the interaction (that is digital communication) between different actors (G2C, G2B). The last two stages, vertical and horizontal integration support digitisation across departments and also across different government sectors and other private bodies.

Such models have so far served as a blueprint for eGovernment adopters. However, experiences gained by Pakistan and other developing countries have been varied (see Chapter 2 and Chapter 7). To qualify the stages mentioned above, prerequisites for any eGovernment plan and implementation include among others, a well-established digital infrastructure, outreach and digital literacy. However, Pakistan's problem, as this study has pointed out, is not just the digital divide but more importantly the information divide. Even though the employment of various ICTs (fibre, co-axial cable and wireless local loop) through different user-platforms (desktop-computers, mobiles, kiosks) have led to the transaction and integration level eServices in Pakistan, it does not rule out the fact that the ICTs have not helped heal the information divide in Pakistan. The possible venue to purge the information divide within eGovernment scenario are government websites/web portals with which Pakistan's experience remains limited, as this study has pointed out.

The web-content analysis of the flagship eGovernment website of Pakistan, www.pakistan.gov.pk developed by the EGD, shows that although most of the initiatives and services are listed, a vast majority of them do not work or lead to broken URL links (see Chapter 4). The absence of local language content on government's official portal leaves the notion of inclusion and culturally-aware content high, dry and cold. When basic information is not available, there is no easy way for citizens to find out the services available to them, the options to avail them or even knowing simple facts like the requirements and timelines for acquiring these services. Similarly, it also prevents the closure of older methods of public service provision as citizens who never learn about the new initiative or even know of its existence, find no reason to migrate to eGovernment initiatives. The fact that most of Pakistani citizens lack fixed internet connection and rely more on mobile telephony, shows that for disseminating government information, the choice of platform needs to be revisited. As the analysis in this study has showed, most of the government websites have the same problem. Very few of them have complete, authentic and regularly updated information and still a miniscule number offer mobile version of their site.

In the current eGovernment scenario, technology adoption (banks, ATMs, unassisted kiosks, mobile services) can be observed in Pakistan. However, the fact remains that Pakistan's current eGovernment scenario does not completely qualify for the catalogue stage (see Figure 41). Such a scenario will eventually be a paradox to the eGovernment implementation in Pakistan. While Pakistan may continue to work on eliminating digital divide, the information divide will continue to prevail. NADRA's kiosk machines or the evolution and adoption of eService like eSahulat, are evident that demand for these facilities exists. These options present a workable communication channel which has the potential to also serve the catalogue stage, that is, government information available through kiosk or eSahulat type of platforms. These experiments may require careful engineering to deconstruct and market it appropriately to ensure its uptake. Jumping across the catalogue stage, however, will not ensure this ubiquitous uptake and nor will it help in efficiently streamlining parallel streams of electronic and non-electronic public service delivery and information.

The use of mobile phone SIM card's built-in application (SIM Toolkit, popularly known as STK), also provides a good opportunity to be used as a possible extension of information dissemination through mobile phones. The STK are currently used to offer value-added services by private mobile operators and can be used to get information such as news, weather forecast, sports updates or prayer timing alerts. This option requires further exploration to create and disseminate at least the catalogue stage information of eGovernment services. Public-private partnerships as discussed in this study have the potential to undertake such information divide oriented case studies to develop citizen-centric services. This exercise may also be a positive way to simulate the local domestic IT market which already has a strong portfolio and can be used to create new stakeholders in this model and increase the exposure of common people.

This study has also highlighted the fact that since its inception, NADRA's portfolio has continued to grow in the field of eGovernment in Pakistan. It has taken over a number of high-profile eGovernment projects not only in but also outside of Pakistan. However, NADRA's kiosk and eSahulat programme would be remembered for redefining government's interaction with citizens within the eGovernment scenario. An important aspect that this study tried to underscore is the emergence of digital communication channels triggered by NADRA. These alternative channels of communication which provide local/indigenous technological solutions such as eSahulat are becoming more dominant. These solutions on the one hand can be projected as socially fit cases but on the other hand they present a number of considerations. For example, a huge number of people in Pakistan leapfrogged from landline phones to the mobile phone subscription. This leapfrogging may have worked for the mobile phone industry, partly because of its oral communication potential, but as far as the human-machine interaction is concerned, this research documents that leapfrogging to the kiosk machine was not realised. People, as of now, still prefer and rely on an assisted technology or its abridgement. In both cases (mobile phone and eSahulat) people have preferred to use a technological option that falls within their existent communication and social patterns.

Similarly, NADRA is also positioning itself as a major public collaborator for electronic public service delivery. It is collaborating with banks and mobile phone companies in order to provide citizen verification mechanism for transactions like opening an account or buying a subscription. It is also taking part in inter-agency collaborations to offer services that are not only favoured for the socio-economic uplift but also the political transformation in Pakistan. In this regard, Election Commission of Pakistan has collaborated with NADRA to provide a SMS-based voter verification system with the central database authority in order to inform over 85 million voters of their electoral registration.

This study has also tried to analyse the emerging and changing eGovernment scenario in Pakistan. It is observed that between 2000 and 2012 the eGovernment scenario has gone through two parallel phases. First is the evolution phase which was backed by the government. This phase saw the launch of institutions (NADRA and EGD) as well as the policies (IT Policy (2000), ETO (2002), eGovernment Strategy (2005)). Along with that, one can observe impact of globalisation and private sector shape the adoption which is backed by market forces, digital and knowledge divide, eReadiness and ICT adoption factors. This phase witnesses the public-sector entities such as NADRA acknowledging the digital divide and presence of limited internet users with the launching of kiosk machine. With the government's legal cover (Branchless Banking regulation) the mobile network operators use NADRA's experience and also enter the G2C market. Following the kinds of eServices and the technological channels used, this study has also tried to understand the nature of eGovernment currently practised in Pakistan. This model is hybrid in nature and utilises both manual and automated procedures. With such a model in place, this study has also observed the growth of multiple channels electronic public service delivery. These channels of eGovernment communication are entirely different from the eGovernment models which are practised particularly in the digitally advanced countries where web/internet is main channel of communication. The multiple channels in Pakistan range from kiosk machines to mobile phones and using third-party retailer networks. Similarly, this study has also argued that technological solutions that are similar to social culture and customs may find more recipients if the technologies are socially embedded. That is, technological solutions correspond to historical patterns of communication, take indigenous knowledge into account and communicate in local languages.

Such a model of eGovernment may continue to evolve and mature over time and at the same time will be influenced by rapidly changing IT industry and global trends in technological developments. Several countries are looking into the potential benefits that increased computer processing power, big data storage and widespread networking of databases can bring. Artificial Intelligence (AI), which is the training or programming of computer modules to do tasks of a repetitive but simpler decision-making nature, that had typically required human presence, could also change service delivery. It is expected to enter the eGovernment domain particularly in citizen-centric services, reducing service time and cost of delivery. In fact, a number of national governments such as the US, Mexico, Japan and Singapore have already launched AI-based citizen platforms that respond to known queries. Imagine an SMS-based service through which user could enquire about the procedure, cost and processing time of an ID-card or other legal document. Based on person's input AI-led application talks to different databases of the public agencies and responds back via SMS or kiosk machine. Such a scenario has the potential to help curb the information divide in a country like Pakistan. However, it might require more legal and technical frameworks for its implementation but is a plausible way to bridge a diverse population with varied technical, linguistic and literacy levels. On the other hand, considering the weak technology infrastructure, as this study has also pointed out, AI-based services could also lead to precarious results in case of data breach.

Nevertheless, the communication paradigm in ICT scenario is rapidly changing in Pakistan. Kiosk machines, eSahulat, SMS-based applications and most of the technological solutions discussed here were launched with a focus on developing strategies that would lead to adoption by a majority of the population. Citizen-centric or citizen-focused approaches are considered as important selling points for any ICT solution. The utilisation of traditional communication patterns seems to be given more importance as observed by the increasing number of options for electronic public service delivery. The research conducted in this study has shown that for a developing country like Pakistan, while there is a scope for eGovernment applications, it must be taken into account that there are several political, societal, social and technological factors that play a role in setting a direction for government to citizen electronic services.

Within the realm of eGovernment, another important aspect that this thesis has reflected upon is the style of governance and origins of civil administration in Pakistan. For that matter, we have dwelled on historical waves of development of Pakistan's bureaucracy and civil service which is a by-product of British colonial legacy. The public services and the concept of service delivery in Pakistan, as pointed out in Chapter 3, is based on rules and regulations that were developed by the British during the colonial period. Elitist and centrist approach towards civil administration along with constant experiments with democracy frequently punctured by military rule has tilted the focus of Pakistan's political preferences more towards democracy vs martial law debate and less towards public service. As a consequence, social tribulations such as pandemic corruption, lax availability and an absence of service oriented approach in public facilitation has diminished the public's trust. These factors among others have also immensely contributed in creating the information divide.

Access to government information and/or clarity about legal procedures and government processes has further shrouded the working of public service. The day-to-day civil and administrative procedures such as registration of birth, death, marriage, divorce, property had been manually handled in Pakistan. By manual it means that there were no standardised practices and every city administration issued legal documents either in handwritten or typewritten form. Forging of these manually issued documents by criminal elements were practices that have been regularly reported in media and known within the government circles[928]. NADRA's Verisys and FRC (Family Registration Certificate) are some G2C eServices that try to address malpractices of identity frauds. Any person undergoing a legal or commercial trade with a fellow citizen can use NADRA's Verisys citizen authentication system via kiosk, eSahulat or mobile phones. Similarly, the FRC service provides a pictorial hierarchy of a citizen in a family tree format which authenticates a person's familial and immediate blood relations printable in A4 size paper format. In several family or legal disputes NADRA's FRC service has the potential to help eliminate counterfeiting. This practice has been declared as a necessary prerequisite not only by bureaucratic and legal system in Pakistan but also by several foreign embassies functioning in Pakistan[929].

The major developments in ICTs, deregulation of telecommunications and formation of new public-sector organisations such as NADRA in Pakistan have also coincided with global and regional developments. Events such 9/11 in US,

---

928   Computerization and the 'patwari culture' (2015, January 28), The Nation. Retrieved 04.05.2015 from https://nation.com.pk/28-Jan-2015/computerization-and-the-patwari-culture

929   Several European embassies and foreign missions of different countries have now made NADRA issued documents such as FRC or CRC a mandatory condition for the issuance of visa. See for example: German Embassy Islamabad (2018). Schengen Visa Requirements. Retrieved 28.12.2018   from https://pakistan.diplo.de/blob/2163332/93695879eadf8da9fea01ecee31f27ed/visa-requirements---schengen-data.pdf

followed by US-led mission in Afghanistan and the spurning of the global "war on terror" brought fragility to Pakistan's eastern and western borders. In the post 9/11 world, Pakistan has struggled with a number of existential threats posed by rising militancy, weak democratic set-up and incidents of sectarian violence with fluctuating frequency. Particularly between 2007 and 2013, Pakistan's security detail was repeatedly compromised and throughout these years several suicide bomb attacks took place in Pakistan. One such attack occurred in one of Islamabad's five-star hotel, just a few hundred meters from NADRA's headquarters[930]. The massive digital database of Pakistani population hosted at NADRA's headquarters in Islamabad puts Pakistan in a vulnerable and questionable situation in the presence of weak security apparatus. The same applies to telecommunication towers of several mobile network operators which remain under constant threat particularly in geographical terrains where militant groups have gained a foothold. Many a time attacks on tower sites had been reported[931]. Amid threats of physical security and virtual threats such as hacking, a redefinition of sovereignty and of the nation-state in a digital world is required.

An additional problem carried over into the new millennium in Pakistan is the acute electricity/power crisis. For G2C services to grow and digitisation to hold value, constant provision of power is a necessity. However, in Pakistan's 70 years of independent history, the power crisis has remained an unsolvable challenge[932]. Electricity shortfalls between 2005 and 2015 have been frequent. Up to half of national power requirements are not being met. On the one hand Pakistan is gradually moving towards multi-channels electronic services but on the other hand it is ironic that the major metropolises in a nuclear armed state suffer electricity shortfall up to 15-18 hours a day[933]. These issues will continue to hinder the growth of not just electronic public service delivery but also IT industry as a whole.

---

930 Both NADRA and the targeted hotel in question are located in the secure so-called Red Zone area of Islamabad and at 5-minute distance from each other. See: Google Maps (2015). "NADRA's location in Islamabad's Red Zone district". Retrieved 04.05.2015 from https://goo.gl/maps/r85vySnqf5QWtgBn8.; Also see: Islamabad Marriott hotel bomb killed 52, says Pakistan (2008, September 21), The Telegraph. Retrieved 04.05.2015 from https://www.telegraph.co.uk/news/worldnews/asia/pakistan/3041148/Islamabad-Marriott-hotel-bomb-killed-52-says-Pakistan.html

931 Kardon, I. B. (2011). China and Pakistan: Emerging Strains in the Entente Cordiale. Retrieved 04.05.2015 from https://project2049.net/wp-content/uploads/2018/05/china_pakistan_emerging_strains_in_the_entente_cordiale_kardon.pdf ; Bid to blow up mobile phone tower foiled (2015d, October 31), Daily Dawn. Retrieved 02.11.2015 from https://www.dawn.com/news/1216554; Militant dies of `heart attack` (2011b, April 23), Daily Dawn. Retrieved 02.11.2015 from https://www.dawn.com/news/623414/newspaper/newspaper/column

932 Kessides, I. N. (2013). Chaos in power: Pakistan's electricity crisis. Energy policy, 55, 271-285; Pakistan's power crisis may eclipse terrorist threat (2012, May 27), The Washington Post. Retrieved 04.05.2015 from https://www.washingtonpost.com/world/asia_pacific/pakistans-power-crisis-may-eclipse-terrorist-threat/2012/05/27/gJQAPhOSuU_story.html

933 Power shortfall exceeds 6,000MW (2014b, April 29), Daily Dawn. Retrieved 04.05.2015 from https://www.dawn.com/news/1102945

The three hypotheses (Chapter 1) that were set out in this study are presented as observations here:

Growth and development in technology and specifically Information and Communication Technology have indeed heralded an era for reinventing public service delivery which has the potential to evolve better governance, introduce accountability, trim down bloated budgets and increase efficiency. Building on the promise of higher rates of data fidelity and standardised digital record management, these enhanced services also assure a tailored and specific experience at the time and convenience of the citizen. These theoretical benefits are indeed grounded in the presence and uptake of a number of eGovernment initiatives in Pakistan including the NADRA-ECP voter verification SMS service or NADRA Kiosk and eSahulat. These have also been extended by the private sector as was demonstrated by the case study of the branchless banking by Mobilink. This mobile banking service is now in fact available through multiple banks and telecom operators and has found a ready market in rural and far-flung areas making mobile usage all the more pertinent.

Another factor that has been highlighted in the analysis of the eGovernment amenities is the uptake of web-based services. As discussed in the eGovernment portal index (Chapter 4), the relative quality of these services has not gone beyond the interaction stage. Majority of websites still curate information, updates and procedures for their respective agencies but do not offer provisions for completing tasks online. This points to absence of both the push and pull factors in adoption. Despite having passed a national ICT policy and harmonising an eGovernment plan, implementation and realisation of these aspirations is late in coming from the government's side. Simple tasks still require visits to offices of government agencies and interaction with bureaucracy. Absence of procedures and simple user guidelines make it difficult for the first-time service seekers to know the operational procedures. Even experienced citizens, who might have done the same task offline, are not aware of resources on sitemaps that can help them adopt eService channels. Absence of user-guides or how-to-do tips trigger mistrust. It reinforces the impression of ineffectuality of eGovernment services. This leads firstly to added disenchantment with this paradigm-changing avenue and secondly to its eventual decline from daily usage and adoption.

Similarly, the government needs to move beyond the web-based services and actually employ channels, which are more user-accessible and preferred. The largest and lowest hanging fruit here is the mobile-services section which has a much higher penetration rate than internet and web for most people. Similarly, applications may need to be scaled to work on normal mobile phones with GPRS services and SMS and not for smartphones alone[934]. In the rising smartphone segment, which picked up with the auctioning of 3G and 4G licenses- mobile applications, websites and gateways may be designed that can associate these

---

934   Davison, R., et al.: 2000

services with the SIM identification and thus allow people to use mobile phones as virtual wallets (as seen in the case of Mobilink in Chapter 6). In developing countries, this by far is the most promising trend that can be forecast.

The case of comparing kiosks and eSahulat also raises a similar observation, where eSahulat soon took over with its popularity. The only difference here is the factor of human assistance which was lacking in kiosks (although present in local language as well). This proved to be the catalyst for its adoption. Human interaction, when seen with societal interaction patterns, shows that social human capital and prevalent communication patterns are important factors that should be considered when designing technology. However, this does not mean that machine interfaces or independent transactions may never succeed in such societies but could be delayed by high threshold level of trust while using machines or inexperience in using them properly[935]. On the flip side, this opens a completely new market which is competitive enough to start public-private partnerships which generate income for last-mile service providers who in turn would be agents of change in hard-to-reach communities and can in fact be their first introduction point for digitisation.

---

935 Howcroft, D., Mitev, N. and Wilson, M. (2004). What We May Learn from the Social Shaping of Technology Approach. In Mingers and Willcocks (Eds.), Social theory and philosophy for information systems (pp. 329-371). Chichester: John Wiley & Sons Ltd

Appendix I

List of Interview and Conversation Partners

- Zubair Saleem, Project Manager, Electronic Government Directorate, Ministry of Information Technology, Government of Pakistan. 18 March, 2009

- Faiz Bashir Malik, Project Manager, Electronic Government Directorate, Ministry of Information Technology, Government of Pakistan. 18 March, 2009

- Bilal Manzoor, Project Manager, Electronic Government Directorate, Ministry of Information Technology, Government of Pakistan. 19 March, 2009

- Muhammad Nasir, Director Database, Electronic Government Directorate, Ministry of Information Technology, Government of Pakistan. 19 March, 2009

- Naveed ul Haq, Assistant Director (ICT), Pakistan Telecommunication Authority, Cabinet Division, Government of Pakistan. 15 December, 2009. 25 September, 2010.

- Usman Mobin, Chief Technology Officer, National Database Registration Authority, Ministry of Interior, Government of Pakistan. 25 August, 2008, 15 February, 2009

- Usman Javed, Manager Technology & Development, National Database Registration Authority, Ministry of Interior, Government of Pakistan. 15 February, 2009

- Ahmed Hussain Naseem, Database Administrator, National Database Registration Authority, Ministry of Interior, Government of Pakistan. 3-10 March, 2009

- Irteza Haider, Senior Program Officer, National Rural Support Program, Islamabad, Pakistan. 22 September, 2010

- Deirdre Williams, Information Specialist, Saint Lucia, West Indies. 4 Dec, 2008

- Saif ul Islam, Group Head, Allied Bank of Pakistan, Islamabad, Pakistan. 7 Feb, 2009

- Mehdi Raza Zaidi, Foreign Trade Officer, Bank Alfalah Limited, Islamabad, Pakistan, 6 Feb, 2009

- Muhammad Ahsan Mansoor, Valued Added Services, Mobilink, Islamabad. 7-15 Feb, 2009. 20-24 Sep, 2010

- Fouad Bajwa, ICT Consultant, Lahore, Pakistan. 18 Nov, 2009

- Shaheryar Idrees, Valued Added Services, Telenor, Pakistan, 22 Sep, 2010

- Dr. Anita Greenhill, Manchester Business School, Manchester University, UK. 15 May, 2010

- Dr. Richard Heeks, Centre for Development Informatics, Manchester University, UK, 20 May, 2010

Appendix II

Interview questions for NADRA

1) What was the main driving force behind constituting NADRA?
2) Why does it function under Ministry of Interior and not under Ministry of Information Technology?
3) Where was the technical equipment developed for NADRA?
4) Which software programmes are currently used by NADRA?
5) Are the software programmes for CNIC, kiosk and eSahulat locally developed or imported from the foreign vendors?
6) What was the inspiration and underlying goals behind developing kiosk?
7) What features are currently offered by the kiosk platform?
8) Is it going to include other languages than English or Urdu? If so what other languages are under consideration?
9) Is it possible to count number of users based on their choice of menu language?
10) What communication technologies does the kiosk use when a transaction takes place?
11) Are there considerations given or a strategy adopted for rural population?
12) Will it always remain a touch screen or would speech-to-text (voice recognition) feature also be introduced later?
13) How many kiosk machines have been produced and installed?
14) Which services are most frequently used by the kiosk?
15) What geographical parameters were followed for distribute kiosk machines all over Pakistan?
16) Is it possible to get provincial overview of kiosk transactions?
17) Does NADRA create any monthly, quarterly, six monthly or yearly review of kiosk transactions and revenue?
18) How do you see NADRA's future in terms of eGovernment projects and its primary concern maintenance citizen database?
19) To what level is citizen database shared with private entities such as banks, telecom providers and other government institutions?
20) What security features are currently in place to ensure the data privacy?
21) How does NADRA place itself in current eGovernment scenario
22) How does NADRA approach the concerns raised in the press about the misuse?
23) How does NADRA position itself as a parallel entity in eGovernment domain in the presence of EGD as a primary agency?

Appendix III

NADRA kiosk/eSahulat sample data

Total Number of Transactions	City	Gender	Amount Generated
3219	ryk	m	1674544
6193	khi	f	5687670
16	fsl	f	15505
2	ryk	f	904
660	rwp	f	425390
16	kri	m	10017
40	shk	f	24531
6	lrk	f	3824
192	mul	f	93337
4	khs	f	701
41749	mul	m	27203698
169	gur	f	138530
5295	gur	m	3773153
2168	suk	m	1956993
210	abt	m	142630
76	pwr	f	111349
4	wah	f	1008
12416	isl	m	13454199
802	hrd	m	608477
82	sgh	f	58201
278	wah	m	160797
80	gjr	m	61245
26	gjr	f	35526
23403	rwp	m	21805357
291	chi	m	316855
80	khs	m	51118
3416	shk	m	2068598
2915	slk	m	2450771
1423	bha	m	1060332
18	hrd	f	10027
2	bha	f	320
6443	sgh	m	5546323
3	slk	f	3884

Appendix IV

Letter of consent by the BatchGeo website in order to use its geographical information system for creation of maps.

Subject:	Re: use of maps for my thesis
From:	BatchGeo Support (admin@batchgeo.com)
To:	hasnain.bokhari@yahoo.com;
Date:	Monday, April 8, 2013 5:12 PM

Hi Hasnain,

No formal approval needed. Free maps are there for you to use.

Thank you,
Tracy

On Sun, Apr 7, 2013 at 4:23 PM, Hasnain Bokhari <hasnain.bokhari@yahoo.com> wrote:
> Dear Sir/Madam,
>
> I am writing my thesis in which I would like to show a graphical view of certain areas through maps. I wanted to know if I need to have a formal approval from your side to use the maps made in batchgeo?
>
> Looking forward to hearing from you at the earliest possible convenience.
>
> Many thanks,
> Hasnain Bokhari

--
BatchGeo - http://batchgeo.com/

Appendix V

Notification for the initiation of Electronic Government Directorate which is no longer available in any Government website.

<u>NOTIFICATION</u>

No.3-16/2001/Coord. The Government of Pakistan is pleased to abolish the Information Technology Commission established vide Finance Division No.1(20)DS/TR.97 dated 9th August, 1997 and simultaneously establish an Electronic Government Directorate to be headed by a Programme Coordinator.

2.   The Business/Functions of the Electronic Government Directorate will be as under:-

. (i) Implementation of different projects related to the Electronic Government (E-Govt.) Programme;

. (ii) Provide technical advice & guidelines for implementation of E-Govt. projects at the Federal, Provincial and District levels;

. (iii) Plan and prepare electronic government projects;

. (iv) Provide standards for software and infrastructure in the field of E-Govt; and

. (v) To undertake any other assignment/matter that the government may assign to the directorate.

3. The Electronic Government Directorate will work under the Administrative/Financial control of the Information Technology & Telecom Division.

Source: http://pakistan.gov.pk/e-government-directorate/about/notification1.pdf accessed 10.10.2007

Bibliography

Books

Abbate, J. (1999). *Inventing the Internet.* Cambridge, Mass: MIT Press

Ahmed, S. (2001). The Fragile Base of Democracy in Pakistan. In Shastri, A. and Wilson, A. J. (Eds.), *The Post-Colonial States of South Asia: Democracy, Development and Identity* (pp. 41-68). New York: Palgrave Macmillan

Akbar, M. J. (2012). *Tinderbox : the past and future of Pakistan.* New York: Harper Perennial

Al Ajeeli, A. and Al-Bastaki, Y. A. L., (Eds.) (2011). *Handbook of research on E-services in the public sector : E-government strategies and advancements.* Hershey: Information Science Reference

Al-Hakim, L., (Ed.) (2007). *Global e-government theory, applications and benchmarking.* Hershey, PA: Idea Group Publishing

Alam, M. (2013). Pakistan's Devolution of Power Plan 2001: A Brief Dawn for Local Democracy? In Sansom, G. and McKinlay, P. (Eds.), *New Century Local Government: Commonwealth Perspectives* (pp. 44-57). London: Commonwealth Secretariat

Alavi, H. (1986). Ethnicity, Muslim society, and the Pakistan ideology. In Weiss, A. M. (Ed.), *Islamic reassertion in Pakistan: The application of Islamic laws in a modern state* (pp. 21-47). Syracuse: Syracuse University Press

Alhassan, A. and Chakravartty, P. (2011). Postcolonial Media Policy Under the Long Shadow of Empire. In Mansell, R. and Raboy, M. (Eds.), *The handbook of global media and communication policy* (pp. 366-382). West Sussex: Blackwell Publishing

Amin-Khan, T. (2012). *The post-colonial state in the era of capitalist globalization: Historical, political and theoretical approaches to state formation.* New York: Routledge

Anderson, B. (2007). *Information and communication technologies in society : e-living in a digital Europe.* London: Routledge

Archibugi, D. and Michie, J., (Eds.) (1997). Technology, globalisation and economic performance. Cambridge: Cambridge University Press

Asif, M. (2011). *Energy crisis in Pakistan : origins, challenges and sustainable solutions.* Karachi: Oxford University Press

Avgerou, C., et al., (Eds.) (2008). Social Dimensions of Information and Communication Technology Policy. Boston: Springer

Aziz, M. (2008). *Military control in Pakistan : the parallel state.* London; New York: Routledge

Aziz, M. A. (1979). *A history of Pakistan : past and present.* Lahore: Sang-e-Meel Publications

Badshah, A., et al. (2005). *Connected for development : information kiosks and sustainability.* New York: UN Information and Communication Technologies Task Force

Baecker, R. M. and Grudin, J., (Eds.) (2014). Readings in Human-Computer Interaction: toward the year 2000. San Francisco: Morgan Kaufmann Publishers

Bakardjieva, M. (2005). *Internet society : the Internet in everyday life.* London: SAGE

Baqir, M. N. and Iyer, L. (2010). E-government maturity over 10 Years: A comparative analysis of e-government maturity in select countries around the world. *Comparative E-Government* (pp. 3-22). New York: Springer

Basit, M. A. (2012). *Pakistan Citizenship & NADRA Laws.* Lahore: Federal Law House

Bawden, D. (2008). Origins and Concepts of Digital Literacy. In Lankshear, C. and Knobel, M. (Eds.), *Digital literacies: Concepts, policies and practices* (pp. 17-32). New York: Peter Lang

Bayly, C. A. (2000). *Empire and Information: Intelligence gathering and social communication in India, 1780-1870.* Cambridge: Cambridge University Press

Bergman, M. M., (Ed.) (2008). Advances in mixed methods research: Theories and applications. London: Sage

Bhagwandas, R. (2015). Pakistan: Federal Public Service Commission and its Functions. In Chalam, K. S. (Ed.), *Governance in South Asia : State of the Civil Services* (pp. 239-256). New Delhi: SAGE Publications

Bhatnagar, S. (2004). *E-government : from vision to implementation: a practical guide with case studies.* New Delhi: Sage publications

Bhatnagar, S. (2009). *Unlocking E-Government potential : concepts, cases and practical insights.* Los Angeles: Sage Publications

Bijker, W. E., et al., (Eds.) (2012). Social Construction of Technological Systems: New Directions in the Sociology and History of Technology. Cambridge: MIT Press

Bracey, B. and Culver, T., (Eds.) (2005). Harnessing the potential of ICT for education: a multistakeholder approach: Proceedings from the Dublin Global Forum of the United Nations ICT Task Force. New York: United Nations Publications

Braibanti, R. (1966). *Research on the bureaucracy of Pakistan: a critique of sources, conditions, and issues, with appended documents.* Durham: Duke University Press

Brandt, R. L. (2011). *One click : Jeff Bezos and the rise of Amazon.com.* London: Viking

Bratich, J. Z., et al., (Eds.) (2003). Foucault, cultural studies, and governmentality. Albany: State University of New York Press

Burki, J. (2008). *Changing perceptions and altered reality : Pakistan's economy under Musharraf, 1999-2007.* Oxford: Oxford University Press

Castells, M. (2009). *Communication power.* Oxford; New York: Oxford University Press

Castells, M. (2010a). *The power of identity.* Oxford: Wiley-Blackwell

Castells, M. (2010b). *The rise of the network society.* Chichester, West Sussex: Wiley-Blackwell

Castells, M., et al. (2007). *Mobile Communication and Society: A Global Perspective.* Cambridge: Mit Press

Chadwick, A. and Howard, P. N. (2010). *Routledge handbook of Internet politics.* London: Taylor & Francis

Chaudry, A. (2011). *Political Administrators: The Story of the Civil Service of Pakistan.* Oxford: Oxford University Press

Chen, H., et al., (Eds.) (2008). Digital government e-government research, case studies and implementation. Berlin: Springer

Chen, W. and Wellman, B. (2005). Minding the cyber-gap: the Internet and social inequality. In Romero, M. and Margolis, E. (Eds.), *The Blackwell companion to social inequalities* (pp. 523-545). Malden, MA: Blackwell Publishing

Chong, A. and Inter-American Development, B. (2011). *Development connections : unveiling the impact of new information technologies.* New York: Palgrave Macmillan

Choudhury, D. K. L. (2010a). The Telegraph and the Uprisings of 1857. *Telegraphic Imperialism: Crisis and Panic in the Indian Empire, c. 1830* (pp. 31-49). Basingstoke: Palgrave Macmillan

Choudhury, D. K. L. (2010b). The Discipline of Technology. *Telegraphic Imperialism: Crisis and Panic in the Indian Empire, c. 1830* (pp. 50-78). London: Palgrave Macmillan

Chrissis, M. B., et al. (2011). *CMMI for Development: Guidelines for Process Integration and Product Improvement.* New Jersey: Addison-Wesley

Chung, Y.-I. (2007). *South Korea in the fast lane: Economic development and capital formation.* Oxford: Oxford University Press

Cohen, S. P. (2006). *The idea of Pakistan.* Washington (D.C.): Brookings Institution Press

Cohen, S. P. (2011). *The future of Pakistan.* Washington, D.C.: Brookings Institution Press

Coleman, S. (2005). *African e-governance-Opportunities and Challenges.* Oxford University: Oxford Internet Institute

Creswell, J. W. (2003). *Research design: qualitative, quantitative, and mixed method approaches.* Thousand Oaks: Sage Publications

Creswell, J. W. and Plano Clark, V. L. (2007). *Designing and conducting mixed methods research.* Thousand Oaks: SAGE Publications

Daechsel, M. (2006). *The politics of self-expression the Urdu middle-class milieu in mid-twentieth century India and Pakistan.* London; New York: Routledge

David, S. (2003). *The Indian Mutiny: 1857.* London: Penguin Books

Drüke, H., (Ed.) (2004). Local Electronic Government: A comparative study. London: Routledge

Durrant, F. (2002). e-Government and the Internet in the Caribbean: An initial assessment. In Traunmüller, R. and Lenk, K. (Eds.), *Electronic Government. First International Conference, EGOV 2002 Aix-en-Provence, France, September 2–6, 2002 Proceedings* (pp. 101-104). Berlin: Springer

Dwivedi, O. P. and Nef, J. (2004). From Development Administration to New Public Management in Postcolonial Settings. In Mudacumura, G. M. and Shamsul, M. (Eds.), *Handbook of Development Policy Studies* (pp. 153-176). New York: Marcel Dekker

Elahi, M. and Zia, A. (2008). Pakistan. In Banerjee, I. and Logan, S. (Eds.), *Asian Communication Handbook 2008* (pp. 369-404). Singapore: Asian Media Information and Communication Centre

Feenberg, A. (1995). *Alternative modernity : the technical turn in philosophy and social theory*. Berkeley: University of California Press

Feenberg, A. (2002). *Transforming technology: a critical theory revisited*. New York: Oxford University Press

Ferneding, K. A. (2003). *Questioning technology : electronic technologies and educational reform*. New York: Peter Lang

Flyverbom, M. (2011). *The power of networks: Organizing the global politics of the Internet*. Cheltenham: Edward Elgar Publishing

Fuglerud, K. S. (2009). Universal Design in ICT Services. In Vavik, T. (Ed.), *Inclusive Buildings, Products & Services: Challenges in Universal Design* (pp. 244-267). Trondheim: Tapir Academic Press

Fuglsang, L. (2001). Three perspectives in STS in the policy context. In Cutcliffe, S. H. and Mitcham, C. (Eds.), *Visions of STS: Counterpoints in science, technology, and society studies* (pp. 33-49). Albany: State University of New York Press

Germanakos, P., et al. (2007). Multi-channel delivery of e-services in the light of m-government challenge. In Kushchu, I. m. (Ed.), *Mobile Government: An Emerging Direction in e-Government* (pp. 292-317). Hershey: IGI Global

Gillies, J. and Cailliau, R. (2000). *How the Web was born : the story of the World Wide Web*. Oxford: Oxford University Press

Golding, P. (2011). *Connected services : a guide to the Internet technologies shaping the future of mobile services and operators*. Chichester, West Sussex: Wiley

Gordon, C. (1991). Governmental Rationality: An Introduction. In Burchell, G. et al. (Eds.), *The Foucault effect : studies in governmentality - with two lectures by and an interview with Michel Foucault* (pp. 1-51). Chicago: University of Chicago Press

Grönlund, Å. (2010). Ten Years of E-Government: The 'End of History' and New Beginning. In Wimmer, M. A. et al. (Eds.), *Electronic Government: 9th IFIP WG 8.5 International Conference, EGOV 2010, Lausanne, Switzerland, August 29 - September 2, 2010. Proceedings* (pp. 13-24). Berlin: Springer

Hagen, M. and Kubicek, H. (2000). One Stop Government in Europe: An Overview. In Hagen, M. (Ed.), *One-stop-government in Europe: Results from 11 National Surveys* (pp. 1-36). Bremen: University of Bremen

Hague, B. N. and Loader, B. D., (Eds.) (2005). *Digital democracy: Discourse and decision making in the information age.* London: Routledge

Hanna, N. K. (2010). *Transforming government and building the information society: Challenges and opportunities for the developing world.* New York: Springer

Haq, K. (2009). *Technology and human development in South Asia.* Oxford: Oxford Univ. Press

Heeks, R. (1999). *Reinventing government in the information age : international practice in IT-enabled public sector reform.* London: Routledge

Heeks, R. (2006b). *Implementing and managing eGovernment: An International Text.* London: SAGE

Henman, P. (2010). *Governing Electronically: E-Government and the Reconfiguration of Public Administration, Policy and Power.* Hampshire: Palgrave Macmillan

Hoff, J., et al. (2003). Technology and Social Change. *Democratic governance and new technology* (pp. 28-48). London: Routledge

Hoffmann, B. (2004). *The politics of the Internet in Third World development : challenges in contrasting regimes with case studies of Costa Rica and Cuba.* New York: Routledge

Hönke, J. (2013). *Transnational companies and security governance: Hybrid practices in a postcolonial world.* Oxon: Routledge

Howcroft, D., et al. (2004). What We May Learn from the Social Shaping of Technology Approach. In Mingers, J. and Willcocks, L. (Eds.), *Social theory and philosophy for information systems* (pp. 329-371). Chichester: John Wiley & Sons Ltd

Imam, A. and Dar, E. A. (2014). *Democracy and Public Administration in Pakistan.* New York: CRC Press

Jaffrelot, C. (2004). *A history of Pakistan and its origins.* London: Anthem

Jalal, A. (1993). The State and Political Privilege in Pakistan. In Weiner, M. and Banuazizi, A. (Eds.), *The Politics of social transformation in Afghanistan, Iran, and Pakistan* (pp. 152-184). Syracuse, N.Y.: Syracuse University Press

Jalal, A. (2014). *The struggle for Pakistan : a Muslim homeland and global politics.* Cambridge (Mass.); London: The Belknap Press of Harvard University Press

James, J. (2013). *Digital interactions in developing countries: an economic perspective.* London: Routledge

Karagiannis, T. (2014). The New face of the Internet. In Harper, R. H. (Ed.), *Trust, Computing and Society* (pp. 38-67). New York: Cambridge University Press

Kayani, D., & Rafi, F (2013). *Pakistan Votes.* Islamabad: Coffee Communications

Kennedy, C. H. (1987). *Bureaucracy in Pakistan*. Karachi: Oxford University Press

Kenney, M. and Florida, R. (2000). Venture Capital in Silicon Valley: Fueling New Firm Formation. In Kenney, M. (Ed.), *Understanding Silicon Valley : the anatomy of an entrepreneurial region* Stanford, Calif.: Stanford Univ. Press

Kerr, I. J. (2007). *Engines of Change: The Railroads That Made India*. Westport: Praeger

Kervenoael, R. d. and Kocoglu, I. (2012). E-Government Strategy in Turkey: A Case for m-Government? In Bwalya, K. J. and Zulu, S. (Eds.), *Handbook of Research on E-Government in Emerging Economies: Adoption, E-Participation, and Legal Frameworks. Vol.1* (pp. 351-373). Hershey (PA): Information Science Reference

Khan, I. H. (2011). *Electoral malpractices during the 2008 elections in Pakistan*. Karachi: Oxford University Press

Khan, M. M. (2002). Resistance to administrative reforms in South Asian civil bureaucracies. In Farazmand, A. (Ed.), *Administrative reform in developing nations* (pp. 73-88). Westport, Conn.: Praeger

Kirkman, G. S., et al., (Eds.) (2002). The global information technology report, 2001-2002: Readiness for the networked world. New York: Oxford University Press

Knayar, P. (2007). *The great uprising: India, 1857*. New Delhi: Penguin Books

Kochanek, S. A. (1974). *Business and politics in India*. Berkeley: Univ. of California Press

Kuppuram, G. and Kumudamani, K. (1990). *History of science and technology in India*. Delhi: Sundeep Prakashan

Kurbalija, J. (2008). The World Summit on Information Society and the Development of Internet Diplomacy. In Cooper, A. F. et al. (Eds.), *Global Governance and Diplomacy* (pp. 180-207). London: Palgrave Macmillan

Lemos, R. and Varon Ferraz, J. (2013). Information and communication technologies for development. In Currie-Alder, B. et al. (Eds.), *International Development: Ideas, Experience, and Prospects* Oxford: Oxford University Press

Lenk, K. and Traunmüller, R. (2002). Electronic Government: Where Are We Heading? In Lenk, K. and Traunmüller, R. (Eds.), *Electronic Government: First international conference, EGOV 2002, Aix-en-Provence, France, September 2-6, 2002: Proceedings* (pp. 1-9). Berlin: Springer

Lerner, D. (1968). *The passing of traditional society*. New York: Free Press

Llewellyn-Jones, R. (2007). *The great uprising in India, 1857-58 : untold stories, Indian and British*. Woodbridge: Boydell & Brewer

Loudon, M. and Rivett, U. (2013). Enacting openness in ICT4D research. In Smith, M. L. and Reilly, K. M. A. (Eds.), *Open development: networked innovations in international development.* (pp. 53-78). Cambridge MA: MIT Press

Lyon, D. (2002). *Surveillance society : Monitoring everyday life*. Buckingham: Open University Press

Madon, S. (2009). *E-governance for Development: a focus on rural India*. Basingstoke: Palgrave Macmillan

Mahboob, A. (2002). No English, no future! Language Policy in Pakistan. In Obeng, S. and Hartford, B. (Eds.), *Political independence with linguistic servitude: The politics about languages in the developing world* (pp. 15-39). New York: Nova Science Publishers

Mahmood, S. (2009). *Reform of the public services in Pakistan*. New York: Nova Science Publishers

Malik, I. H. (1997). The Supremacy of the Bureaucracy and the Military in Pakistan. *State and Civil Society in Pakistan: Politics of Authority, Ideology and Ethnicity* (pp. 57-80). London: Palgrave Macmillan

Malik, J. (2008). *Madrasas in South Asia : teaching terror?* London: Routledge

Mathiason, J. (2008). *Internet governance: the new frontier of global institutions*. Oxon: Routledge

McCartney, M. (2011). *Pakistan--the political economy of growth, stagnation and the state, 1951-2009*. Abingdon, Oxon ; New York: Routledge

McPhail, T. L., (Ed.) (2009). Development communication : reframing the role of the media. Chichester, West Sussex: Wiley-Blackwell

Meijer, A., et al., (Eds.) (2009). ICTs, citizens and governance: After the hype! Amsterdam: IOS Press

Misra, A. (2010). *India-Pakistan : coming to terms*. New York: Palgrave Macmillan

Niaz, I. (2010). *The Culture of Power and Governance of Pakistan: 1947 - 2008*. Oxford: Oxford University Press

Norris, D. F. (2008). *E-government research : policy and management*. Hershey: IGI Pub.

Norris, P. (2001). *Digital divide : civic engagement, information poverty, and the Internet worldwide*. Cambridge: Cambridge University Press

O'Hara, K. S. D. (2006). *inequality.com : power, poverty and the digital divide*. Oxford: Oneworld Publicationss

Ogura, T. (2006). Electronic government and surveillance-oriented society. In Lyon, D. (Ed.), *Theorizing surveillance: The panopticon and beyond* (pp. 270-295). London: Routledge

Papadopoulos, T., (Ed.) (2012). Public sector reform using information technologies transforming policy into practice. Hershey, Pa.: IGI Global

Pinch, T. (2009). The social construction of technology (SCOT): The old, the new, and the nonhuman. In Vannini, P. (Ed.), *Material culture and technology in everyday life: ethnographic approaches.* (pp. 45-58). New York: Peter Lang

Putnis, P. (2013). International press and the Indian uprising. In Carter, M. and Bates, C. (Eds.), *Mutiny at the Margins: New Perspectives on the Indian*

Uprising of 1857: Vol 3: Global Perspectives (pp. 1-17). London, New Delhi: Sage Publications

Qadeer, M. A. (2006). *Pakistan : social and culural transformations in a Muslim nation*. London: Routledge

Rahman, A. (1999). *Women and microcredit in rural Bangladesh: anthropological study of the rhetoric and realities of Grameen Bank lending*. Boulder, CO: Westview Press, Inc.

Rahman, T. (2003). Language policy, multilingualism and language vitality in Pakistan. In Saxena, A. and Borin, L. (Eds.), *Lesser-Known Languages of South Asia: Status and Policies, Case Studies, and Applications of Information Technology* (pp. 73-104). Berlin: Walter de Gruyter

Rahman, T. (2008). Language policy and education in Pakistan. In Hornberger, N. H. (Ed.), *Encyclopedia of language and education* (pp. 383-392). Boston, MA: Springer

Raine, L. and Wellman, B. (2012). *Networked: The New Social Operating System*. Massachussets: Massachusetts Institute of Technology

Rand, G. (2013). Reconstructing the Imeprial Military after Rebellion. In Rand, G. and Bates, C. (Eds.), *Mutiny at the Margins: New Perspectives on the Indian Uprising of 1857: Volume IV: Military Aspects of the Indian Uprising* (pp. 93-112). London, New Delhi: Sage Publications

Raseroka, K. (2006). Access to information and knowledge. In Jørgensen, R. F. (Ed.), *Human rights in the global information society* (pp. 91-105). Cambridge: The MIT Press

Reddick, C. G., (Ed.) (2010). Citizens and e-government: Evaluating policy and management: Evaluating policy and Management. Hershey: IGI Global

Reddick, C. G. (2010). *Comparative e-government*. New York: Springer

Reed, T. V. (2014). *Digitized lives: culture, power, and social change in the internet era*. London: Routledge

Resnick, P. (2005). Impersonal sociotechnical capital, ICTs, and collective action among strangers. In Dutton, W. et al. (Eds.), *Transforming enterprise: the economic and social implications of information technology* (pp. 399-412). Cambridge: The MIT Press

Reza, Y. (1999). *History of Pakistan Telecommunications*. Islamabad: Pakistan Telecommunication Authority

Rizvi, H.-A. (2000). *Military, State and Society in Pakistan*. London: Macmillan Press

Roy, P., et al. (1969). *The Impact of communication on rural development : an investigation in Costa Rica and India. Report of a research project initiated by Unesco*. Paris: UNESCO

Ryan, J. (2010). *A History of the Internet and the Digital Future*. London: Reaktion Books

Schell, B. H. (2007). *The Internet and society : a reference handbook*. Santa Barbara: ABC-CLIO

Servaes, J. (2008). *Communication for development and social change.* New Delhi, India: Sage Publications

Servaes, J. (2013). Comparing development communication. In Esser, F. and Hanitzsch, T. (Eds.), *The handbook of comparative communication research* (pp. 86-102). New York: Routledge

Servaes, J. (2014). *Technological determinism and social change: Communication in a Tech-Mad World.* London: Lexington Books

Servaes, J., et al., (Eds.) (1996). Participatory communication for social change. New Delhi: Sage

Servaes, J. and Malikhao, P. Development Communication Approaches in an International Perspective. In Servaes, J. (Ed.), *Approaches to Development Communication* (pp. 214-251). Paris: UNESCO

Servon, L. J. (2008). *Bridging the digital divide: Technology, community and public policy.* Oxford: Blackwell Publishing

Shaikh, F. (2009). *Making sense of Pakistan.* New York: Columbia University Press

Shaw, J. K. (2001). *Telecommunications deregulation and the information economy.* Boston; London: Artech House

Siddiqa-Agha, A. (2007). *Military Inc. : inside Pakistan's military economy.* London: Pluto Press

Siddiqui, S. (2012). *Education, Inequalities, and Freedom: A Sociopolitical Critique.* Islamabad: Narratives Publication

Slevin, J. (2000). *Internet and society.* Cambridge: Polity Press

Sparks, C. (2007). *Globalization, development and the mass media.* London: Sage

Steyn, J., et al., (Eds.) (2010). ICTs for Global Development and Sustainability: Practice and Applications: Practice and Applications. Hershey: IGI Global

Steyn, J. and Johanson, G. (2011). *ICTs and sustainable solutions for the digital divide: theory and perspectives.* Hershey, PA: IGI Global

Storey, D. (1999). Popular Culture, Discourse and Development. In Jacobson, T. L. and Servaes, J. (Eds.), *Theoretical approaches to participatory communication* (pp. 337-358). Cresskill, NJ: Hampton Press

Susanto, T. D. and Goodwin, R. (2006). Opportunity and overview of SMS-based e-government in developing countries. In Morgan, K. et al. (Eds.), *Advances in Education, Commerce and Governance second International Conference on the Internet Society Advances in Education, Commerce and Governance* (pp. 255-264). South Hampton: WIT Press

Susanto, T. D. and Goodwin, R. (2011a). An SMS-Based e-Government Model: What Public Services can be Delivered through SMS? In Al Ajeeli, A. and Al-Bastaki, Y. A. L. (Eds.), *Handbook of Research on e-Services in the Public Sector: eGovernment Strategies and Advancements* (pp. 137-146). Hershey PA: Information Science Reference

Susanto, T. D. and Goodwin, R. (2011b). An SMS-Based e-Government Model: What Public Services can be Delivered through SMS? In Al Ajeeli, A. T. and Al-Bastaki, Y. A. L. (Eds.), *Handbook of Research on E-Services in the*

Public Sector: E-Government Strategies and Advancements (pp. 137-146). Hershey: IGI Global

Talbot, I. (2002). Does the Army Shape Pakistan's Foreign Policy? In Jaffrelot, C. (Ed.), *Pakistan, nationalism without a nation* (pp. 311-333). London: Zed Books

Talbot, I. (2010). India and Pakistan. In Brass, P. R. (Ed.), *Routledge Handbook of South Asian Politics* (pp. 43-56). Oxon: Routledge

Talbot, I. (2012). *Pakistan : a new history*. New York: Columbia University Press

Thomas, M., (Ed.) (2011). Deconstructing Digital Natives: Young People, Technology, and the New Literacies. New York: Routledge

Tian, Y. and Stewart, C. (2006). History of E-Commerce. In Khosrow-Pour, M. (Ed.), *Encyclopedia of e-commerce, e-government, and mobile commerce* (pp. 559-564). Hershey: IGI Global

Torero, M. and Von Braun, J. (2006). *Information and communication technologies for development and poverty reduction: The potential of telecommunications*. Baltimore: Johns Hopkins University Press

United Nations Economic and Social Commission for Asia and the Pacific (2003). *Initiatives for e-commerce capacity-building of small and medium enterprises : proceedings and papers presented at the Regional Consultative Meeting on Initiatives for E-Commerce Capacity-Building of Small and Medium Enterprises, Seoul, 13-15 November 2002*. New York: United Nations

Unwin, T. (2009). *ICT4D : information and communication technology for development*. Cambridge: Cambridge University Press

Wang, Y.-J., et al. (2009). *Rethinking e-government services : user-centred approaches*. Paris: OECD

Warschauer, M. (2003b). *Technology and social inclusion rethinking the digital divide*. Cambridge: MIT Press

Weerakkody, V. and Reddick, C. G. (2012). *Public sector transformation through e-government: experiences from Europe and North America*. New York: Routledge

West, D. M. (2005). *Digital government: Technology and public sector performance*. Princeton: Princeton University Press

West, D. M. (2007). Global perspectives on e-government. In Mayer-Schönberger, V. and Lazer, D. (Eds.), *Governance and information technology: From electronic government to information government* (pp. 17-32). Cambridge: The MIT Press

William McIver, J. (2003). A Community Informatics for the Information Society. In Girard, B. and Siochrú, S. Ó. (Eds.), *Communicating in the Information Society* (pp. 33-64). Geneva: United Nations Research Institute for Social Development (UNRISD)

Wolpert, S. (2004). *A new history of India*. New York: Oxford University Press

Worth, A. (2014). *Imperial media : colonial networks and information technologies in the British literary imagination, 1857-1918*. Columbus: The Ohio State University Press

Ziring, L. (1997). *Pakistan in the twentieth century : a political history*. Karachi: Oxford Univ. Press

Conference Papers/Proceedings

Alvi, A. M. (1993). *National Registration System in Pakistan*. In East and Sourth Asian Workshop on Strategies for Accelerating the Improvement of Civil Registration and Vital Statistics Systems. 29 November - 3 December 1993, Beijing, China. (pp. 1-19). Retrieved 04.05.2014 from https://unstats.un.org/unsd/demographic/meetings/wshops/1993_China_CR VS/docs/1993_Doc.11_Pakistan.pdf

Bokhari, H. and Khan, M. (2012). *Digitisation of electoral rolls: analysis of a multi-agency e-government project in Pakistan*. In Proceedings of the 6th International Conference on Theory and Practice of Electronic Governance. (pp. 158-165). Albany: ACM Press.

Dyson, L. E. (2004). *Cultural issues in the adoption of information and communication technologies by Indigenous Australians*. In Proceedings cultural attitudes towards communication and technology. (pp. 58-71). Murdoch University Perth. Retrieved 04.05.2014 from http://citeseerx.ist.psu.edu/viewdoc/download?doi=10.1.1.198.7661&rep=re p1&type=pdf

Flor, A. G. (2001). *ICT and poverty: The indisputable link.* . Paper presented at the Third Asian Development Forum on "Regional Economic Cooperation in Asia and the Pacific", organised by Asian Development Bank.Retrieved 09.10.2009 from http://ftp.unpad.ac.id/orari/library/library-ref-ind/ref-ind-1/application/poverty-reduction/! ICT4PR/ICT and poverty - the link (WB Report).pdf

Fountain, J. E. (2002). *Building a deeper understanding of e-government*. In The Future of e-Governance Workshop (September 10, 2002) at the Campbell Public Affairs Institute in Maxwell School of Syracuse University. Retrieved 04.05.2014 from https://www.maxwell.syr.edu/uploadedFiles/campbell/events/fountain.pdf

Gaby, S. and Henman, P. (2004). *E-Health: transforming doctor-patient relationships with a dose of technology*. In Paper presented at: Aus- tralian Electronic Governance Conference; 14th and 15th April 2004. Centre for Public Policy, University of Melbourne Victoria. Retrieved 08.10.2008 from http://www.public-policy.unimelb.edu.au/egovernance/ConferenceContent.html

Gronlund, A. (2005). *What's In a Field - Exploring the eGoverment Domain*. In Proceedings of the 38th Annual Hawaii International Conference on System Sciences. (pp. 3-6). New York: IEEE.

Hafkin, N. J. (2002). *Gender issues in ICT policy in developing countries: An overview.* Expert Group Meeting organised by United Nations Division for the Advancement of Women (DAW) on "Information and communication technologies and their impact on and use as an instrument for the advancement and empowerment of women".Retrieved 03.05.2014 from citeseerx.ist.psu.edu/viewdoc/download?doi=10.1.1.463.3492&rep=rep1&type=pdf

Homburg, V. (2004). *E-government and NPM: a perfect marriage?* In Proceedings of the 6th international conference on Electronic commerce. (pp. 547-555). Delft: ACM Press.

Hornung, H. and Baranauskas, M. C. C. (2007). *Interaction Design in eGov systems: challenges for a developing country.* In Proceedings of the 34th Seminário Integrado de Software e Hardware, Porto Alegre, Brazil. (pp. 2217-2231).

Malik, S. J. (1993). *Poverty in Pakistan, 1984-85 to 1987-88.* In Lipton, M. and Gaag, J. v. d. (Eds.), Including the Poor: Proceedings of a Symposium Organized by the World Bank and the International Food Policy Research Institute. (pp. 487-519). Retrieved 02.07.2011 from http://documents.worldbank.org/curated/en/649751468764364351/pdf/multi-page.pdf

Meijer, A. and Lofgren, K. (2010). *Selling technology to the policy sciences: Marketing strategies for specialized scholars.* Conference: Internet, Politics, Policy 2010: An Impact Assessment.Retrieved 04.05.2014 from http://blogs.oii.ox.ac.uk/ipp-conference/sites/ipp/files/documents/IPP2010_Meijer_Lofgren_Paper.pdf

Ojo, A., et al. (2005). *Determining Progress Towards e-Government-What are the Core Indicators?* In 5th European Conference on on e-Government (ECEG2005). (pp. 313-322). Retrieved 09.11.2009 from http://collections.unu.edu/eserv/UNU:3052/Adegboyega-ECEG2005.pdf

Rafiq, A. and Gao, P. (2008). *The transformation of mobile telecommunications industry in Pakistan.* In The Pacific Asia Conference on Information Systems. (pp. 95). Retrieved 04.12.2010 from https://aisel.aisnet.org/cgi/viewcontent.cgi?article=1188&context=pacis2008

Raza, F., et al. (2005). *Land records information management system.* In 25th Annual ESRI International User Conference, San Diego, California. Retrieved 04.05.2014 from http://proceedings.esri.com/library/userconf/proc05/papers/pap1279.pdf

Razak, N. A. and Malek, J. A. (2009). *Education for All: Meeting Millennium Development Goals via the Telecenters.* In Proceedings of the 5th WSEAS/IASME International Conference on Educational Technologies (EDUTE' 09). (pp. 218-220). Retrieved 04.05.2014 from http://www.wseas.us/e-library/conferences/2009/lalaguna/EDUTE/EDUTE-33.pdf

Rorissa, A., et al. (2010). *A tale of two continents: Contents of African and Asian e-government Websites.* In 43rd Hawaii International Conference on System Sciences. (pp. 1-9). IEEE. Retrieved 04.05.2014 from https://ieeexplore.ieee.org/document/5428303

Dictionary

Facility. [Def. 2]. *Oxford Dictionary Online.* In Oxford Dictionary. Retrieved from https://ur.oxforddictionaries.com/translate/urdu-english

Kiosk. [Def. 2]. *Cambridge Dictionary Online.* In Cambridge Dictionary. Retrieved from https://dictionary.cambridge.org/dictionary/english/kiosk?q=kiosk

Electronic Resources

Ansari, S. and Saleem, S. (2009b). .pk: Pakistan. In Akhtar, S. and Arinto, P. B. (Eds.). *Digital Review of Asia Pacific 2009-2010.* (pp. 294-301). Ottawa: International Development Research Centre. https://idl-bnc-idrc.dspacedirect.org/bitstream/handle/10625/38550/IDL-38550.pdf?sequence=1&isAllowed=y

Arch, A. and Hardy, B. (2005). E-Government: Accessible to All. *Future Challenges for E-government.* (pp. 54-74). Retrieved 04.05.2014 from https://www.finance.gov.au/sites/default/files/Future-Challenges-for-Egovernment-Volume02.pdf

Baer, W. S. (1998). Will the Internet Transform Higher Education? https://www.rand.org/content/dam/rand/pubs/reprints/2005/RP685.pdf

Bessette, G. (2004). Involving the community: a guide to participatory development communication. Ottawa: International Development Research Centre (IDRC). Retrieved 29.10.2009 from https://idl-bnc-idrc.dspacedirect.org/bitstream/handle/10625/31476/IDL-31476.pdf?sequence=33&isAllowed=y

Deloitte & Touche LLP (2008). Economic impact of mobile communications in Serbia, Ukraine, Malaysia, Thailand, Bangladesh, and Pakistan: A report prepared for Telenor ASA. Retrieved 05.05.2014 from https://www.telenor.rs/media/TelenorSrbija/fondacija/economic_impact_of_mobile_communications.pdf

Deloitte LLP (2015). Digital inclusion and mobile sector taxation in Pakistan. Retrieved 03.05.2015 from https://www.gsma.com/publicpolicy/wp-content/uploads/2015/02/GSMA2015_Report_DigitalInclusionAndMobileSectorTaxationInPakistan.pdf

European Country of Origin Information Network (2013). Bericht zur Fact Finding Mission Pakistan. Retrieved 03.06.2014 from https://www.ecoi.net/en/file/local/1068808/1729_1374674206_ffm-bericht-pakistan-2013-06.pdf

Felix Librero (Ed.) (2008). Digital Review of Asia Pacific 2007-2008. Retrieved 28.10.2010 from http://www.digital-review.org/uploads/files/pdf/2007-2008/intro.pdf

Gallup Pakistan (2013). Use of Mobile Money in Pakistan: Findings from FITS Study. Retrieved 04.05.2014 from http://gallup.com.pk/wp-content/uploads/2016/02/Aug-03-20131.pdf

Gallup Pakistan (2014). Pakistan ICT Indicators Survey - 2014. Retrieved 04.05.2014 from http://gallup.com.pk/wp-content/uploads/2017/01/Pakistan-ICT-Indicators-Survey-2014-1.pdf

Gelb, A. (2014). Technology in the Service of Development: The NADRA Story. Retrieved 28.12.2014 from https://www.cgdev.org/sites/default/files/CGD-Essay-Malik_NADRA-Story_0.pdf

German Embassy Islamabad (2018). Schengen Visa Requirements. Retrieved 28.12.2018 from https://pakistan.diplo.de/blob/2163332/93695879eadf8da9fea01ecee31f27ed/visa-requirements---schengen-data.pdf

Gerster, R. and Zimmermann, S. (2003). Information and Communication Technologies (ICTs) for Poverty Reduction?: Discussion Paper. Retrieved 18.09.2008 from http://www.gersterconsulting.ch/docs/ICT_for_poverty_reduction.pdf

Goodwin, P. (1998). Pakistan prepares to redraw population map. Retrieved 04.05.2014 from http://news.bbc.co.uk/2/hi/despatches/61407.stm

Haider, Z. (2007). Bhutto challenges Pakistani voter list in court. Retrieved 04.05.2014 from https://www.reuters.com/article/us-pakistan-bhutto-idUSISL18463420070626

Hameed, T. (2007). ICT as an enabler of socio-economic development. Retrieved December 17, 2011 from http://unpan1.un.org/intradoc/groups/public/documents/un-dpadm/unpan043799.pdf

Hasan, A. and Raza, M. (2009). Migration and small towns in Pakistan: Working Paper Series on Rural-Urban Interactions and Livelihood Strategies (Working Paper 15). Retrieved 13.01.2012 from https://pubs.iied.org/pdfs/10570IIED.pdf

Heeks, R. (2005a). ICTs and the MDGs: On the Wrong Track? Retrieved 30.11.2010 from http://hummedia.manchester.ac.uk/institutes/gdi/publications/workingpapers/di/di_sp07.pdf

Houreld, K. (2012). Pakistan ID cards remove ghost voters, target poor for aid. Retrieved 04.05.2014 from https://uk.reuters.com/article/uk-pakistan-identity/pakistan-id-cards-remove-ghost-voters-target-poor-for-aid-idUKBRE8AL0Y120121122

Hussain, N. and Tahir, A. (2014). Financial Inclusion's Catalytic Role in the Urbanization of Pakistan's Rural Poor. In Kugelman, M. (Ed.). *Pakistan's Runaway Urbanization: What Can Be Done.* (pp. 135-139). Retrieved

13.11.2014 from https://www.wilsoncenter.org/publication/pakistans-runaway-urbanization

International Foundation for Electoral Systems (2013b). Using a Gender Lens to Examine Pakistan's Historic Election. Retrieved 04.05.2014 from https://www.ifes.org/news/using-gender-lens-examine-pakistans-historic-election

International Telecommunication Union (2011a). Mobile Banking. Retrieved 12.05.2013 from https://www.itu.int/net/itunews/issues/2011/07/32.aspx

International Telecommunication Union (2013c). Universal Service Fund and Digital Inclusion For All Study. Retrieved 04.05.2014 from https://www.itu.int/en/ITU-D/Conferences/GSR/Documents/ITU USF Final Report.pdf

Kamran, T. (2008). Democracy and governance in Pakistan. Lahore: South Asia Partnership-Pakistan. Retrieved 30.04.2019 from http://sappk.org/wp-content/uploads/publications/eng_publications/Democracy_and_Governance.pdf

Kardon, I. B. (2011). China and Pakistan: Emerging Strains in the Entente Cordiale. Retrieved 04.05.2015 from https://project2049.net/wp-content/uploads/2018/05/china_pakistan_emerging_strains_in_the_entente_cordiale_kardon.pdf

Keogh, D. and Wood, T. (2005). Village Phone Replication Manual: Creating sustainable access to affordable telecommunications for the rural poor. New York: United Nations Information and Communication Technologies Task Force. http://www.infodev.org/infodev-files/resource/InfodevDocuments_14.pdf

Lamb, R. D. and Hameed, S. (2012). Subnational Governance, Service Delivery, and Militancy in Pakistan: A Report of the CSIS Program on Crisis, Conflict, and Cooperation. Retrieved 04.05.2014 from https://csis-prod.s3.amazonaws.com/s3fs-public/legacy_files/files/publication/120610_Lamb_SubnatGovernPakistan_web.pdf

Mas, I. and Kumar, K. (2008). Banking on Mobiles: Why, How, for Whom? Retrieved 10.12.2012 from https://www.cgap.org/sites/default/files/researches/documents/CGAP-Focus-Note-Banking-on-Mobiles-Why-How-for-Whom-Jul-2008.pdf

Mas, I. and Radcliffe, D. (2010). Mobile payments go viral: M-PESA in Kenya. In Chuhan-Pole, P. and Angwafo, M. (Eds.), *Yes Africa Can: Success Stories from a Dynamic Continent* (pp. 353-369). Retrieved from http://siteresources.worldbank.org/AFRICAEXT/Resources/258643-1271798012256/YAC_Consolidated_Web.pdf

Masood, J. and Malik, S. (2008). .pk. In Librero, F. and Arinto, P. B. (Eds.). *Digital Review of Asia Pacific 2007–2008.* (pp. 263 - 267). Retrieved 04.05.2014 from http://www.digital-review.org/uploads/files/pdf/2007-2008/intro.pdf

McCarty, M. Y. and Bjaerum, R. (2013). Easypaisa: Mobile Money Innovation in Pakistan. Retrieved 03.05.2014 from https://www.gsma.com/mobilefordevelopment/wp-content/uploads/2013/07/Telenor-Pakistan.pdf

Michaelsen, M. (2011). New Media vs. Old Politics. Retrieved 22.02.2012 from http://www.fes-asia.org/media/publication/2011_NewMediaVsOldPolitics_Pakistan_fesmediaAsiaSeries_Michaelsen.pdf

Ming, A., et al. (2013). e-Governance in Small States. London: Commonwealth Secretariat. Retrieved 04.05.2014 from https://sscoe.thecommonwealth.org/wp-content/uploads/2018/11/eGovernanceinSmallStates.pdf

Permal, A. J. (2014). Going native. Retrieved 04.05.2015 from https://aurora.dawn.com/news/1140683/going-native

State Bank of Pakistan (2012b). Branchless Banking Newsletter: Leverging Technologies and Partnerships to Promote Financial Inclusion. Retrieved 04.05.2014 from http://www.sbp.org.pk/publications/acd/2012/BranchlessBanking-Jul-Sep-2012.pdf

Unwin, T. (2005). Partnerships in development practice: evidence from multi-stakeholder ICT4D partnership practice in Africa. Retrieved 09.09.2011 from https://unesdoc.unesco.org/ark:/48223/pf0000142982

Vaid, M. and Tourangbam, M. (2014). Pakistan's Energy Crisis. Retrieved 30.08.2014 from https://foreignpolicy.com/2014/08/13/pakistans-energy-crisis/

VOA Urdu (2011). ڈاکیا : ماضی کا مضبوط معاشرتی کردار زوال پذیر ہے [Postman: The strong social character of the past is on the decline]. Retrieved 04.05.2014 from https://www.urduvoa.com/a/pakistan-postal-service-21apr11-120382089/1132690.html

Wolcott, P. and Goodman, S. E. (2000). The internet in Turkey and Pakistan: a comparative analysis. Retrieved 04.05.2014 from https://www.researchgate.net/publication/264999624_The_Internet_in_Turkey_and_Pakistan_A_Comparative_Analysis

World Bank. (2008b). Finance For All? Policies and Pitfalls in Expanding Access. Washington DC: World Bank. Retrieved 09.12.2010 from https://siteresources.worldbank.org/INTFINFORALL/Resources/4099583-1194373512632/FFA_book.pdf

Yusuf, H. (2013). Mapping Digital Media: Pakistan: A Report by the Open Society Foundations. Retrieved 02.01.2014 from https://www.opensocietyfoundations.org/uploads/4d1a6626-d6d7-41b5-befc-d8f5a78d545c/mapping-digital-media-pakistan-20130902.pdf

Yusuf, H. and Schoemaker, E. (2013). The media of Pakistan: Fostering inclusion in a fragile democracy? (Policy Briefing #9). Retrieved 01.10.2013 from

http://downloads.bbc.co.uk/mediaaction/pdf/bbc_media_action_pakistan_p
olicy_briefing.pdf

Government Documents

Cabinet Division (2006). *Year Book of Cabinet Division – 2005-06*. Retrieved
04.05.2014   from
http://cabinet.gov.pk/cabinet/userfiles1/file/Publications/year-book-2005-
06.pdf

Deutscher Bundestag (German Federal Parliament) (2014). *Deutscher Bundestag
(18. Wahlperiode): Antwort der Bundesregierung; Drucksache 18/1271*.
Retrieved   14.05.2014   from
http://dipbt.bundestag.de/doc/btd/18/012/1801271.pdf

Election Commission of Pakistan (2008). *Delegation to Observe the
Parliamentary Elections in the Islamic Republic of Pakistan (14 - 21
February 2008)*. Retrieved 04.05.2014 from
http://www.epgencms.europarl.europa.eu/cmsdata/upload/b7862f31-0376-
4959-a4b7-5b64e74a421c/Election_report_Pakistan_18_February_2008.pdf

Election Commission of Pakistan (2013a). *Second Five-Year Strategic Plan
2014-2018 (revised)*. Retrieved 04.05.2015 from
https://www.ecp.gov.pk/Documents/sp_revised/Strategic Plan Revised
final.pdf

Election Commission of Pakistan (2015). *ECP Newsletter: Issue I*. Retrieved
04.05.2015 from https://www.ecp.gov.pk/Documents/Newsletter.pdf

Electronic Government Directorate (2002). *Notification: No.3-16/2001/Coord.
(Establishment of Electronic Government Directorate)*. Retrieved
10.10.2007   from   http://pakistan.gov.pk/e-government-
directorate/about/notification1.pdf

Electronic Government Directorate (2005a). *E-Government Strategy and 5-Year
Plan for the Federal Government*. Retrieved 16.12.2010 from
http://202.83.164.28:9080/egdsite05/downloads/E-Government Strategy and
5-Year Plan (19%5B1%5D.06.2005).pdf

Electronic Government Directorate (2005b). *Briefing For Focal Persons About
E-Government Strategy and 5-Year Plan*. Retrieved 28.10.2009 from
http://pakistan.gov.pk/e-government-
directorate/presentations/briefing_for_focal Persons13102005Final.pdf

Electronic Government Directorate (2006a). *Consultancy Study on 20 Most
Important Services Assessment for Pakistan Government Services Inventory
Questionnaire (Phase 2)*. Retrieved 09.08.2008 from
http://pakistan.gov.pk/e-government-
directorate/information/media/GSI_QUESTIONNAIRE_Phase_1_210306.p
df

Electronic Government Directorate (2006b). *Consultancy Study on 20 Most
Important Services Assessment for Pakistan Government Services Inventory*

Questionnaire (Phase 1). Retrieved 09.08.2008 from http://pakistan.gov.pk/e-government-directorate/information/media/GSI_QUESTIONNAIRE_Phase_1_210306.pdf

Electronic Government Directorate (2009b). *Projects Implemented by EGD.* Retrieved 30.11.2010 from http://www.pakistan.gov.pk/e-government-directorate/projects/projects.jsp

Electronic Government Directorate (2012). *Abridged Version: E-Government Strategy for the Federal Government*. Retrieved 01.12.2013 from http://e-government.gov.pk/gop/index.php?q=aHR0cDovLzE5Mi4xNjguNzAuMTM2L2VnZW9uZy9GaW5hbEFicmlkZ2VkJTIwRU dvdmVybm1lbnRTdHJhdGVneSUyMFdlYjEucGRm

Federal Ombudsman Secretariat (2015). *Report on the functioning of National Database Registration Authority*. Retrieved 15.07.2015 from http://www.mohtasib.gov.pk/images/pdfs/NADRA.pdf

Frequency Allocation Board (2004). *Broadband Policy*. Retrieved 14.06.2009 from http://www.fab.gov.pk/images/pdf/Broadband Policy.pdf

Gazette of Pakistan (1960). *Ordinance No. X of 1960: Municipal Administration Ordinance*. Retrieved 04.05.2014 from https://cmsdata.iucn.org/downloads/municipal_administration_ordinance_1960.pdf

Gazette of Pakistan (2009). *Ordinance No. LIX of 2009*. Retrieved 20.11.2011 from http://www.na.gov.pk/uploads/documents/1302654535_961.pdf

Ministry of Finance (2001). *Interim Poverty Reduction Strategy Paper (I-PRSP)*. Retrieved 28.07.2008 from http://www.finance.gov.pk/poverty/iprsp_2.pdf

Ministry of Finance (2015). *Pakistan Economic Survey (2014-15): Transport and Communications*. Retrieved 15.10.2015 from http://www.finance.gov.pk/survey/chapters_15/13_Transport.pdf

Ministry of Information Technology (2006). *National ICT R&D Fund Policy Framework*. Retrieved 04.05.2014 from https://moitt.gov.pk/SiteImage/Misc/files/National ICT R%26D Fund Policy Framework%2C 2006.pdf

Ministry of Planning Development and Reform (2004). *Public Sector Development Projects 2002-03*. Retrieved 03.02.2009 from https://www.pc.gov.pk/uploads/archives/PSDP2002-2003.pdf

Ministry of Science and Technology (2000a). *IT Policy and Action Plan: Including initial steps required for updating and Implementing a modular National IT Policy & Plan*. Retrieved 23.06.2008 from http://anf.gov.pk/library/acts/Pakistan IT Policy and Action Plan.pdf

Ministry of Science and Technology (2000b). *IT Policy and Action Plan*. Retrieved 02.09.2008 from https://moitt.gov.pk/moit/userfiles1/file/policies/Pakistan IT Policy Action Plan 2000.pdf

Ministry of Social Welfare and Special Education (2007). *Pakistan's Consolidated Third and Fourth Periodic Report to the UN Committee on the Rights of the Child on the implementation of the Convention on the Rights of the Child.* Retrieved 04.05.2014 from https://www2.ohchr.org/english/bodies/crc/docs/AdvanceVersions/CRC.C.PAK.4.pdf

National Database Registration Authority (2008b). *NADRA announces another public service(e-Sahulat).* Retrieved 07.02.2009 from http://www.nadra.gov.pk/DesktopModules/top/topmore.aspx?tabID=0&ItemID=53&bID=0&Mid=3026

National Database Registration Authority (2009a). *NADRA's organisational structure.* Retrieved 09.06.2009 from http://nadra.gov.pk/images/organogram.jpg

National Database Registration Authority (2009d). *National Database is Fully Secured: Deputy Chairman NADRA.* Retrieved 09.07.2009 from https://www.nadra.gov.pk/DesktopModules/top/topmore.aspx?tabID=0&ItemID=57&bID=0&Mid=3026

National Database Registration Authority (2010). *NADRA's Services Brochure.* Retrieved 29.01.2010 from http://www.nadra.gov.pk/downloads/solutions/brochure.pdf

National Database Registration Authority (2014a). *Civil Registration Management System.* Retrieved 04.05.2015 from https://www.pc.gov.pk/uploads/crvs/CRVS.pdf

National Information Technology Board (2014). *Advisory Services.* Retrieved 04.05.2014 from http://nitb.gov.pk/newnitb/index.php?section=functions&page=advisoryservice

Pakistan Bureau of Statistics (2017). *Province Wise Provisional Results of Census - 2017.* http://www.pbs.gov.pk/sites/default/files/PAKISTAN TEHSIL WISE FOR WEB CENSUS_2017.pdf

Pakistan Computer Bureau (2009). *Functions of and Trainings provided by PCB* Retrieved 24.10.2009 from http://www.pcb.gov.pk/Trainings.asp

Pakistan Telecommunication Authority (2010b). *Telecom Indicators.* Retrieved 10.11.2010 from https://www.pta.gov.pk/en/telecom-indicators/2

Pakistan Telecommunication Authority (2012). *Annual Reports.* Retrieved 20.01.2012 from https://www.pta.gov.pk/en/data-&-research/publications/annual-reports

Pakistan Telecommunications Authority (2011). *Telecom Indicators.* Retrieved April 15, 2011 from http://www.pta.gov.pk/index.php?option=com_content&task=view&id=269&Itemid=658

Prime Minister's Office (2013). *National Agenda for Real Change: Manifesto (PML-N).* Retrieved 20.04.2013 from https://pmo.gov.pk/documents/manifesto.pdf

Prime Minister's Secretariat (2008). *Report of the National Commission for Government Reform on Reforming the Government in Pakistan. Vol 1.* Retrieved 04.05.2014 from https://www.pc.gov.pk/uploads/report/NCGR_Vol_I-1.pdf

Security and Exchange Commission of Pakistan (2014). *Company Name Search.* Retrieved 19.10.2014 from https://eservices.secp.gov.pk/eServices/NameSearch.jsp

Senate of Pakistan (2013). *Report of The Senate Standing Committee on Cabinet Secretariat and Capital Administration & Development.* Retrieved 04.05.2014 from http://www.senate.gov.pk/uploads/documents/1428589865_112.pdf

State Bank of Pakistan (1956). *The State Bank of Pakistan Act, 1956.* Retrieved 04.05.2014 from http://www.sbp.org.pk/about/act/SBP-Act.pdf

State Bank of Pakistan (2002b). *Electronic Transactions Ordinance.* Retrieved 15.03.2009 from http://www.sbp.org.pk/about/act/ETC202.pdf

State Bank of Pakistan (2007). *Draft: Policy Paper on Regulatory Framework for Mobile Banking in Pakistan.* Retrieved 09.03.2009 from http://www.sbp.org.pk/bprd/2007/Policy_Paper_RF_Mobile_Banking_07-Jun-07.pdf

State Bank of Pakistan (2008a). *Branchless Banking Regulations for Financial Institutions Desirous to undertake Branchless Banking.* Retrieved 15.11.2011 from http://www.sbp.org.pk/bprd/2008/Annex_C2.pdf

State Bank of Pakistan (2008b). *Pakistan 10 Year Strategy Paper for the Banking Sector Reforms.* Retrieved 23.05.2011 from http://www.sbp.org.pk/bsd/10YearStrategyPaper.pdf

State Bank of Pakistan (2011). *Branchless Banking Newsletter (Issue 1).* Retrieved 04.05.2014 from http://www.sbp.org.pk/publications/acd/2011/BranchlessBanking-Jul-Sep-2011.pdf

State Bank of Pakistan (2015). *Branchless Banking Newsletters.* Retrieved 31.12.2015 from http://www.sbp.org.pk/publications/acd/branchless.htm

State Bank of Pakistan (2017). *Quarterly Compendium: Statistics of the Banking System.* http://www.sbp.org.pk/ecodata/fsi/qc/2017/Dec.pdf

Journal Articles

A.Shaikh, A., et al. (2017). Exploring the nexus between financial sector reforms and the emergence of digital banking culture – Evidences from a developing country. *Research in International Business and Finance, 42*(2017), 1030-1039

Ahmed, F. (1996). Pakistan: Ethnic Fragmentation or National Integration? *The Pakistan Development Review, 35*(4), 631-645

Akakandelwa, A. (2011). An exploratory survey of the SADC e-government web sites. *Library Review, 60*(5), 421-431

Akram, M. and Mahmood, A. (2007). The status and teaching of English in Pakistan. *Language in India, 7*(12), 1-7

Al-Hujran, O. (2012). Toward the utilization of m-Government services in developing countries: a qualitative investigation. *International Journal of Business and Social Science, 3*(5), 155-160

Aladwani, A. M. (2013). A contingency model of citizens' attitudes toward e–government use. *Electronic Government, An International Journal, 10*(1), 68-85

Alampay, E. (2006). Beyond access to ICTs: Measuring capabilities in the information society. *International journal of education and development using ICT, 2*(3), 4-22

Alavi, H. (1975). India and the colonial mode of production. *Economic and Political Weekly, 10*(1975), 1235-1262

AlAwadhi, S. and Morris, A. (2009). Factors influencing the adoption of e-government services. *Journal of Software, 4*(6), 584-590

Ali, M. and Mujahid, N. (2015). Electronic Government Re-Inventing Governance: A Case Study of Pakistan. *Public Policy and Administration Research, 5*(2), 1-8

Ali, S. A. (1995). Unicameralism in United Pakistan: Why and How? *Pakistan Horizon, 48*(3), 69-80

Alshawi, S. and Alalwany, H. (2009). E-government evaluation: Citizen's perspective in developing countries. *Information technology for development, 15*(3), 193-208

Alzouma, G. (2005). Myths of digital technology in Africa: Leapfrogging development? *Global Media and Communication, 1*(3), 339-356

Ameen, K. and Gorman, G. (2009). Information and digital literacy: a stumbling block to development? A Pakistan perspective. *Library Management, 30*(1/2), 99-112

Ancarani, A. (2005). Towards quality e-service in the public sector: The evolution of web sites in the local public service sector. *Managing Service Quality: An International Journal, 15*(1), 6-23

Andonova, V. (2006). Mobile phones, the Internet and the institutional environment. *Telecommunications Policy, 30*(1), 29-45

Ansari, S. and Khan, A. (2009). Telecommunication trends in Pakistan. *Market Forces, 4*(4), 213-218

Asgarkhani, M. (2005). Digital government and its effectiveness in public management reform: A local government perspective. *Public Management Review, 7*(3), 465-487

Asiedu, C. (2012). Information communication technologies for gender and development in Africa: The case for radio and technological blending. *International Communication Gazette, 74*(3), 240-257

Asokan, N., et al. (1997). The state of the art in electronic payment systems. *Computer, 30*(9), 28-35

Avgerou, C. (2008). Information systems in developing countries: a critical research review. *Journal of information Technology, 23*(3), 133-146

Baber, Z. (2001). Colonizing nature: scientific knowledge, colonial power and the incorporation of India into the modern world-system. *The British journal of sociology, 52*(1), 37-58

Bakardjieva, M. and Feenberg, A. (2002). Community technology and democratic rationalization. *The information society, 18*(3), 181-192

Bano, M. (2008). Dangerous correlations: Aid's impact on NGOs' performance and ability to mobilize members in Pakistan. *World Development, 36*(11), 2297-2313

Baptista, M. (2005). E-government and state reform: Policy dilemmas for Europe. *The Electronic Journal of e-Government, 3*(4), 167-174

Barima, A. K. and Farhad, A. (2010). Challenges of making donor-driven public sector reform in sub-Saharan Africa sustainable: Some experiences from Ghana. *International Journal of Public Administration, 33*(12-13), 635-647

Barrett, M., et al. (2015). Service innovation in the digital age: key contributions and future directions. *MIS quarterly, 39*(1), 135-154

Barwise, P. (2001). TV, PC, or Mobile? Future Media for Consumer e-Commerce. *Business Strategy Review, 12*(1), 35-42

Basu, S. (2004). E-government and developing countries: an overview. *International Review of Law, Computers & Technology, 18*(1), 109-132

Batool, S. H. and Mahmood, K. (2010). Entertainment, communication or academic use? A survey of Internet cafe users in Lahore, Pakistan. *Information Development, 26*(2), 141-147

Becker, S. A. (2005). E-government usability for older adults. *Communications of the ACM, 48*(2), 102-104

Berners-Lee, T., et al. (1992). World-Wide Web: the information universe. *Internet Research, 2*(1), 52-58

Best, M. L. (2010). Understanding our knowledge gaps: Or, do we have an ICT4D field? And do we want one? *Information Technologies & International Development, 6*(SE), pp. 49-52

Beynon-Davies, P. (2007). Models for e-government. *Transforming Government: People, Process and Policy, 1*(1), 7-28

Bhatnagar, S. C. (2002). E-government: lessons from implementation in developing countries. *Regional Development Dialogue, 23*(2), 164-175

Bijker, W. E. (2009). How is technology made?—That is the question! *Cambridge journal of economics, 34*(1), 63-76

Boudreaux, G. and Sloboda, B. (1999). The fast changing world of the internet. *Management Quarterly, 40*(2), 2

Bowonder, B. and Boddu, G. (2005). Internet kiosks for rural communities: using ICT platforms for reducing digital divide. *International Journal of Services Technology and Management, 6*(3), 356-378

Brown, M. M. (2007). Understanding e-government benefits: An examination of leading-edge local governments. *The American Review of Public Administration, 37*(2), 178-197

Bwalya, K. J. (2009). Factors affecting adoption of e-government in Zambia. *The Electronic Journal of Information Systems in Developing Countries, 38*(1), 1-13

Çam, A. R. (2007). SMS information system: Mobile access to justice. *European Journal of ePractice, 4*(10), 35-41

Carleheden, M. and Borch, C. (2010). Editorial: Successive Modernities. *Journal of Social Theory, 11*(2), 5-8

Cecchini, S. and Raina, M. (2004). Electronic government and the rural poor: The case of Gyandoot. *Information Technologies and International Development, 2*(2), 65-76

Chakravartty, P. (2004). Telecom, national development and the Indian state: a postcolonial critique. *Media, Culture & Society, 26*(2), 227-249

Chaudhry, A. and Ahmad, I. (2011). Examining the relationship of work-life conflict and employee performance (a case from NADRA Pakistan). *International journal of business and management, 6*(10), 170-177

Chen, Y.-C. and Dimitrova, D. V. (2006). Electronic government and online engagement: citizen interaction with government via web portals. *International Journal of Electronic Government Research, 2*(1), 54-76

Choudhury, D. K. L. (2000). 'Beyond the reach of monkeys and men'? O'Shaughnessy and the telegraph in India c. 1836-56. *The Indian Economic & Social History Review, 37*(3), 331-359

Choudhury, D. K. L. (2009). The Telegraph, Censorship, and 'Clemency': Canning during 1857. *Contemporary Perspectives, 3*(1), 34-54

Chun, S. and Cho, J.-S. (2012). E-participation and transparent policy decision making. *Information Polity, 17*(2), 129-145

Chun, S. A., et al. (2010). Government 2.0: Making connections between citizens, data and government. *Information Polity, 15*(1), 1-9

Ciborra, C. and Navarra, D. D. (2005). Good governance, development theory, and aid policy: Risks and challenges of e-government in Jordan. *Information technology for development, 11*(2), 141-159

Clarke, S., et al. (2013). ICT 4 the MDGs? A perspective on ICTs' role in addressing urban poverty in the context of the Millennium Development Goals. *Information Technologies & International Development, 9*(4), 55-70

Cordella, A. (2007). E-government: towards the e-bureaucratic form? *Journal of information Technology, 22*(3), 265-274

Corradini, F., et al. (2007). The e-Government digital credentials. *International Journal of Electronic Governance, 1*(1), 17-37

Cruz-Jesus, F., et al. (2012). Digital divide across the European Union. *Information & Management, 49*(6), 278-291

Cullen, R. (2005). E-Government, A Citizens' Perspective. *Journal of E-Government, 1*(3), 5-28

Curran, J. M. and Meuter, M. L. (2005). Self-service technology adoption: comparing three technologies. *Journal of services marketing, 19*(2), 103-113

Cyan, M. R. (2006). Main issues for setting the civil service reform agenda in Pakistan. *The Pakistan Development Review, 45*(4), 1241-1254

Dada, D. (2006). E-readiness for developing countries: moving the focus from the environment to the users. *The Electronic Journal of Information Systems in Developing Countries, 27*(1), 1-14

Davison, R., et al. (2000). Technology Leapfrogging in Developing Countries-An Inevitable Luxury? *The Electronic Journal of Information Systems in Developing Countries, 1*(1), 1-10

Davison, R. M., et al. (2005). From government to e-government: a transition model. *Information Technology & People, 18*(3), 280-299

Dawes, S. S. (2008). The evolution and continuing challenges of e-governance. *Public administration review, 68*(S1), 86-102

Dawes, S. S., et al. (2009). From "need to know" to "need to share": Tangled problems, information boundaries, and the building of public sector knowledge networks. *Public administration review, 69*(3), 392-402

Dhagamwar, V. (1995). Elections in Pakistan, 1993. *Indian Journal of Gender Studies, 2*(1), 115-120

Donner, J. (2007). The rules of beeping: exchanging messages via intentional "missed calls" on mobile phones. *Journal of Computer-mediated communication, 13*(1), 1-22

Donovan, K. (2012). Mobile money for financial inclusion. *Information and Communications for development, 61*(1), 61-73

Dutton, W., et al. (2005). The cyber trust tension in E-government: Balancing identity, privacy, security. *Information Polity, 10*(1, 2), 13-23

Easterly, W. (2009). How the millennium development goals are unfair to Africa. *World Development, 37*(1), 26-35

Ebrahim, Z. and Irani, Z. (2005). E-government adoption: architecture and barriers. *Business Process Management Journal, 11*(5), 589-611

Englebert, P. (2000). Pre-colonial institutions, post-colonial states, and economic development in tropical Africa. *Political Research Quarterly, 53*(1), 7-36

Fonseca, C. (2010). The digital divide and the cognitive divide: Reflections on the challenge of human development in the digital age. *Information Technologies & International Development, 6*(SE), pp. 25-30

Freund, C. and Weinhold, D. (2002). The Internet and international trade in services. *American Economic Review, 92*(2), 236-240

Fuchs, C. and Horak, E. (2008). Africa and the digital divide. *Telematics and Informatics, 25*(2), 99-116

Gagnon, Y.-C., et al. (2010). Multichannel delivery of public services: A new and complex management challenge. *International Journal of Public Administration, 33*(5), 213-222

Gascó, M. (2005). Exploring the e-government gap in South America. *International Journal of Public Administration, 28*(7-8), 683-701

Gibbs, J., et al. (2003). Environment and policy factors shaping global e-commerce diffusion: A cross-country comparison. *The information society, 19*(1), 5-18

Gichoya, D. (2005). Factors affecting the successful implementation of ICT projects in government. *The Electronic Journal of e-Government, 3*(4), 175-184

Gil-García, J. R. and Pardo, T. A. (2005). E-government success factors: Mapping practical tools to theoretical foundations. *Government Information Quarterly, 22*(2), 187-216

Gorla, N. (2008). Hurdles in rural e-government projects in India: lessons for developing countries. *Electronic Government, An International Journal, 5*(1), 91-102

Gorman, M. (1971). Sir William O'Shaughnessy, Lord Dalhousie, and the Establishment of the Telegraph System in India. *Technology and Culture, 12*(4), 581-601

Gotoh, R. (2009). Critical factors increasing user satisfaction with e-government services. *Electronic Government, An International Journal, 6*(3), 252-264

Govindaraju, P. and Mabel, M. (2010). The status of information and communication technology in a coastal village: A case study. *International Journal of Education and Development using Information and Communication Technology, 6*(1), 128-135

Guma, P. (2012). Public Sector Reform, E-Government and the Search for Excellence in Africa: Experiences from Uganda. *Electronic journal of e-government, 11*(2), 241-253

Gunkel, D. J. (2003). Second thoughts: toward a critique of the digital divide. *New media & society, 5*(4), 499-522

Gupta, M. and Jana, D. (2003). E-government evaluation: A framework and case study. *Government Information Quarterly, 20*(4), 365-387

Hardy, C. A. and Williams, S. P. (2008). E-government policy and practice: A theoretical and empirical exploration of public e-procurement. *Government Information Quarterly, 25*(2), 155-180

Hasnain, Z. (2010). Devolution, accountability, and service delivery in Pakistan. *The Pakistan Development Review, 49*(2), 129-152

Hawari, A. a. and Heeks, R. (2010). Explaining ERP failure in a developing country: a Jordanian case study. *Journal of Enterprise Information Management, 23*(2), 135-160

Headrick, D. (2010). A double-edged sword: communications and imperial control in British India. *Historical Social Research, 35*(1), 51-65

Heeks, R. (2002a). Information systems and developing countries: Failure, success, and local improvisations. *The information society, 18*(2), 101-112

Heeks, R. (2002b). e-Government in Africa: Promise and practice. *Information Polity, 7*(2,3), 97-114

Heeks, R. (2002c). e-Government in Africa: Promise and practice. *Information Polity, 7*(2), 97-114

Heeks, R. (2005a). e-Government as a Carrier of Context. *Journal of Public Policy, 25*(1), 51-74

Heeks, R. (2006a). Theorizing ICT4D Research. *Information Technologies & International Development, 3*(3), 1-4

Heeks, R. (2008). ICT4D 2.0: The next phase of applying ICT for international development. *Computer, 41*(6), 26-33

Heeks, R. (2010). Do information and communication technologies (ICTs) contribute to development? *Journal of International Development, 22*(5), 625-640

Heeks, R. and Bailur, S. (2007). Analyzing e-government research: Perspectives, philosophies, theories, methods, and practice. *Government Information Quarterly, 24*(2), 243-265

Helbig, N., et al. (2009). Understanding the complexity of electronic government: Implications from the digital divide literature. *Government Information Quarterly, 26*(1), 89-97

Henman, P. (2013). Governmentalities of Gov 2.0. *Information, Communication & Society, 16*(9), 1397-1418

Herani (2007). Computerized National Identify Card, NADRA KIOSKs and its prospects. Indian Journal of Science and Technology, 1(2), 1-5

Holden, L. and Chaudhary, A. (2013). Daughters' inheritance, legal pluralism, and governance in Pakistan. *The Journal of Legal Pluralism and Unofficial Law, 45*(1), 104-123

Holmes, P., et al. (1996). International competition policy and telecommunications. Lessons from the EU and prospects for the WTO. *Telecommunications Policy, 20*(10), 755-767

Hughes, N. and Lonie, S. (2007). M-PESA: Mobile Money for the "Unbanked" Turning Cellphones into 24-Hour Tellers in Kenya. *Innovations: Technology, Governance, Globalization, 2*(1-2), 63-81

Hulme, D. and Scott, J. (2010). The political economy of the MDGs: Retrospect and prospect for the world's biggest promise. *New Political Economy, 15*(2), 293-306

Humes, L. L. and Reinhard, N. (2009). E-Government and the Influence of Power in the Development of Information Infrastructure. *Journal of Global Information Technology Management, 12*(2), 61-79

Huque, A. S. (2005). Explaining the myth of public sector reform in South Asia: de-linking cause and effect. *Policy and Society, 24*(3), 97-121

Husain, I. (1973). Mechanics of Development Planning in Pakistan: A Suggested Framework. *Pakistan Economic and Social Review, 11*(4), 454-462

Hussain, M. and Kokab, R. U. (2013). Institutional influence in Pakistan: Bureaucracy, cabinet and parliament. *Asian Social Science, 9*(7), 173-178

Hussain, S., et al. (2009). Demographic transition and economic growth in Pakistan. *European Journal of Scientific Research, 31*(3), 491-499

Ilshammar, L., et al. (2005). Public E-Services in Sweden. *Scandinavian Journal of Information Systems, 17*(2), 11–40

Iqbal, A. J. and Javed, H. (2013). Public Sector Project Management: Perception of Risks in IT. *Journal of Strategy and Performance Management, 1*(1), 24-32

Iqbal, M. and Ahmad, E. (2006). Is Good Governance an Approach to Civil Service Reforms? *The Pakistan Development Review, 45*(4), 621-637

Islam, M. S. and Business, S. (2008). Towards a sustainable e-Participation implementation model. *European Journal of ePractice, 5*(10), 1-12

Islam, N. (1989). Colonial legacy, administrative reform and politics: Pakistan 1947-1987. *Public Administration & Development, 9*(3), 271-285

Islam, N. (2004). Sifarish, sycophants, power and collectivism: administrative culture in Pakistan. *International Review of Administrative Sciences, 70*(2), 311–330

James, J. (2002). Universal access to information technology in developing countries. *Regional Studies, 36*(9), 1093-1097

James, J. (2009). Leapfrogging in mobile telephony: A measure for comparing country performance. *Technological Forecasting and Social Change, 76*(7), 991-998

James, J. and Versteeg, M. (2007). Mobile phones in Africa: how much do we really know? *Social indicators research, 84*(1), 117

Janssen, D., et al. (2004). If you measure it they will score: An assessment of international eGovernment benchmarking. *Information Polity, 9*(3, 4), 121-130

Jhatial, A. A., et al. (2014). Rhetorics and realities of management practices in Pakistan: Colonial, post-colonial and post-9/11 influences. *Business History, 56*(3), 456-484

Jones, C. and Czerniewicz, L. (2010). Describing or debunking? The net generation and digital natives. *Journal of Computer Assisted Learning, 26*(5), 317-320

Jones, G. N. (1997). Pakistan: A civil service in an obsolescing imperial tradition. *Asian journal of public Administration, 19*(2), 321-364

Kamran, S. (2010). Mobile phone: Calling and texting patterns of college students in Pakistan. *International journal of business and management, 5*(4), 26-36

Kamran, T. (2002). Electoral Politics in Pakistan (1955-1969). *Journal of Pakistan Vision, 10*(1), 82-97

Kayani, M. B., et al. (2011). Analyzing barriers in e-government implementation in Pakistan. *International Journal for Infonomics, 4*(3), 494-500

Kazi, S. (1995). Rural Women, Poverty and Development in Pakistan. *Asia-Pacific Journal of Rural Development, 5*(1), 78-92

Kernaghan, K. (2005). Moving towards the virtual state: integrating services and service channels for citizen-centred delivery. *International Review of Administrative Sciences, 71*(1), 119-131

Kessides, I. N. (2013). Chaos in power: Pakistan's electricity crisis. *Energy policy, 55*, 271-285

Khan, F. (2007). Corruption and the Decline of the State in Pakistan. *Asian Journal of Political Science, 15*(2), 219-247

Khan, I. A. (2010). Public sector institutions, politics and outsourcing: Reforming the provision of primary healthcare in Punjab, Pakistan. *Journal of International Development, 22*(4), 424-440

Khan, M. A. and Ahmed, A. (2007). Foreign aid—blessing or curse: Evidence from Pakistan. *The Pakistan Development Review, 46*(3), 215-240

Khan, T. H., et al. (2014). Governance challenges in Public Sector: an Analysis of Governance and Efficiency Issues in Public Sector Institutions in Pakistan. *Governance, 4*(9), 58-67

King, S. and Cotterill, S. (2007). Transformational government? The role of information technology in delivering citizen-centric local public services. *Local Government Studies, 33*(3), 333-354

Kleine, D. and Unwin, T. (2009). Technological Revolution, Evolution and New Dependencies: what's new about ICT4D? *Third World Quarterly, 30*(5), 1045-1067

Kolsaker, A. and Lee-Kelley, L. (2007). G2C e-government: modernisation or transformation? *Electronic Government, An International Journal, 4*(1), 68-75

Koltay, T. (2011). The media and the literacies: Media literacy, information literacy, digital literacy. *Media, Culture & Society, 33*(2), 211-221

Kostopoulos, G. K. (1998). Global delivery of education via the Internet. *Internet Research, 8*(3), 257-265

Kraemer, K. L., et al. (1994). Information technology and transitions in the public service: A comparison of Scandinavia and the United States. *International Journal of Public Administration, 17*(10), 1871-1905

Kumar, R. and Best, M. L. (2007). Social impact and diffusion of telecenter use: A study from the sustainable access in rural India project. *The Journal of Community Informatics, 2*(3), Special Issue: Telecentres

Kunstelj, M. and Vintar, M. (2004). Evaluating the progress of e-government development: A critical analysis. *Information Polity, 9*(3), 131-148

Lahiri, N. (2003). Commemorating and remembering 1857: The revolt in Delhi and its afterlife. *World Archaeology, 35*(1), 35-60

Lai, R. and Haleem, A. (2002). E-governance: an emerging paradigm. *Vision, 6*(2), 99-109

Lange, M. K. (2004). British colonial legacies and political development. *World Development, 32*(6), 905-922

Lassinantti, J., et al. (2014). Shaping Local Open Data Initiatives: Politics and Implications. *Journal of theoretical and applied electronic commerce research, 9*(2), 17-33

Lau, T., et al. (2008). Adoption of e-government in three Latin American countries: Argentina, Brazil and Mexico. *Telecommunications Policy, 32*(2), 88-100

Layne, K. and Lee, J. (2001). Developing fully functional E-government: A four stage model. *Government Information Quarterly, 18*(2), 122-136

Lee, H., et al. (2008). Analysing South Korea's ICT for development aid programme. *The Electronic Journal of Information Systems in Developing Countries, 35*(1), 1-15

Lips, A. M. B., et al. (2009). Managing citizen identity information in E-government service relationships in the UK: the emergence of a surveillance state or a service state? *Public Management Review, 11*(6), 833-856

Lips, M. (2010). Rethinking citizen–government relationships in the age of digital identity: Insights from research. *Information Polity, 15*(4), 273-289

Looney, R. E. (1998). Telecommunications policy in Pakistan. *Telematics and Informatics, 15*(1-2), 11-33

Lu, M.-t. (2001). Digital Divide in Developing Countries. *Journal of Global Information Technology Management, 4*(3), 1-4

Madon, S. (2005). Governance lessons from the experience of telecentres in Kerala. *European Journal of Information Systems, 14*(4), 401-416

Maheshwari, S. (1974). Administrative Reforms in Pakistan. *The Indian Journal of Political Science, 35*(2), 144-156

Mahmood, K. (2005). Multipurpose community telecenters for rural development in Pakistan. *The Electronic Library, 23*(2), 204-220

Majeed, G. (2010). Ethnicity and ethnic conflict in Pakistan. *Journal of Political Studies, 17*(2), 51

Malik, S., et al. (2009). Socio-economic Impact of Cellular Phones Growth in Pakistan: An Empirical Analysis. *Pakistan Journal of Social Sciences, 29*(1), 23-37

Mandarano, L., et al. (2010). Building social capital in the digital age of civic engagement. *Journal of Planning Literature, 25*(2), 123-135

Maniruzzaman, T. (1971). Crises in Political Development" and the Collapse of the Ayub Regime in Pakistan. *The Journal of Developing Areas, 5*(2), 221-238

Mazhar, F., et al. (2012). Consumer trust in e-commerce: A study of consumer perceptions in Pakistan. *African Journal of Business Management, 6*(7), 2516-2528

McIvor, R., et al. (2002). Internet technologies: supporting transparency in the public sector. *International Journal of Public Sector Management, 15*(3), 170-187

Mofleh, S., et al. (2008a). The gap between citizens and e-government projects: the case for Jordan. *Electronic Government, An International Journal, 5*(3), 275-287

Mofleh, S., et al. (2008b). Developing countries and ICT initiatives: Lessons learnt from Jordan's experience. *The Electronic Journal of Information Systems in Developing Countries, 34*(1), 1-17

Mohabbat Khan, M. (1999). Civil service reforms in British India and united Pakistan. *International Journal of Public Administration, 22*(6), 947-954

Morris, M. and Ogan, C. (1996). The Internet as mass medium. *Journal of Communication, 46*(1), 39-50

Mukerji, M. (2008). Telecentres in rural India: emergence and a typology. *The Electronic Journal of Information Systems in Developing Countries, 35*(1), 1-13

Murthy, D. and Longwell, S. A. (2013). Twitter and Disasters. *Information, Communication & Society, 16*(6), 837-855

Mustafa, D. and Sawas, A. (2013). Urbanisation and political change in Pakistan: Exploring the known unknowns. *Third World Quarterly, 34*(7), 1293-1304

Mustafa, R. (2015). Business model innovation: pervasiveness of mobile banking ecosystem and activity system–an illustrative case of Telenor Easypaisa. *Journal of Strategy and Management, 8*(4), 342-367

Mutula, S. M. (2005). Peculiarities of the digital divide in sub-Saharan Africa. *Program: electronic library and information systems, 39*(2), 122-138

Mutula, S. M. and Mostert, J. (2010). Challenges and opportunities of e-government in South Africa. *Electronic Library, The, 28*(1), 38-53

Nair, K. and Prasad, P. (2002). Development through information technology in developing countries: experiences from an Indian state. *The Electronic Journal of Information Systems in Developing Countries, 8*(1), 1-13

Ndou, V. (2004). E-government for developing countries: opportunities and challenges. *The Electronic Journal of Information Systems in Developing Countries, 18*

Ni, A. Y. and Ho, A. T.-K. (2005). Challenges in e-government development: Lessons from two information kiosk projects. *Government Information Quarterly, 22*(1), 58-74

Niaz, I. (2011). Advising the State: Bureaucratic Leadership and the Crisis of Governance in Pakistan, 1952–2000. *Journal of the Royal Asiatic Society, 21*(1), 41-53

Nygren, K. G. (2009). E-governmentality: On electronic administration in local government. *The Electronic Journal of e-Government, 7*(1), 55-64

Nygren, K. G. (2010). "Monotonized administrators" and "personalized bureaucrats" in the everyday practice of e-government: Ideal-typical occupations and processes of closure and stabilization in a Swedish municipality. *Transforming Government: People, Process and Policy, 4*(4), 322-337

Otjacques, B., et al. (2007). Interoperability of e-government information systems: Issues of identification and data sharing. *Journal of management information systems, 23*(4), 29-51

Paoli, A. D. and Leone, S. (2015). Challenging conceptual and empirical definition of e-government toward effective e-governance. *International Journal of Social Science and Humanity, 5*(2), 186

Parent, M., et al. (2005). Building citizen trust through e-government. *Government Information Quarterly, 22*(4), 720-736

Parvez, Z. and Ahmed, P. (2006). Towards building an integrated perspective on e-democracy. *Information, Community & Society, 9*(5), 612-632

Pettit, J., et al. (2009). Citizens' media and communication. *Development in Practice, 19*(4-5), 443-452

Pieterson, W. (2010). Citizens and service channels: Channel choice and channel management implications. *International Journal of Electronic Government Research, 6*(2), 37-53

Pigg, K. E. and Crank, L. D. (2004). Building community social capital: The potential and promise of information and communications technologies. *The Journal of Community Informatics, 1*(1), 58-73

Pinkerton, A. (2008). Radio and the Raj: broadcasting in British India (1920–1940). *Journal of the Royal Asiatic Society, 18*(2), 167-191

Porter, G. (2012). Mobile phones, livelihoods and the poor in Sub-Saharan Africa: Review and prospect. *Geography Compass, 6*(5), 241-259

Power, D. and Power, M. R. (2009). Communication and culture: signing deaf people online in Europe. *Technology and Disability, 21*(4), 127-134

Pradhan, Q. (2007). Empire in the hills: the making of hill stations in colonial India. *Studies in History, 23*(1), 33-91

Prakash, A. and De', R. (2007). Importance of development context in ICT4D projects: A study of computerization of land records in India. *Information Technology & People, 20*(3), 262-281

Prasad, K. (2012). E-governance policy for modernizing government through digital democracy in India. *Journal of Information Policy, 2*, 183-203

Rahman, T. (1997a). The medium of instruction controversy in Pakistan. *Journal of Multilingual and Multicultural Development, 18*(2), 145-154

Rahman, T. (1997b). The Urdu—English Controversy in Pakistan. *Modern Asian Studies, 31*(1), 177-207

Rahman, T. (2005). Passports to privilege: The English-medium schools in Pakistan. *Peace and Democracy in South Asia, 1*(1), 24-44

Rahman, T. (2009). Language ideology, identity and the commodification of language in the call centers of Pakistan. *Language in Society, 38*(2), 233-258

Ramli, R. M. (2012). Hybrid approach of e-government on Malaysian e-government experience. *International Journal of Social Science and Humanity, 2*(5), 366-370

Rao, S. S. (2008). Social development in Indian rural communities: Adoption of telecentres. *International Journal of Information Management, 28*(6), 474-482

Rawan, S. M. (2002). Modern mass media and traditional communication in Afghanistan. *Political communication, 19*(2), 155-170

Reddick, C. G. (2004). A two-stage model of e-government growth: Theories and empirical evidence for US cities. *Government Information Quarterly, 21*(1), 51-64

Reddick, C. G. (2005). Citizen interaction with e-government: From the streets to servers? *Government Information Quarterly, 22*(1), 38-57

Reddick, C. G., et al. (2012). Channel choice and the digital divide in e-government: the case of Egypt. *Information technology for development, 18*(3), 226-246

Rice, M. F. (2003). Information and communication technologies and the global digital divide: Technology transfer, development, and least developing countries. *Comparative Technology Transfer and Society, 1*(1), 72-88

Rocheleau, B. (2007). Whither E-Government? *Public administration review, 67*(3), 584-588

Roman, R. and Colle, R. D. (2003). Content creation for ICT development projects: Integrating normative approaches and community demand. *Information technology for development, 10*(2), 85-94

Roomi, M. A. and Parrott, G. (2008). Barriers to development and progression of women entrepreneurs in Pakistan. *The Journal of Entrepreneurship, 17*(1), 59-72

Rose, R. (2005). A global diffusion model of e-governance. *Journal of Public Policy, 25*(1), 5-27

Rust, R. (2001). The rise of e-service. *Journal of Service Research, 3*(4), 283-284

Sachs, J. D. and McArthur, J. W. (2005). The millennium project: a plan for meeting the millennium development goals. *Lancet, 365*(9456), 347-353

Sagheb-Tehrani, M. (2010). A model of successful factors towards e-government implementation. *Electronic Government, An International Journal, 7*(1), 60-74

Sahay, S. and Avgerou, C. (2002). Introducing the special issue on information and communication technologies in developing countries. *The information society, 18*(2), 73-76

Sale, J. E., et al. (2002). Revisiting the quantitative-qualitative debate: Implications for mixed-methods research. *Quality and quantity, 36*(1), 43-53

Sangwan, S. (1988). Indian response to European science and technology 1757–1857. *The British Journal for the History of Science, 21*(2), 211-232

Sarfo, E. (2011). Offering healthcare through radio: An analysis of radio health talk by medical doctors. *Journal of Language, Discourse and Society, 1*(1), 104-125

Sarkar, S. (2010). Technological momentum: Bengal in the nineteenth century. *Indian Historical Review, 37*(1), 89-109

Sawyer, S. and Eschenfelder, K. R. (2002). Social informatics: Perspectives, examples, and trends. *Annual review of information science and technology, 36*(1), 427-465

Saxena, K. (2005). Towards excellence in e-governance. *International Journal of Public Sector Management, 18*(6), 498-513

Schmidt, P. (2003). New model, old barriers: remaining challenges to African civil society participation. *Information Technologies and International Development, 1*(3-4), 100-103

Schuppan, T. (2009). E-Government in developing countries: Experiences from sub-Saharan Africa. *Government Information Quarterly, 26*(1), 118-127

Schware, R. and Deane, A. (2003). Deploying e-government programs: The strategic importance of "I" before "E". *info, 5*(4), 10-19

Schwester, R. (2009). Examining the barriers to e-government adoption. *The Electronic Journal of e-Government, 7*(1), 113 - 122

Selwyn, N. (2002). 'E-stablishing'an inclusive society? Technology, social exclusion and UK government policy making. *Journal of social Policy, 31*(1), 1-20

Servaes, J. and Malikhao, P. (2005). Participatory communication: The new paradigm. *Media & global change. Rethinking communication for development*, 91-103

Shafique, F. and Mahmood, K. (2008). Indicators of the emerging information society in Pakistan. *Information Development, 24*(1), 66-78

Shah, A. (2002). South Asia faces the future: Democracy on hold in Pakistan. *Journal of Democracy, 13*(1), 67-75

Shah, M. (2007). E-governance in India: Dream or reality? *International Journal of Education and Development using Information and Communication Technology, 3*(2), 125-137

Shah, S. A. H., et al. (2006). Convergence Model of Governance: A Case Study of the Local Government System of Pakistan. *The Pakistan Development Review, 45*(4), 855-871

Sharma, V. and Seth, P. (2011). Effective Public Private Partnership through E-Governance Facilitation. *Computer & Communications Sciences, 34*(1), 15-25

Shaukat, M. (2009). Developments of Information Technology, Telecom and E-Commerce in Business Environment of Pakistan: An Analysis of Banking and Manufacturing Sectors. *Pakistan Journal of Social Sciences, 29*(2), 259-278

Shea, T., et al. (2007). Putting the world in the world wide web: the globalisation of the internet. *International Journal of Business Information Systems, 2*(1), 75-98

Shin, D.-H. (2007). A critique of Korean National Information Strategy: Case of national information infrastructures. *Government Information Quarterly, 24*(3), 624-645

Shipps, B. (2013). Social networks, interactivity and satisfaction: Assessing socio-technical behavioral factors as an extension to technology acceptance. *Journal of theoretical and applied electronic commerce research, 8*(1), 35-52

Singh, A. K. and Sahu, R. (2008). Integrating Internet, telephones, and call centers for delivering better quality e-governance to all citizens. *Government Information Quarterly, 25*(3), 477-490

Siriginidi, S. R. (2009). Achieving millennium development goals: Role of ICTS innovations in India. *Telematics and Informatics, 26*(2), 127-143

Slotten, H. R. (2002). Satellite communications, globalization, and the Cold War. *Technology and Culture, 43*(2), 315-350

Solymar, L. (2000). The effect of the telegraph on law and order, war, diplomacy, and power politics. *Interdisciplinary Science Reviews, 25*(3), 203-210

Spence, R. and Smith, M. L. (2010). ICT, development, and poverty reduction: Five emerging stories. *Information Technologies & International Development, 6*(SE), 11-17

Sreekumar, T. T. (2007). Decrypting e-governance: Narratives, power play and participation in the Gyandoot intranet. *The Electronic Journal of Information Systems in Developing Countries, 32*(1), 1-24

Steinmueller, W. E. (2002). Knowledge-based economies and information and communication technologies. *International Social Science Journal, 54*(171), 141-153

Stemberger, M. I. and Jaklic, J. (2007). Towards E-government by business process change—A methodology for public sector. *International Journal of Information Management, 27*(4), 221-232

Subrahmanyam, G. (2006). Ruling continuities: Colonial rule, social forces and path dependence in British India and Africa. *Commonwealth & Comparative Politics, 44*(1), 84-117

Suoranta, J. (2003). The world divided in two: Digital divide information and communication technologies, and the'youth question'. *Journal for Critical Education Policy Studies, 1*(2), 1-31

Susanto, T. and Goodwin, R. (2013). User acceptance of SMS-based e-government services: Differences between adopters and non-adopters. *Government Information Quarterly, 30*(4), 486-497

Susanto, T. D. and Goodwin, R. (2010). Factors Influencing Citizen Adoption of SMS-Based e-Government Services. *The Electronic Journal of e-Government, 8*(1), 55-71

Syed, A. H. (1971). Bureaucratic Ethic & Ethos in Pakistan. *Polity, 4*(2), 159-194

Taj, A. and Baker, K. (2018). Multi-level Governance and Local Government Reform in Pakistan. *Progress in Development Studies, 18*(4), 267-281

Tat-Kei Ho, A. (2002). Reinventing local governments and the e-government initiative. *Public administration review, 62*(4), 434-444

Taylor, J. and Lips, A. (2008). The citizen in the information polity: Exposing the limits of the e-government paradigm. *Information Polity, 13*(3), 139-152

Taylor, J. A. and Lips, A. M. B. (2008). The citizen in the information polity: Exposing the limits of the e-government paradigm. *Information Polity, 13*(3,4), 139-152

Thomas, P. (2009). Bhoomi, Gyan Ganga, e-governance and the right to information: ICTs and development in India. *Telematics and Informatics, 26*(1), 20-31

Thompson, M. (2004). Discourse, 'Development' & the 'digital divide': ICT & the World Bank. *Review of African Political Economy, 31*(99), 103-123

Thompson, M. (2008). ICT and development studies: Towards development 2.0. *Journal of International Development, 20*(6), 821-835

Torres, L., et al. (2005). E-government and the transformation of public administrations in EU countries: Beyond NPM or just a second wave of reforms? *Online Information Review, 29*(5), 531-553

Trimi, S. and Sheng, H. (2008). Emerging trends in M-government. *Communications of the ACM, 51*(5), 53-58

Unegbu, V. E. and Haliso, Y. (2012). Local Government Financial Information Communication, Citizenry Awareness and Empowerment in Nigeria. *Journal of Economics and Sustainable Development, 3*(12), 116-128

van Velsen, L., et al. (2009). Requirements engineering for e-Government services: A citizen-centric approach and case study. *Government Information Quarterly, 26*(3), 477-486

Vance, A., et al. (2008). Examining trust in information technology artifacts: the effects of system quality and culture. *Journal of management information systems, 24*(4), 73-100

Vandemoortele, J. (2008). Making sense of the MDGs. *Development, 51*(2), 220-227

Vandemoortele, J. (2011). A fresh look at the MDGs. *Journal of the Asia Pacific Economy, 16*(4), 520-528

Vassilakis, C., et al. (2007). A knowledge-based approach for developing multi-channel e-government services. *Electronic Commerce Research and Applications, 6*(1), 113-124

Venkatesh, V., et al. (2014). Understanding e-Government portal use in rural India: role of demographic and personality characteristics. *Information Systems Journal, 24*(3), 249-269

Venkatraman, S. and Hughes, S. (2009). The development of an information systems strategic plan: an e-government perspective. *International Journal of Business Excellence, 2*(1), 50-64

Verdegem, P. and Verleye, G. (2009). User-centered E-Government in practice: A comprehensive model for measuring user satisfaction. *Government Information Quarterly, 26*(3), 487-497

Warkentin, M., et al. (2002). Encouraging citizen adoption of e-government by building trust. *Electronic markets, 12*(3), 157-162

Warren, M. (2007). The digital vicious cycle: Links between social disadvantage and digital exclusion in rural areas. *Telecommunications Policy, 31*(6), 374-388

Warschauer, M. (2003a). Demystifying the digital divide. *Scientific American, 289*(2), 42-47

Waseem, M. (2002). Causes of Democratic Downslide. *Economic and Political Weekly, 37*(44-45), 4532-4538

Waseem, M. (2011). Pakistan:A Majority-Constraining Federalism. *India Quarterly, 67*(3), 213-228

Watling, S. (2011). Digital exclusion: coming out from behind closed doors. *Disability & Society, 26*(4), 491-495

Weerakkody, V., et al. (2012). E-government implementation strategies in developed and transition economies: A comparative study. *International Journal of Information Management, 32*(1), 66-74

Welch, E. W., et al. (2004). Linking citizen satisfaction with e-government and trust in government. *Journal of public administration research and theory, 15*(3), 371-391

Wells, M. G. (2000). Business process re-engineering implementations using Internet technology. *Business Process Management Journal, 6*(2), 164-184

Wescott, C. G. (2001). E-Government in the Asia Pacific region. *Asian Journal of Political Science, 9*(2), 1-24

West, D. M. (2004a). E-Government and the Transformation of Service Delivery and Citizen Attitudes. *Public administration review, 64*(1), 15-27

West, D. M. (2004b). E-government and the transformation of service delivery and citizen attitudes. *Public administration review, 64*(1), 15-27

Wigand, R. T. (1997). Electronic commerce: Definition, theory, and context. *The information society, 13*(1), 1-16

Wilder, A. (2009). The politics of civil service reform in Pakistan. *Journal of International Affairs, 63*(1), 19-37

Williams, R. and Edge, D. (1996). The social shaping of technology. *Research policy, 25*(6), 865-899

Yildiz, M. (2007). E-government research: Reviewing the literature, limitations, and ways forward. *Government Information Quarterly, 24*(3), 646-665

Yu, F.-L. T. (2008). Uncertainty, human agency and e-government. *Transforming Government: People, Process and Policy, 2*(4), 283-296

Yu, L. (2006). Understanding information inequality: Making sense of the literature of the information and digital divides. *Journal of Librarianship and Information Science, 38*(4), 229-252

Zahra, K., et al. (2008). Telecommunication infrastructure development and economic growth: A panel data approach. *The Pakistan Development Review, 47*(4-II), 711-726

Zaidi, S. A. (2005). State, Military and Social Transition: Improbable Future of Democracy in Pakistan. *Economic and Political Weekly, 40*(49), 5173-5181

Zhao, J. J. and Zhao, S. Y. (2010). Opportunities and threats: A security assessment of state e-government websites. *Government Information Quarterly, 27*(1), 49-56

Zhou, Z. and Gao, F. (2007). E-government and Knowledge Management. *International Journal of Computer Science and Network Security, 7*(6), 285-289

Ziring, L. and Robert LaPorte, J. (1974). The Pakistan Bureaucracy: Two Views. *Asian Survey, 14*(12), 1086-1103

Newspaper/Magazine Articles

ECP signs contract with NADRA for verification (2011, June 27), *Aaj News*. Retrieved 04.05.2014 from http://www.aaj.tv/english/national/ecp-signs-contract-with-nadra-for-verification/

Arshad, A. (2014, November 01). Banking by app. *Aurora. Dawn Media Group*. Retrieved 15.01.2015 from https://aurora.dawn.com/news/1140685

The mysterious powers of Microsoft Excel (2013, 21 April), *BBC News*. Retrieved 04.05.2014 from https://www.bbc.com/news/magazine-22213219

[New facility to pay utility bills] سہولت نئی کی کرنے ادا بل یوٹلٹی (2006, October 12), *BBC Urdu*. Retrieved 04.05.2014 from https://www.bbc.com/urdu/pakistan/story/2006/10/061012_utility_bills_kiosks_uk.shtml

Alternate channels for E-Billing (2004a, January 15), *Business Recorder*. Retrieved 10.10.2009 from https://fp.brecorder.com/2004/01/20040115191221

Islamabad vehicle registration system to be computerised (2004b, May 02), *Business Recorder*. Retrieved 04.05.2014 from https://fp.brecorder.com/2004/05/20040502122986/

SBP stresses photo identity for bank account (2005a, August 31), *Business Recorder*. Retrieved 04.05.2014 from https://fp.brecorder.com/2005/08/20050831319362/

Nadra bill payment machines to start work from next week (2005b, November 28), *Business Recorder*. Retrieved 04.05.2014 from https://fp.brecorder.com/2005/11/20051128358037

Nadra signs accord for KESC bills payment (2005c, September 09), *Business Recorder*. Retrieved 04.05.2014 from https://fp.brecorder.com/2005/09/20050909323033

Nadra likely to sign MoUs with utility companies (2005d, April 16), *Business Recorder*. Retrieved 04.05.2014 from https://fp.brecorder.com/2005/04/20050416233198

MoU for online payment of utility bills signed (2005e, August 10), *Business Recorder*. Retrieved 04.05.2014 from https://fp.brecorder.com/2005/08/20050810310524/

Nadra advises gas consumers to get registered (2005f, August 30), *Business Recorder*. Retrieved 04.05.2014 from https://fp.brecorder.com/2005/08/20050830319049/

Mobilink continues to grow despite increasing competition (2005g, June 17), *Business Recorder*. Retrieved 03.04.2009 from https://fp.brecorder.com/2005/06/20050617282489/

Nadra and PTCL sign contract (2006, March 03), *Business Recorder*. Retrieved 04.05.2014 from https://fp.brecorder.com/2006/03/20060303393678/

Nadra to participate in card technology conference in Egypt from May 6 (2007a, May 01), *Business Recorder*. Retrieved 04.05.2014 from https://fp.brecorder.com/2007/05/20070501558069

Nadra dispersing valuable data to government departments (2007b, April 30), *Business Recorder*. Retrieved 04.05.2014 from https://fp.brecorder.com/2007/04/20070430558432/

PTA launches 'Rabta Ghar' (2007c, October 30), *Business Recorder*. Retrieved March 29, 2010 from https://fp.brecorder.com/2007/10/20071030645727

E-electoral rolls soon: ECP (2007d, January 28), *Business Recorder*. Retrieved 12.04.2014 from https://fp.brecorder.com/2007/01/20070128522704/

1LINK building the e-Banking infrastructure in Pakistan (2008a, April 23), *Business Recorder*. Retrieved 04.05.2014 from https://fp.brecorder.com/2008/04/20080423726993

Digital signature certificates: SECP, NIFT sign agreement (2008b, October 31), *Business Recorder*. Retrieved 23.07.2009 from https://fp.brecorder.com/2008/10/20081031828456/

Nadra's e-Sahulat project woos huge response (2008c, March 14), *Business Recorder*. Retrieved 04.05.2014 from https://fp.brecorder.com/2008/03/20080314708668

Several ATM machines go out of order (2009, November 28), *Business Recorder*. Retrieved 04.05.2014 from https://fp.brecorder.com/2009/11/20091128991571

e-Sahulat Branchless Banking: Nadra facilitates financial institutions, telecom operators (2010a, December 23), *Business Recorder*. Retrieved 04.05.2014 from https://fp.brecorder.com/2010/12/201012231137422/

Data Democratisation – ensuring Transparency in Disaster Relief (2010b, October 31), *Business Recorder*. Retrieved 04.05.2014 from https://fp.brecorder.com/2010/10/201010311118592/

2007 electoral rolls: 37.18 million unverified voters removed (2011, September 25), *Business Recorder*. Retrieved 04.05.2014 from https://fp.brecorder.com/2011/09/201109251234878/

Nadra carries out data entry: ECP meeting reviews situation (2012a, February 10), *Business Recorder*. Retrieved 04.05.2014 from https://fp.brecorder.com/2012/02/201202101152327/

EGD projects: National Assembly panel detects Rs 1 billion embezzlement (2012b, November 08), *Business Recorder*. Retrieved 04.05.2014 from https://fp.brecorder.com/2012/11/201211081256384

Consistant policies can tap Pakistan's economic potential (2013a, March 16), *Business Recorder*. Retrieved 04.05.2014 from https://www.brecorder.com/2013/03/16/110972/consistant-policies-can-tap-pakistans-economic-potential

Mobile banking with 1.4mn accounts showing remarkable surge (2013b, April 14), *Business Recorder*. Retrieved 04.05.2014 from https://www.brecorder.com/2013/04/14/115161

Mobile Banking is a growth story in Pakistan (2013c, March 14), *Business Recorder*. Retrieved 04.05.2014 from https://fp.brecorder.com/2013/03/201303141163171/

Pakistan's e-banking – past, present and future (2013d, April 23), *Business Recorder*. Retrieved 04.05.2014 from https://fp.brecorder.com/2013/04/201304251178401/

'The Nadra story' (2014a, November 12), *Business Recorder*. Retrieved 04.05.2015 from https://epaper.brecorder.com/2014/11/12/7-page/466533-news.html

Telecom sector attracts US$ 12bn investments in seven years (2014b, June 17), *Business Recorder*. Retrieved 30.09.2014 from https://www.brecorder.com/2014/06/17/179565/

'Smart cards to greatly benefit private sector,' DG Operations, Nadra Technologies Limited (2014c, June 27), *Business Recorder*. Retrieved 04.05.2015 from https://fp.brecorder.com/2014/06/201406271197131/

ICT Review / 2014 (2014d, December 16), *Business Recorder*. Retrieved 04.05.2015 from http://www.brecorder.com/pdf/ict2014.pdf

"National IT Board" being set up: PCB-EGD merger approved (2014e, July 19), *Business Recorder*. Retrieved https://fp.brecorder.com/2014/07/201407191204134

eGovernment: SECP's eServices greatly help business entities (2015, January 11), *Business Recorder*. Retrieved 04.05.2015 from https://fp.brecorder.com/2015/01/201501111141316/

Pakistan ends cheap late-night mobile phone deals to protect young from 'vulgarity' (2012 *The Telegraph*. Retrieved https://www.telegraph.co.uk/news/worldnews/asia/pakistan/9692675/Pakistan-ends-cheap-late-night-mobile-phone-deals-to-protect-young-from-vulgarity.html

Immigration identity system at airport (2001, December 09), *Daily Dawn*. Retrieved 11.05.2014 from https://www.dawn.com/news/9830/karachi-immigration-identity-system-at-airport

PTCL losing customers to cheap cell phones (2002a, January 16), *Daily Dawn*. Retrieved 15.04.2009 from https://www.dawn.com/news/15176

Legal cover soon for e-signatures, -documents (2002b, September 03), 23.08.2009 from https://www.dawn.com/news/55507/legal-cover-soon-for-e-signatures-documents

Electronic Transactions Ordinance promulgated (2002c, September 12), *Daily Dawn*. Retrieved 04.05.2014 from http://www.dawn.com/news/56846/electronic-transactions-ordinance-promulgated

Seminar calls for freedom of electronic media (2002d, February 01), *Daily Dawn*. Retrieved 30.08.2008 from https://www.dawn.com/news/17381/karachi-seminar-calls-for-freedom-of-electronic-media

Software park, IT board inaugurated (2002e, February 17), *Daily Dawn*. Retrieved 04.05.2014 from https://www.dawn.com/news/22096

Nadra to compile data of 140m people (2002f, August 11), *Daily Dawn*. Retrieved 04.05.2014 from https://www.dawn.com/news/52097

Waiting for the NIC: Dateline Islamabad (2002g, February 12), *Daily Dawn*. Retrieved 04.05.2014 from https://www.dawn.com/news/1062686/dawn-features-february-12-2002

Sindh IT board constituted (2002h, November 22), *Daily Dawn*. Retrieved 04.05.2014 from https://www.dawn.com/news/67931

E-governance will help in eradication of corruption: Ata (2002i, November 05), *Daily Dawn*. Retrieved 04.05.2014 from https://www.dawn.com/news/65117

E-banking be made secure, says SBP chief (2003a, July 27), *Daily Dawn*. Retrieved 04.05.2014 from https://www.dawn.com/news/132128/karachi-e-banking-be-made-secure-says-sbp-chief

No extension in Dec 31 deadline for ID cards (2003b, November 15), *Daily Dawn*. Retrieved 04.05.2014 from https://www.dawn.com/news/125001

Nadra to introduce new verification facility (2003c, July 28), *Daily Dawn*. Retrieved 04.05.2014 from https://www.dawn.com/news/132294

Nadra's hi-tech system opens today (2003d, July 29), *Daily Dawn*. Retrieved 04.05.2014 from https://www.dawn.com/news/132510

Nadra set to introduce online verification system (2003e, August 11), *Daily Dawn*. Retrieved 04.05.2014 from https://www.dawn.com/news/134572

Nadra refuses to own its mistakes in CNICs (2003f, January 09), *Daily Dawn*. Retrieved 04.05.2014 from https://www.dawn.com/news/76685

E-commerce and its future in Pakistan (2003g, April 21), *Daily Dawn*. Retrieved 26.07.2009 from https://www.dawn.com/news/96984

Incentives to private TV channels stressed (2003h, August 02), *Daily Dawn*. Retrieved 01.09.2008 from https://www.dawn.com/news/133754/karachi-incentives-to-private-tv-channels-stressed

Rs452 million data centre project approved: Communication within depts (2004a, June 09), *Daily Dawn*. Retrieved 04.05.2014 from https://www.dawn.com/news/361312

Tax imposed on surcharge: Late payment of power bills (2004b, September 29), *Daily Dawn*. Retrieved 04.05.2014 from https://www.dawn.com/news/372024

Machine readable passports in six months (2004c, January 11), *Daily Dawn*. Retrieved 14.05.2014 from https://www.dawn.com/news/390878

MRP facility handed over to passport dept (2004d, December 28), *Daily Dawn*. Retrieved 14.05.2014 from https://www.dawn.com/news/378252

Old passports will be replaced by June 2006 (2004e, October 27, 2004), *Daily Dawn*. Retrieved 14.05.2014 from https://www.dawn.com/news/398323

New passports from September (2004f, March 31), *Daily Dawn*. Retrieved 13.05.2014 from https://www.dawn.com/news/355229

Issuing of manual passports to stop by June 30 (2005a, May 07), *Daily Dawn*. Retrieved 04.05.2014 from https://www.dawn.com/news/138212

Nadra signs MoU for payment of bills (2005b, August 10), *Daily Dawn*. Retrieved 04.05.2014 from https://www.dawn.com/news/151636

Nadra bill machines start functioning next week (2005c, November 28), *Daily Dawn*. Retrieved 04.05.2014 from https://www.dawn.com/news/167475

Immigration takes over passport project (2005d, January 01), *Daily Dawn*. Retrieved 04.05.2014 from https://www.dawn.com/news/378681/immigration-takes-over-passport-project

Karachi: Nadra terms MRP system perfect (2006a, April 2008), *Daily Dawn*. Retrieved 14.05.2014 from https://www.dawn.com/news/189724/karachi-nadra-terms-mrp-system-perfect

PTA cuts bandwidth rates (2006b, June 24), *Daily Dawn*. Retrieved 06.04.2009 from https://www.dawn.com/news/198300

Hospitals to be computerised (2006c, June 13), *Daily Dawn*. Retrieved 04.05.2014 from https://www.dawn.com/news/196753/hospitals-to-be-computerised

Process to issue passport, visa being revamped (2006d, October 20), *Daily Dawn*. Retrieved 04.05.2014 from https://www.dawn.com/news/215619/process-to-issue-passport-visa-being-revamped

Nadra franchises kiosks (2007a, March 10), *Daily Dawn*. Retrieved 13.12.2009 from https://www.dawn.com/news/236682

Nadra to register births, marriages (2007b, August 13), *Daily Dawn*. Retrieved 04.05.2014 from https://www.dawn.com/news/260982

New act to regulate payment systems (2007c, June 16), *Daily Dawn*. Retrieved 04.05.2014 from https://www.dawn.com/news/251898

Computerisation comes at a price: Birth, death certificates (2007d, August 11), *Daily Dawn*. Retrieved 04.05.2014 from https://www.dawn.com/news/260696

Nadra employees caught in fake ID cards racket (2007e, November 17), *Daily Dawn*. Retrieved 04.05.2014 from https://www.dawn.com/news/276183

Cellphone users jump to 52m from 1.2m in 2002 (2007f, May 13), *Daily Dawn*. Retrieved 02.09.2009 from https://www.dawn.com/news/246666

400 telecentres for villages (2007g, February 16), *Daily Dawn*. Retrieved 09.03.2009 from https://www.dawn.com/news/233033

Over 100 more get Nadra's 'e-Sahulat' licence (2008, July 06), *Daily Dawn*. Retrieved 04.05.2014 from https://www.dawn.com/news/310440

The great digital divide (2009a, September 21), *Daily Dawn*. Retrieved 07.10.2011 from https://www.dawn.com/news/838989

The outreach of mobile banking (2009b, November 02), *Daily Dawn*. Retrieved 04.05.2014 from https://www.dawn.com/news/838971/the-outreach-of-mobile-banking

Classified data stolen from PM`s secretariat (2010a, June 09), *Daily Dawn*. Retrieved 04.05.2014 from https://www.dawn.com/news/857783/classified-data-stolen-from-pm-s-secretariat

Need for electoral reforms in Pakistan (2010b, August 11), *Daily Dawn*. Retrieved 04.05.2014 from https://www.dawn.com/news/813373

Demand to strengthen election commission (2010c, August 11), *Daily Dawn*. Retrieved 04.05.2014 from https://www.dawn.com/news/919073/demand-to-strengthen-election-commission

What are you doing on Facebook? (2011a, February 22), *Daily Dawn*. Retrieved 23.10.2012 from https://www.dawn.com/news/608010

Still far from the target (2011b, October 05), *Daily Dawn*. Retrieved 02.06.2012 from https://www.dawn.com/news/664142

E-sahulat — a nightmare for Nadra franchisees (2011c, August 05), *Daily Dawn*. Retrieved 04.05.2014 from https://www.dawn.com/news/649717

Militant dies of `heart attack` (2011d, April 23), *Daily Dawn*. Retrieved 02.11.2015 from https://www.dawn.com/news/623414/newspaper/newspaper/column

Nadra conducts its own probe into 'UK visa scam' (2012a, July 24), *Daily Dawn*. Retrieved 04.05.2014 from https://www.dawn.com/news/736956

Cartels in telecom sector (2012b, October 11), *Daily Dawn*. Retrieved 15.12.2012 from https://www.dawn.com/news/755950

Over 1.67m people used SMS service to verify voter list (2012c, March 02), *Daily Dawn*. Retrieved 04.05.2014 from https://www.dawn.com/news/699639

Long hours for little pay (2013a, June 02), *Daily Dawn*. Retrieved 04.05.2014 from https://www.dawn.com/news/1015552

Nadra chief briefs EU team on steps to ensure fair polls (2013b, April 24), *Daily Dawn*. Retrieved 05.05.2014 from https://www.dawn.com/news/793880

Accidental deaths: Nadra, State Life join hands to give insurance cover to people (2013c, January 24), *Daily Dawn*. Retrieved 04.05.2014 from https://www.dawn.com/news/781098

Online FIR registration procedure announced (2013d, July 06), *Daily Dawn*. Retrieved 04.05.2014 from https://www.dawn.com/news/1023147

Power shortfall exceeds 6,000MW (2014a, April 29), *Daily Dawn*. Retrieved 04.05.2015 from https://www.dawn.com/news/1102945

Branchless banking: Branching out (2014b, June 18), *Daily Dawn*. Retrieved 23.06.2014 from https://www.dawn.com/news/1113550

Telecom operators celebrate one million 3G, 4G users (2014c, November 05), 05.05.2015 from https://www.dawn.com/news/1142409

Easypaisa launches funds transfer facility to and from banks (2014d, June 13), *Daily Dawn*. Retrieved 04.05.2014 from https://www.dawn.com/news/1112521

Analysis: Banking of the future (2015a, May 18), *Daily Dawn*. Retrieved 02.06.2015 from https://www.dawn.com/news/1182626

Offline transactions for online business (2015b, February 22), *Daily Dawn*. Retrieved 03.05.2015 from https://www.dawn.com/news/1164656/offline-transactions-for-online-business

Strategic media cell (2015c, November 02), *Daily Dawn*. Retrieved 15.11.2015 from https://epaper.dawn.com/DetailImage.php?StoryImage=02_11_2015_005_005

Bid to blow up mobile phone tower foiled (2015d, October 31), *Daily Dawn*. Retrieved 02.11.2015 from https://www.dawn.com/news/1216554

Terrorists to 'lose big weapon' as Pakistan tightens mobile phone control (2015e, February 26), *Daily Dawn*. Retrieved 02.06.2015 from https://www.dawn.com/news/1166052

[Faded stands the redness of rose; raped lies its fragnance!] گل بوئے اڑا اڑا گل رنگِ لٹی لٹی (2010, March 18), *Daily Nawaiwaqt*. Retrieved 04.05.2014 from https://www.nawaiwaqt.com.pk/18-Mar-2010/106629

The importance of birth registration (2014, October 21), *Daily Times*. Retrieved 04.05.2015 from https://dailytimes.com.pk/102786/the-importance-of-birth-registration/

Nadra to set up more centres in Hyderabad (2003, December 20), *Daily Dawn*. Retrieved 03.05.2014 from https://www.dawn.com/news/130311

Certification authority formed (2004, September 24), *Daily Dawn*. Retrieved 04.05.2014 from https://www.dawn.com/news/371664

The futility of Indo-Pak cyber wars (2011, July 28), 04.05.2014 from https://www.dawn.com/news/647571

Citizens use SMS to verify their votes (2012, February 29), *Daily Dawn*. Retrieved 04.05.2014 from https://www.dawn.com/news/699224

NADRA launches Revolutionized Pension Disbursement Project (2013, April 13), *Dunya News*. Retrieved 04.05.2014 from https://dunyanews.tv/en/Pakistan/168968-NADRA-launches-Revolutionized-Pension-Disbursement

Haider, H. (2014, March 10). Branchless Banking Section: Banking Review 2013. *Business Recorder*. Retrieved 04.05.2014 from https://www.brecorder.com/pdf/banking-review-2013.pdf

Kugelman, M. (2011, July 11). Pakistan's demographic dilemma. *Foreign Policy*. Retrieved 30.06.2013 from https://foreignpolicy.com/2011/07/11/pakistans-demographic-dilemma/

From an idea to reality: Pakistan's smart card (2012, November 11), *The News*. Retrieved 25.01.2013 from **Error! Hyperlink reference not valid.**

Get your expired IDs renewed, NADRA asks citizens (2011, June 15), *Pakistan Today*. Retrieved 14.08.2013 from https://www.pakistantoday.com.pk/2011/06/15/get-your-expired-ids-renewed-nadra-asks-citizens/

NADRA information security system fully secured (2012a, September 10), *Pakistan Today*. Retrieved 04.05.2014 from https://www.pakistantoday.com.pk/2012/09/10/nadra-information-security-system-fully-secured/

Out-of-order ATMs creating problems for customers (2012b, April 6), *Pakistan Today*. Retrieved 04.05.2014 from https://www.pakistantoday.com.pk/2012/04/06/out-of-order-atms-creating-problems-for-customers/

Mobile market not saturated: Mobilink chief (2009, 16 November), *Business Recorder*. Retrieved 05.10.2011 from http://fp.brecorder.com/2009/11/20091116987256/

Universal Service Fund: A success story (2012a, May 18), *Business Recorder*. Retrieved 02.02.2013 from https://fp.brecorder.com/2012/05/201205181190823/

ECP directs NADRA to update FERs every month (2012b, September 04), *Business Recorder*. Retrieved 04.05.2014 from https://fp.brecorder.com/2012/09/201209041233785/

NADRA issues 200,000 SNIC (2012c, December 17), *Business Recorder*. Retrieved 04.05.2014 from https://www.brecorder.com/2012/12/17/96294/

Schumpeter (2013, August 17). Marketing a missed call. Retrieved from https://www.economist.com/schumpeter/2013/08/17/marketing-a-missed-call

E-Resistance Blooms in Pakistan (2007, November 13), *Spiegel Online*. Retrieved 02.02.2009 from https://www.spiegel.de/international/business/cyber-demonstrations-e-resistance-blooms-in-pakistan-a-517023.html

Corruption: Nadra staff caught making fake IDs (2010, December 10), *The Express Tribune*. Retrieved 04.05.2014 from https://tribune.com.pk/story/88181/corruption-nadra-staff-caught-making-fake-ids/

As mobile banking grows, Zong, Askari Bank join the race (2012a, November 20), *The Express Tribune*. Retrieved 04.05.2014 from https://tribune.com.pk/story/468247/as-mobile-banking-grows-zong-askari-bank-join-the-race/?amp=1

Voter verification: ECP launches SMS service (2012b, February 29), *The Express Tribune*. Retrieved 04.05.2014 from https://tribune.com.pk/story/343403/voter-verification-ecp-launches-sms-service/

Voter list keeps 7m CNIC holders out (2012c, September 04), *The Express Tribune*. Retrieved 04.05.2014 from https://tribune.com.pk/story/431030/voter-list-keeps-7m-cnic-holders-out/

ECP 'SMS voters' verification service may break World Record. (2012d, March 25), *The Express Tribune*. Retrieved https://tribune.com.pk/story/350452/ecp-sms-voters-verification-service-may-break-world-record/

Poll preparations: ECP sets up display centres for electoral rolls (2012e, February 20), *The Express Tribune*. Retrieved 04.05.2014 from https://tribune.com.pk/story/338951/poll-preparations-ecp-sets-up-display-centres-for-electoral-rolls/

Preparing transparent and error-free electoral rolls (2012f, October 19), *The Express Tribune*. Retrieved 04.05.2014 from https://tribune.com.pk/story/453924/preparing-transparent-and-error-free-electoral-rolls/

5-year roundup: By-polls scorecard puts PML-N ahead of PPP (2013a, March 22), *The Express Tribune*. Retrieved 04.05.2014 from https://tribune.com.pk/story/524655/5-year-roundup-by-polls-scorecard-puts-pml-n-ahead-of-ppp/

End of an era: Pindi's Imperial Market is going cellular (2013b, July 21), *The Express Tribune*. Retrieved 04.05.2014 from https://tribune.com.pk/story/580119/end-of-an-era-pindis-imperial-market-is-going-cellular/

Computerising land records: Contract awarded in violation of rules (2013c, October 06), *The Express Tribune*. Retrieved 04.05.2014 from https://tribune.com.pk/story/614128/computerising-land-records-contract-awarded-in-violation-of-rules/

Election Day: Expired CNIC holders would be allowed to vote (2013d, May 08), *The Express Tribune*. Retrieved 03.05.2014 from https://tribune.com.pk/story/546150/election-day-expired-cnic-holders-would-be-allowed-to-vote/

QMobile: Conquering the Pakistani market, one phone at a time (2013e, April 7), *The Express Tribune*. Retrieved 04.05.2014 from https://tribune.com.pk/story/532133/qmobile-conquering-the-pakistani-market-one-phone-at-a-time/

Pakistan's next media revolution (2013f, February 14), *The Express Tribune*. Retrieved 04.05.2014 from https://tribune.com.pk/story/506846/pakistans-next-media-revolution/

Branchless banking – an effort to control cash, document economy (2013g, April 02), *The Express Tribune*. Retrieved 04.05.2014 from https://tribune.com.pk/story/529689/branchless-banking-an-effort-to-control-cash-document-economy/

30m internet users in Pakistan, half on mobile: Report (2013h, June 24), *The Express Tribune*. Retrieved 05.07.2013 from https://tribune.com.pk/story/567649/30m-internet-users-in-pakistan-half-on-mobile-report/

When land grabbers rule (2014a, January 28), *The Express Tribune*. Retrieved 04.05.2014 from https://tribune.com.pk/story/664487/when-land-grabbers-rule/

IT embezzlement: 13 E-services projects referred to FIA (2014b, March 21), *The Express Tribune*. Retrieved 04.05.2014 from https://tribune.com.pk/story/685362/it-embezzlement-13-e-services-projects-referred-to-fia/

Operational problems: Affiliate dept draining IT ministry, says Anusha (2014c, January 22), *The Express Tribune*. Retrieved 04.05.2014 from https://tribune.com.pk/story/661785/operational-problems-affiliate-dept-draining-it-ministry-says-anusha/

Technology merger: Govt merges bloated tech departments to form National IT Board (2014d, July 18), *The Express Tribune*. Retrieved 04.05.2015 from https://tribune.com.pk/story/737597/technology-merger-govt-merges-bloated-tech-departments-to-form-national-it-board/

Governance: E-office implementation reviewed (2014e, November 08), *The Express Tribune*. Retrieved 05.12.2014 from https://tribune.com.pk/story/787411/governance-e-office-implementation-reviewed/

Pakistan to have 100 million bank accounts by 2025 (2015, November 27), *The Express Tribune*. Retrieved 03.12.2015 from https://tribune.com.pk/story/999299/one-minute-bank-account-pakistan-to-have-100-million-bank-accounts-by-2025/

Telenor buys Tameer Bank (2016, March 17), *The Express Tribune*. Retrieved https://tribune.com.pk/story/1067643/telenor-buys-tameer-bank

In just 25 years, the mobile phone has transformed the way we communicate (2010, January 01), *The Guardian*. Retrieved 04.05.2014 from https://www.theguardian.com/business/2010/jan/01/25-years-phones-transform-communication

Chairman Nadra resigns (2008a, August 08), *The Nation*. Retrieved 04.05.2014 from https://nation.com.pk/08-Aug-2008/chairman-nadra-resigns

NADRA ready to launch e-Sahulat (2008b, May 22), *The Nation*. Retrieved 04.05.2014 from https://nation.com.pk/22-May-2008/nadra-ready-to-launch-esahulat

Ufone launches UPayments (2010, December 03), *The Nation*. Retrieved 03.04.2013 from https://nation.com.pk/03-Dec-2010/ufone-launches-upayments

Harnessing technology for electoral transparency (2011, October 17), *The Nation*. Retrieved 04.05.2014 from https://nation.com.pk/17-Oct-2011/harnessing-technology-for-electoral-transparency

Indian hackers hack http://www.multan.gov.pk/ (2012, May 22), *The Nation*. Retrieved 04.05.2014 from https://nation.com.pk/22-May-2012/indian-hackers-hack-www-multan-gov-pk

Nadra shelves project despite Malik's opposition (2013a, February 03), *The Nation*. Retrieved 04.05.2014 from https://nation.com.pk/03-Feb-2013/nadra-shelves-project-despite-maliks-opposition

HBL partners with Nadra for branchless banking (2013b, November 01), *The Nation*. Retrieved 04.05.2014 from https://nation.com.pk/01-Nov-2013/hbl-partners-with-nadra-for-branchless-banking

Govt steps up efforts to introduce e-govt (2013c, September 02), *The Nation*. Retrieved 23.09.2013 from https://nation.com.pk/02-Sep-2013/govt-steps-up-efforts-to-introduce-e-govt

CCoR approves IT entities merger (2014, July 19), *The Nation*. Retrieved 04.05.2015 from https://nation.com.pk/19-Jul-2014/ccor-approves-it-entities-merger

Computerization and the 'patwari culture' (2015, January 28), *The Nation*. Retrieved 04.05.2015 from https://nation.com.pk/28-Jan-2015/computerization-and-the-patwari-culture

Coup in Pakistan: The Overview; Pakistan Army seizes power hours after Prime Minister Dismisses his Military Chief (1999, October 13), *The New York Times*. Retrieved 09.10.2010 from https://www.nytimes.com/1999/10/13/world/coup-pakistan-overview-pakistan-army-seizes-power-hours-after-prime-minister.html

Nadra plans to franchise kiosks (2007, March 10), *The News*. Retrieved 04.05.2014 from https://www.thenews.com.pk/archive/print/46012-nadra-plans-to-franchise-kiosks

Electoral rolls are full of errors (2009, August 03), *The News*. Retrieved 04.05.2014 from https://www.thenews.com.pk/archive/print/665969-electoral-rolls-are-full-of-errors

NADRA, Mobilink join hands for utility bills payment (2011a, January 13), *The News*. Retrieved 12.11.2013 from https://www.thenews.com.pk/archive/print/279693

Nadra completes 80m draft electoral rolls (2011b, August 19), *The News*. Retrieved 04.05.2014 from https://www.thenews.com.pk/archive/print/616008-nadra-completes-80m-draft-electoral-rolls

ECP one step closer to 'one CNIC, one vote' goal (2011c, August 19), *The News*. Retrieved 04.05.2014 from https://tribune.com.pk/story/234741/ecp-one-step-closer-to-one-cnic-one-vote-goal/

Nadra, ECP working on electoral reforms (2013a, November 25), *The News*. Retrieved 04.05.2014 from https://www.thenews.com.pk/archive/amp/468792

55m text messages sent to ECP, companies make billions (2013b, May 15), *The News*. Retrieved 04.05.2014 from https://www.thenews.com.pk/archive/print/630483-55m-text-messages-sent-to-ecp,-companies-make-billions

Nadra ready to verify thumb impressions (2013c, May 16), *The News*. Retrieved 04.05.2014 from https://www.thenews.com.pk/archive/print/630507-nadra-ready-to-verify-thumb-impressions

Islamabad Marriott hotel bomb killed 52, says Pakistan (2008, September 21), *The Telegraph*. Retrieved 04.05.2015 from https://www.telegraph.co.uk/news/worldnews/asia/pakistan/3041148/Islamabad-Marriott-hotel-bomb-killed-52-says-Pakistan.html

Pakistan's power crisis may eclipse terrorist threat (2012, May 27), *The Washington Post*. Retrieved 04.05.2015 from https://www.washingtonpost.com/world/asia_pacific/pakistans-power-crisis-may-eclipse-terrorist-threat/2012/05/27/gJQAPhOSuU_story.html

25 fastest growing companies announced (2011a, January 20), *The Express Tribune*. Retrieved 26.02.2012 from https://tribune.com.pk/story/106200/allworld-network-announces-pakistan-fast-growth-25-winners/

Pakistan to replace 'insecure' US border watch software (2011b, June 08), *The Express Tribune*. Retrieved 04.05.2014 from https://tribune.com.pk/story/184568/pakistan-to-replace-insecure-us-border-watch-software/

Warraich, H. (2011, March 18). Pakistan's social media landscape. *Foreign Policy*. Retrieved 04.05.2014 from https://foreignpolicy.com/2011/03/18/pakistans-social-media-landscape/

Press Releases

Telenor Easypaisa. (2009a, October 15). *Telenor Pakistan & Tameer Bank launch 'easypaisa'.* [Press release]. Retrieved from https://www.easypaisa.com.pk/pressview/ep-launch.html

Pakistan Telecommunication Authority (PTA). (2009b, November 19). *PTA and SBP to formulate third party mobile banking regulations.* [Press release]. Retrieved 28.08.2012 from https://www.pta.gov.pk/index.php/en/media-center/single-media/pta-and-sbp-to-formulate-third-party-mobile-banking-regulations

Election Commission of Pakistan (ECP). (2012, March 25). *ECP inaugurates SMS facility tomorrow.* [Press release]. Retrieved from http://www.ecp.gov.pk/ViewPressReleaseNotific.aspx?ID=1503

Election Commission of Pakistan. (2012, February 28). *ECP inaugurates SMS facility tomorrow* [Press release]. Retrieved 04.05.2014 from https://www.ecp.gov.pk/PrintDocument.aspx?PressId=11053&type=Text

Mobilink. (2008a, July 08). *Mobilink Genie pioneers Mobile Commerce in Pakistan.* [Press release]. Retrieved from https://www.jazz.com.pk/media-center/press-releases/mobilink-genie-pioneers-mobile-commerce-in-pakistan/

Mobilink. (2008b, November 28). *Mobilink and Pakistan Post Office to launch Mobile Money Order Service.* [Press release]. Retrieved 28.10.2011 from https://www.jazz.com.pk/media-center/press-releases/mobilink-and-pakistan-post-officeto-launch-mobile-money-order-service/

Mobilink. (2011a, February 22). *Mobilink Launches Utility Bill Payment Solution.* [Press release]. Retrieved from https://www.jazz.com.pk/media-center/press-releases/mobilink-launches-utility-bill-payment-solution/

Mobilink. (2011b, April 13). *Mobilink & Ufone Pen Agreement for Tower Sharing.* [Press release]. Retrieved 03.05.2014 from https://www.jazz.com.pk/media-center/press-releases/mobilink-ufone-pen-agreement-for-tower-sharing/

Pakistan telecommunication Authority (PTA). (2007a, June 14). *Pakistan Ahead of Regional Countries in Telecom Regulartory Environment.* [Press release]. Retrieved 09.09.2009 from https://www.pta.gov.pk/en/media-center/single-media/pakistan-ahead-of-regional-countries-in-telecom-regulatory-environment

Pakistan Telecommunication Authority. (2009b, December 16). *All Banks to be Connected with Mobile Operators for Mobile Banking Soon.* [Press release]. Retrieved 04.05.2014 from https://www.pta.gov.pk/en/media-center/single-media/all-banks-to-be-connected-with-mobile-operators-for-mobile-banking-soon

Pakistan Telecommunication Authority. (2014b, November 14). *New sims are being issued through biometric verification system.* [Press release]. Retrieved 03.05.2015 from https://www.pta.gov.pk/en/media-center/single-media/new-sims-are-being-issued-through-biometric-verification-system-

Telenor Group. (2008, July 03). *Telenor Pakistan connects the unconnected.* [Press release]. Retrieved 08.09.2010 from https://www.telenor.com/media/press-release/telenor-pakistan-connects-the-unconnected

Virtual University of Pakistan. (2015, August 05). *Virtual University Signed MoU With NADRA Technologies.* [Press release]. Retrieved 30.08.2015 from http://vu.edu.pk/NewsDetails.aspx?type=&NewsId=2383

Reports, Online Databanks

Ahmed, N. (2006). An Overview of e-Participation Models. A publication of the UN Department of Economic and Social Affairs (UNDESA). Retrieved 04.05.2014 from http://unpan1.un.org/intradoc/groups/public/documents/UN/UNPAN023622.pdf

Asian Development Bank (ADB). (2014). Regional: Financial Sector Development in Central and West Asia: Making Mobile Financial Services Work for Central and West Asian Countries. Retrieved 04.06.2015 from https://www.adb.org/sites/default/files/project-document/173360/43359-012-tacr-07.pdf

Chawla, R. and Bhatnagar, S. (2004). Online Delivery of Land Titles to Rural Farmers in Karnataka, India. Retrieved 04.05.2014 from http://unpan1.un.org/intradoc/groups/public/documents/un-dpadm/unpan042925.pdf

Commonwealth Secretariat. (2013). Pakistan General Elections: Report of the Commonwealth Observer Mission. Retrieved 04.05.2014 from https://thecommonwealth.org/sites/default/files/project/documents/Pakistan General Elections 2013 Commonwealth Observer Mission Report.pdf

Cullen, R. and Hernon, P. (2004). Wired for well-being: Citizens' response to e-government. Retrieved 28.10.2010 from http://unpan1.un.org/intradoc/groups/public/documents/APCITY/UNPAN0 17975.pdf

Election Commission of Pakistan. (2013a). Report on the General Elections - 2013. Retrieved 03.05.2014 from https://www.ecp.gov.pk/Documents/General Elections 2013 report/Election Report 2013 Volume-I.pdf

Election Observation and Democracy Support. (2008). Islamic Republic of Pakistan: Final Report - National and Provincial Assembly Elections. Retrieved 04.05.2014 from http://www.eods.eu/library/FR PAKISTAN 16.04.2008_en.pdf

Free and Fair Elections Network. (2010). National Assembly Elections in Pakistan 1970-2008: A compendium of elections related facts and statistics. Retrieved 04.05.2014 from http://fafen.org/wp-content/uploads/2010/08/Compendium-National-Assembly-Elections-1970-2008-Pakistan.pdf

German Development Agency. (2011). The transformative role of Mobile Financial Services and the role of German Development Cooperation. Retrieved 09.12.2011 from http://www2.gtz.de/dokumente/bib-2011/giz2011-0068en-mobile-financial-services.pdf

GIZ. (2008). eGovernment for development: The Promise and the Practice (A Conference Documentation). Retrieved 04.10.2010 from http://www2.gtz.de/dokumente/bib-2009/gtz2009-0504en-e-government-conference.pdf

Hakeem, A. A. (2009). Smart National Identity Card in Pakistan. Presentation Slides. Retrieved 04.05.2014 from http://siteresources.worldbank.org/EXTSAFETYNETSANDTRANSFERS/Resources/Smart-National-ID-cards-in-Pakistan.pdf

infoDev and Center for Democracy & Technology. (2002). The E-Government Handbook For Developing Countries: A Project of InfoDev and The Center for Democracy & Technology. Retrieved 04.05.2014 from http://unpan1.un.org/intradoc/groups/public/documents/apcity/unpan00746 2.pdf

International Finance Corporation. (2014). Alternative Delivery Channels and Technology. Retrieved 04.05.2015 from https://openknowledge.worldbank.org/bitstream/handle/10986/25980/11134 4-WP-ADC-Handbook-2014-PUBLIC.pdf?sequence=1&isAllowed=y

International Foundation for Electoral Systems. (2013b). Electoral Rolls: Pakistan Factsheet. Retrieved 04.05.2014 from https://www.ifes.org/sites/default/files/the_electoral_rolls.pdf

International Foundation for Electoral Systems. (2013c). IFES Pakistan Fact Sheet: IFES in Pakistan. Retrieved 05.05.2014 from https://www.ifes.org/sites/default/files/ifes_in_pakistan.pdf

International Institute for Vital Registration and Statistics. (1995). Organization of National Civil Registration and Vital Statistics Systems: An Update. Retrieved 04.05.2014 from https://www.cdc.gov/nchs/data/isp/063_organization_of_the_national_civil _registration_and_vital_stat_systm_an_update.pdf

International Monetary Fund. (2004). Pakistan: Poverty Reduction Strategy Paper - IMF Country Report No. 04/24. Retrieved 02.02.2009 from https://www.imf.org/external/pubs/ft/scr/2004/cr0424.pdf

International Telecommunication Union. (2007). Foreign Direct Investment in Pakistan Telecommunication Sector. Paper presented by PTA staff member Mr. Asif Inam at Regional Seminar on Costs and Tariffs for Member Countries of the Tariff Group for Asia and Oceania (TAS) in Seol, South Korea. Retrieved 03.08.2011 from https://www.itu.int/ITU-D/finance/work-cost-tariffs/events/tariff-seminars/Korea-07/presentations/FDI_Aasif_Inam.pdf

International Telecommunication Union. (2010). ICT Statistics. Retrieved 12.09.2011 from https://www.itu.int/en/ITU-D/Statistics/Pages/stat/default.aspx

International Telecommunication Union. (2011b). ICT success stories: Innovation and sustainability. Retrieved 04.05.2014 from https://www.itu.int/net/itunews/issues/2011/04/pdf/201104_32.pdf

International Telecommunication Union. (2013a). ITU ICT Statistics. Retrieved 04.05.2014 from http://www.itu.int/en/ITU-D/Statistics/Pages/stat/default.aspx

International Telecommunication Union. (2013b). ICT Statistics. Retrieved 20.05.2013 from http://www.itu.int/en/ITU-D/Statistics/Pages/stat/default.aspx

International Telecommunication Union. (2013c). Contribution by Ethiopia: E-Government Strategy. Paper Presented at the Ministerial Roundtable at the WSIS Forum. Retrieved 04.05.2014 from https://unctad.org/meetings/en/Presentation/CSTD_2013_WSIS_Ethiopia_E-Gov_Strategy.pdf

International telecommunication Union. (2015). m-Powering Development Initiative: A report by the m-Powering Development Initiative Advisory Board. Retrieved 04.05.2015 from https://www.itu.int/en/ITU-D/Initiatives/m-Powering/Documents/m-PoweringDevelopmentInitiative_Report2015.pdf

International Telecomunication Union. (2013). Universal Service Fund and Digital Inclusion for All Retrieved 15.12.2013 from https://www.itu.int/en/ITU-D/Digital-Inclusion/Documents/USF_final-en.pdf

Jafri, S. K. (2012). 3G Mobile Spectrum – Issues & Prospects. In *SBP (State Bank of Pakistan) Research Bulletin Volume 8, Number 1*.Retrieved 13.01.2013 from http://www.sbp.org.pk/research/bulletin/2012/Vol-8-1/opinion2sabina.pdf

Jalali, A. (2006). Socio Economic Impacts of Rural Telecenters in Iran. In *World Bank Seminar on Women's Economic Empowerment and the Role of ICT [presentation slides]*.Retrieved 24.10.2012 from http://siteresources.worldbank.org/INTGENDER/Resources/AliJalali.pdf

Malik, M. T. (2010). National Identity Management in Pakistan. Retrieved 04.05.2014 from http://siteresources.worldbank.org/INTEDEVELOPMENT/Resources/Malik-1.pdf

Mefalopulos, P. (2008). Development communication sourcebook broadening the boundaries of communication. Retrieved 28.09.2010 from http://siteresources.worldbank.org/EXTDEVCOMMENG/Resources/DevelopmentCommSourcebook.pdf

Ministry of Information Technology (Government of Pakistan). (2014). Pakistan IT Policy and Action Plan 2000. Retrieved 04.05.2014 from http://www.moit.gov.pk/gop/index.php?q=aHR0cDovLzE5Mi4xNjguNzAuMTM2L21vaXQvZnJtRGV0YWlscy5hc3B4P2lkPTEwJmFtFtcDtvcHQ9cG9saWNpZXM%3D

National Assembly of Pakistan. (2009). Questions for Oral Answers and their Replies: National Assembly Secretariat (11th Session). Retrieved 04.05.2014 from http://www.na.gov.pk/uploads/documents/questions/1303078749_855.pdf

National Institute of Population Studies. (2013). Pakistan Demographic and Health Survey. Retrieved 30.01.2014 from https://www.nips.org.pk/abstract_files/PDHS Final Report as of Jan 22-2014.pdf

NCS. (2009). e-Government Consulting – Shaping your e-Government Blueprint: Mapping the Future. Retrieved 09.07.2011 from https://www.ncs.com.sg/documents/20184/73726/eGov+Consulting+%28062009%29.pdf/941bf5b9-c760-4054-a6f2-995aae69ae0a

Organisation for Economic Co-operation and Development. (2001). Understanding the Digital Divide. Retrieved 04.05.2014 from http://dx.doi.org/10.1787/236405667766

Organisation for Economic Co-operation and Development. (2003). OECD E-Government Flagship Report "The E-Government Imperative". Retrieved 04.05.2014 from http://www.oecd.org/officialdocuments/publicdisplaydocumentpdf/?cote=GOV/PUMA(2003)6/ANN&docLanguage=En

Pacific Council On International Policy. (2002). Roadmap for E-government in the Developing World: 10 Questions E-Government Leaders Should Ask Themselves (The Working Group on E-Government in the Developing

World). Retrieved 04.05.2014 from http://unpan1.un.org/intradoc/groups/public/documents/caricad/unpan008526.pdf

Pakistan Poverty Alleviation Fund. (2012). Branchless Banking and Savings. Retrieved 04.05.2014 from http://www.ppaf.org.pk/doc/microfinance/27-Session on Branchless Banking and Savings.pdf

Pakistan Telecommunication Authority. (2003). Annual Report 2002-03. Retrieved 03.10.2008 from https://www.pta.gov.pk/media/annual_report.pdf

Pakistan Telecommunication Authority. (2005). Annual Report (2004-05). Retrieved 17.12.2010 from https://www.pta.gov.pk/annual-reports/ann-rep-05.pdf

Pakistan Telecommunication Authority. (2007b). Industry Analysis Report. Retrieved 17.12.2010 from https://www.pta.gov.pk/media/industry_report_2007_1.pdf

Pakistan Telecommunication Authority. (2008). Annual Report. Retrieved 17.12.2010 from https://www.pta.gov.pk/annual-reports/annrep0708/ch_03.pdf

Pakistan Telecommunication Authority. (2009a). Telecom Quarterly Review. Retrieved 19.10.2010 from https://www.pta.gov.pk/media/tqr_dec_09_1.pdf

Pakistan Telecommunication Authority. (2010b). Paradigm Technologies - Broadband Subscribers Survey: Estimating Broadband End-Users and their Experience of Service and its Performance. Retrieved 17.10.2011 from https://www.pta.gov.pk/media/bb_sub_sur_report_10.pdf

Pakistan Telecommunication Authority. (2010c). Annual Report - 2009-10. Retrieved 03.05.2011 from https://www.pta.gov.pk/annual-reports/pta_ann_rep_2010.pdf

Pakistan Telecommunication Authority. (2011). Annual Report 2010-11. Retrieved 03.07.2012 from https://www.pta.gov.pk/annual-reports/pta_ann_rep_11.pdf

Pakistan Telecommunication Authority (PTA). (1996). The Gazette of Pakistan. Act No. XVII OF 1996. Retrieved 08.03.2009 from https://www.pta.gov.pk/media/telecom_act_170510.pdf

Pakistan Telecommunication Authority (PTA). (2012). Proliferation of SMS & MMS in Pakistan with emphasis on Premium Rate SMS services. Retrieved 05.05.2014 from https://www.pta.gov.pk/media/sms_mms_report_130313_1.pdf

Pakistan Telecommunication Authority (PTA). (2014). Annual Report 2013-14. Retrieved 04.05.2015 from https://www.pta.gov.pk/annual-reports/ptaannrep2013-14.pdf

Pinon, R. and Haydon, J. (2010). English Language Quantitative Indicators: Cameroon, Nigeria, Rwanda, Bangladesh and Pakistan. In *A custom report compiled by Euromonitor International for the British Council.*Retrieved 04.05.2014 from http://teachingenglish.britishcouncil.org.cn/sites/teacheng/files/Euromonitor Report A4.pdf

Sardar, A. (2007). Information System Reforms for Improving Governance: Case Study NADRA. Presentation Slides. Retrieved 04.05.2014 from http://siteresources.worldbank.org/PSGLP/Resources/InformationSystemReformsforImprovingGovernance.pdf

Siegmann, K. A. (2010). The Gender Digital Divide in Rural Pakistan-To Measure and to Bridge it. Retrieved 17.08.2012 from https://repub.eur.nl/pub/22393/SiegmannGenderDigitalDivide.pdf

SMEDA. (2009). SME Baseline Survey. Retrieved 09.02.2011 from https://smeda.org/phocadownload/Publicatoins/SME Baseline Survey.pdf

State Bank of Pakistan. (2002a). An Ordinance: Electronic Transaction Ordinance (ETO). Retrieved 23.08.2009 from http://www.sbp.org.pk/about/act/ETC202.pdf

State Bank of Pakistan. (2003). Banking System Review. Retrieved 20.08.2009 from http://www.sbp.org.pk/publications/bsr/bkg_system_review(2003).pdf

State Bank of Pakistan. (2010). The Size of Informal Economy in Pakistan. In *SBP Working Paper Series. No. 33*.Retrieved 23.08.2011 from http://www.sbp.org.pk/repec/sbp/wpaper/wp33.pdf

State Bank of Pakistan. (2012a). Payment Systems Quarterly Reports (2012 - 2013). Retrieved 02.12.2012 from http://www.sbp.org.pk/psd/reports/2012/Third-Quarterly-Review-FY11-12.pdf

Telenor. (2013). Reach. Retrieved 04.05.2014 from https://www.telenor.com/wp-content/uploads/2013/02/TG_Mag_2013-02-20_72dpi_spreads.pdf

The Gazzette of Pakistan. (2013). Pakistan Telecommunication (Re-organization) Act. Retrieved 03.04.2013 from https://www.ecac.org.pk/assets/front/files/TelecomAct.pdf

UNICEF. (2010). Mobiles for Development. Retrieved 23.03.2013 from https://www.unicef.org/cbsc/files/Mobiles4DeReport.pdf

UNICEF. (2014). Annual Report 2013 - Pakistan. Retrieved 04.05.2015 from https://www.unicef.org/about/annualreport/files/Pakistan_COAR_2013.pdf

United Nation Development Programme. (2012). Mobile Technologies and Empowerment: Enhancing human development through participation and innovation. Retrieved 04.05.2014 from https://www.undp.org/content/dam/undp/library/Democratic Governance/Access to Information and E-governance/Mobile Technologies and Empowerment_EN.pdf

United Nations. (2001). Benchmarking E-government: A Global Perspective (Assessing the Progress of the UN Member States). Retrieved 04.05.2014 from https://publicadministration.un.org/egovkb/Portals/egovkb/Documents/un/English.pdf

United Nations. (2003). Global E-government Survey 2003. Retrieved 04.05.2014 from https://publicadministration.un.org/egovkb/portals/egovkb/Documents/un/2003-Survey/unpan016066.pdf

United Nations. (2005). Designing e-government for the poor. Retrieved 04.05.2014 from https://www.adb.org/sites/default/files/publication/159381/adbi-e-gov-poor.pdf

United Nations. (2008). Global E-Government Survey 2008: From E-Government to Connected Governance Retrieved 04.05.2014 from https://publicadministration.un.org/egovkb/en-us/Reports/UN-E-Government-Survey-2008

United Nations. (2012a). Preparatory Process of the United Nations E-Government Survey. Retrieved 04.05.2014 from https://publicadministration.un.org/egovkb/en-us/About/-eGov-Surveys

United Nations. (2012b). E-Government Survey 2012: E-Government for the People. Retrieved 12.12.2013 from https://publicadministration.un.org/egovkb/Portals/egovkb/Documents/un/2012-Survey/Complete-Survey.pdf

United Nations. (2014). UN E-Government Knowledgebase. Retrieved 04.05.2014 from https://publicadministration.un.org/egovkb/en-us/About

United Nations. (2015). Universal Service in Pakistan-The Broadband Perspective. Country Perspective presented by Mr. Mudassar Hussain (Ministry of Information Technology, Pakistan) at the Capacity building workshop on improving broadband connectivity in Republic of Korea (1-2 September 2015). Retrieved 02.12.2015 from https://www.unescap.org/sites/default/files/Item 7 Universal Service in Pakistan.pdf

United Nations Conference on Trade and Development. (2007). The Least Developed Countries Report. Retrieved 04.05.2014 from https://unctad.org/en/Docs/ldc2007_en.pdf

United Nations Development Programme. (2008). The Role of Information Communication Technologies (ICTs) in achieving the Millennium Development Goals. In Mozambique: National Human Development Report. Retrieved 04.05.2014 from http://hdr.undp.org/sites/default/files/mozambique_nhdr_2008_ict.pdf

United Nations Development Programme. (2009). Communication for Development: A glimpse at UNDP's practice. Retrieved 27.08.2011 from https://www.undp.org/content/dam/aplaws/publication/en/publications/democratic-governance/oslo-governance-center/civic-engagement/communication-for-development-a-glimpse-at-undps-practice-/Communication for Development A Glimpse at UNDP%C2%92s Practice (2009).pdf

Universal Service Fund. (2009). Universal Service Fund: Telecom for All. Annual Report 2008-09. Retrieved 25.05.2014 from https://usf.org.pk/assets/publication-pdf/annual-report2008-09.pdf

Universal Service Fund (USF). (2013). Universal Service Fund: Telecom for All. Annual Report 2012-13. Retrieved 03.04.2010 from https://usf.org.pk/assets/publication-pdf/annual-report.pdf

World Bank. (1995). Telecommunications Regulation and Privatization Support Project. Staff Appraisal Report (Nr. 13298-PAK). Retrieved 04.05.2014 from http://documents.worldbank.org/curated/en/879571468774973666/pdf/multi0page.pdf

World Bank. (2009). Bringing Finance to Pakistan's Poor: A Study on Access to Finance for the Underserved and Small Enterprises. Retrieved 12.03.2011 from http://documents.worldbank.org/curated/en/358531468057878762/pdf/486720WP0Bring10Box338916B01PUBLIC1.pdf

World Bank. (2011). The World Bank Annual Report 2011: Year in Review. Retrieved 12.11.2012 from http://siteresources.worldbank.org/EXTANNREP2011/Resources/8070616-1315496634380/WBAR11_YearInReview.pdf

World Bank. (2013a). DataBank: World Development Indicators. Retrieved 02.06.2013 from https://databank.worldbank.org/

World Bank. (2013b). Leveraging Mobile Phones for Innovative Governance Solutions in Pakistan. Retrieved 04.05.2014 from https://www.worldbank.org/en/news/feature/2013/12/11/leveraging-mobile-phones-for-innovative-governance-solutions-in-Pakistan

World Bank. (2015). Pakistan Development Update. Retrieved 05.06.2015 from http://documents.worldbank.org/curated/en/531691468098388002/pdf/957510WP00PUBL0pril20150edited0fina.pdf

World Economic Forum. (2001). The Global Information Technology Report (2001–2002): Readiness for the Networked World. Retrieved 04.05.2014 from http://unpan1.un.org/intradoc/groups/public/documents/un/report.pdf

Theses

Chini, I. (2012). *Governmentality and the information society: ICT policy practices in Greece under the influence of the European Union* (Doctoral Dissertation, London School of Economics and Political Science, UK). Retrieved 04.05.2014 from http://etheses.lse.ac.uk/847/1/Chini_Governmentality_and_the_information_society.pdf

Hussain, E. (2010). *Military agency, politics and the state the case of Pakistan* (Doctoral Dissertation. Faculty of Economics and Social Sciences, University of Heidelberg, Germany). Retrieved 23.08.2012 from http://archiv.ub.uni-heidelberg.de/volltextserver/10947/1/Thesis.Online_Pub.5.8.10.pdf

Web Pages

Atta-ur-Rahman (2002). "Science and Technology in Pakistan: The Way Forward". Retrieved 02.12.2008 from https://www.sciencemag.org/careers/2002/09/science-and-technology-pakistan-way-forward.

Barany, Z. (2009). "Authoritarianism in Pakistan". In Policy Review. Retrieved 02.06.2014 from https://www.hoover.org/research/authoritarianism-pakistan.

Batchgeo (2013). "About batchgeo". Retrieved 29.11.2013 from https://batchgeo.com/about/.

BBC News (2012). "Pakistan's experience with identity management". Retrieved 04.05.2014 from https://www.bbc.com/news/world-asia-18101385.

Capital Development Authority (2008). "Complaint & Enquiry Management System". Retrieved 28.08.2008 from http://www.cda.gov.pk/cda-latest/files/file.asp?var=cems.

Daily Dawn (2012d). "ECP starts Rolls display, SMS service for voters verification. ". Retrieved 25.12.2012 from http://www.dawn.com/2012/02/29/ecp-starts-rolls-display-sms-service-for-voters-verification.html.

Election Commission of Pakistan (2012a). "Progress Report on Implementation of the Five Year Strategic Plan". Retrieved 24.03.2012 from http://www.ecp.gov.pk/Reports/SPStrategicPlanProgress.pdf.

Electronic Government Directorate (2009b). "Web Standards and Guidelines". Retrieved 27.03.2009 from http://www.pakistan.gov.pk/e-government-directorate/standards/postimplement.jsp.

Ghosh, P. (2013). "Talk Dirty? Pakistan To Ban Youths From Obtaining Mobile Phone Package Deals To Prevent 'Immoral Behavior'". In International Business Times. Retrieved 14.12.2013 from https://www.ibtimes.com/talk-dirty-pakistan-ban-youths-obtaining-mobile-phone-package-deals-prevent-immoral-behavior-1401933.

Google Maps (2015). "NADRA's location in Islamabad's Red Zone district". Retrieved 04.05.2015 from https://goo.gl/maps/r85vySnqf5QWtgBn8.

Government of Pakistan (2006a). "Vision and Objectives: About Pakistan.Gov". Retrieved 01.11.2007 from http://www.pakistan.gov.pk/AboutPakGov.jsp - vis.

Government of Pakistan (2006b). "Government of Pakistan's official webportal ". Retrieved November 01 2007, from http://www.pakistan.gov.pk/.

Hasnain, Z. (2008). "Devolution, Accountability, and Service Delivery: Some Insights from Pakistan". World Bank Policy Research Working Paper (4610). Retrieved 02.04.2011 from

https://openknowledge.worldbank.org/bitstream/handle/10986/6713/wps46
10.pdf;sequence=1.

India Filings (2015). "Government Initiatives: eSeva". Retrieved 04.05.2015 from https://www.indiafilings.com/learn/eseva/.

International Telecommunication Union (2003). "Building the Information Society: a global challenge in the new Millennium". Retrieved 03.02.2008 from https://www.itu.int/net/wsis/docs/geneva/official/dop.html.

International Telecommunication Union (2005). "Tunis Agenda for the Information Society". Retrieved 03.03.2008 from https://www.itu.int/net/wsis/documents/doc_multi.asp?lang=en&id=2266|2 267.

Internet Governance Forum (2006). "About the IGF". Retrieved 03.03.2008 from https://www.intgovforum.org/multilingual/tags/about.

Internet World Stats (2012). "Internet World Users by Language". Retrieved 04.05.2014 from https://www.internetworldstats.com/stats7.htm.

Jamil, F. (2013). "Pressured by lawmakers, Pakistan bans SMS and chat packages". In Aaj News. Retrieved 13.12.2013 from https://www.aaj.tv/english/technology/pressured-by-lawmakers-pakistan-bans-sms-and-chat-packages/.

Microsave Consulting (2015). "Over the Counter (OTC) in Pakistan: Why It Works". Retrieved 30.06.2015 from https://www.microsave.net/2015/06/10/over-the-counter-otc-in-pakistan-why-it-works/.

Ministry of Interior (Government of Pakistan) (2014). "Rules of Business". Retrieved 04.05.2015 from https://www.interior.gov.pk/index.php/business-rules.

Ministry of Justice (Republic of Turkey) (2008). "UYAP SMS Information System". Retrieved 10.12.2012 from http://sms.uyap.gov.tr/smseng/engnedir.html.

Mir, A. (2010). "Local Web Content: Mobile Operators Sign MOUs, Telenor has an Urdu Website". Retrieved 04.05.2014 from http://telecompk.net/2010/12/29/local-web-content-mobile-operators-sign-mous-telenor-has-an-urdu-website/.

Mobilink (2013). "Money Transfer". Retrieved 04.05.2014 from http://mobicash.com.pk/money-transfer.

National Database Registration Authority (2005). "President of Pakistan General Pervez Musharraf inaugurated the "NADRA Kiosk"". Retrieved 28.08.2008 from http://www.nadra.gov.pk/DesktopModules/top/topmore.aspx?tabID=0&Ite mID=39&bID=0&Mid=2925.

National Database Registration Authority (2006). "Objectives Achieved by NADRA". Retrieved 09.12.2006 from http://www.nadra.gov.pk/objectives.html.

National Database Registration Authority (2007). "History of NADRA". Retrieved 17.08.2008 from http://nadra.gov.pk/site/351/default.aspx.

National Database Registration Authority (2008b). "Kiosk: Franchise Model (Terms and Conditions)". Retrieved 09.03.2009 from http://www.nadra.gov.pk/Franchise_Model_terms_and_conditions.html.

National Database Registration Authority (2008c). "NADRA has franchised the kiosk machines". Retrieved 09.03.2009 from http://www.nadra.gov.pk/DesktopModules/top/topmore.aspx?tabID=0&ItemID=46&bID=0&Mid=3026.

National Database Registration Authority (2008d). "History of NADRA". Retrieved 18.08.2008 from http://www.nadra.gov.pk/site/351/default.aspx.

National Database Registration Authority (2008e). "Goals and Achievements ". Retrieved 10.09.2009 from http://www.nadra.gov.pk/site/353/default.aspx.

National Database Registration Authority (2009a). "About e-Sahulat". Retrieved 14.09.2009 from http://www.esahulat.com.pk/subpages/about_esahulat.php.

National database Registration Authority (2009c). "Introduction to Kiosk: Frequently Asked Questions ". Retrieved 31.08.2009 from http://www.nadra.gov.pk/site/396/default.aspx - Frequently%20asked%20questions.

National Database Registration Authority (2012a). "NADRA Kiosk Brochure". Retrieved 06.09.2012 from http://www.nadra.gov.pk/downloads/solutions/other-kiosk.pdf.

National Database Registration Authority (2012b). "e-Sahulat: Individual Franchise Scheme". Retrieved 27.08.2012 from http://www.esahulat.com.pk/subpages/pop_up_individual_franchisee_scheme.php.

National Database Registration Authority (2012c). "Family Registration Certificate". Retrieved 21.04.2012 from http://www.nadra.gov.pk/index.php?option=com_content&view=article&id=11&Itemid=14.

National Database Registration Authority (2012d). "e-Sahulat: Frequently Asked Questions". Retrieved 03.02.2012 from http://www.esahulat.com.pk/subpages/faqs.php

National Database Registration Authority (2013a). "e-Sahulat: Franchise Application Form". Retrieved 03.09.2013 from http://www.esahulat.com.pk/files_download/esahulat_application_form_urdu.pdf

National Database Registration Authority (2013b). "e-Commerce Platform". Retrieved 13.01.2013 from http://www.nadra.gov.pk/index.php?option=com_content&view=article&id=51&Itemid=102.

National Database Registration Authority (2014b). "NADRA's International Clients". Retrieved 04.05.2014 from http://nadra.gov.pk/index.php/clients-a-partners/international-clients.

National Database Registration Authority (2014c). "About Ministry of Interior". Retrieved 04.05.2014 from http://nadra.gov.pk/index.php/about-us/ministry-of-interior/about-moi.

Pakistan Telecommunication Authority (2014a). " [Urdu Website of Pakistan Telecommunication Authority]پاکستان ٹیلی نیکیشنکمیو اتھارٹی اردو ویب سائٹ : ہی-ٹی۔اے.حکومت پاکستان". Retrieved 04.05.2014 from https://www.pta.gov.pk/ur.

Potohar Organisation for Development Advocacy (2008). "Facilitating Rural Citizens in CNIC & Voter Registration Process". Retrieved 04.05.2014 from http://www.poda.org.pk/projects/15/?title=Facilitating-Rural-Citizens-in-CNIC-%26-Voter-Registration-Process.

Punjab Information Technology Board (2011). "History". Retrieved 15.10.2011 from http://pitb.gov.pk/?q=history.

Radio Pakistan (2014). "Chronology of Pakistan Broadcasting Corporation (PBC): Radio Pakistan in the light of history". Retrieved 22.05.2014 from http://www.radio.gov.pk/chronology-of-pbc.

Sustainable Development Policy Institute (2014). "Profile: Prof Ahsan Iqbal (Board of Governors, SDPI)". 02.06.2014, from https://www.sdpi.org/about_sdpi/bog_details82.html.

Telenor (2014). "About Easypaisa". Retrieved 04.05.2014 from http://easypaisa.com.pk/index.php/en/about/about-easypaisa.

The Economist (2009). "The power of mobile money". Retrieved 15.04.2011 from http://www.economist.com/node/14505519?story_id=14505519.

Urdu Point (2006). "نادرا کیاسک کی بڑھتی ہوئی مقبولیت، 10لاکھ صارفین نے یوٹیلیٹی بلوں کی ادائیگی کے لئے رجسٹریشن کروالی [Increasing popularity of NADRA kiosk: 1 million subscribers registered themselves for the payment of utility bills] ". Retrieved 04.05.2014 from https://www.urdupoint.com/weather/news-detail/live-news-17018.html.

World Bank (2008b). "Develoment Communication (DevComm)". Retrieved 27.08.2011 from http://web.worldbank.org/archive/website01216/WEB/0__MENUP.HTM.

Yasir, M. (2011). "Interview with Tariq Malik, Deputy Chairman, NADRA". Retrieved 04.05.2014 from https://www.codeweek.pk/2011/06/interview-with-deputy-chairman-nadra/.